Novel Food and Feed Safety

Safety Assessment of Foods and Feeds Derived from Transgenic Crops, Volume 1

This work is published under the responsibility of the Secretary-General of the OECD. The opinions expressed and arguments employed herein do not necessarily reflect the official views of OECD member countries.

This document and any map included herein are without prejudice to the status of or sovereignty over any territory, to the delimitation of international frontiers and boundaries and to the name of any territory, city or area.

Please cite this publication as:
OECD (2015), *Safety Assessment of Foods and Feeds Derived from Transgenic Crops, Volume 1*, Novel Food and Feed Safety, OECD Publishing, Paris.
http://dx.doi.org/10.1787/9789264180147-en

ISBN 978-92-64-18013-0 (print)
ISBN 978-92-64-18014-7 (PDF)

Series: Novel Food and Feed Safety
ISSN 2304-9499 (print)
ISSN 2304-9502 (online)

The statistical data for Israel are supplied by and under the responsibility of the relevant Israeli authorities. The use of such data by the OECD is without prejudice to the status of the Golan Heights, East Jerusalem and Israeli settlements in the West Bank under the terms of international law.

Photo credits: Cover ©danyssphoto/Shutterstock.com; ©withGod/Shutterstock.com; © Aleksandar Mijatovic - Fotolia.com.

Corrigenda to OECD publications may be found on line at: *www.oecd.org/about/publishing/corrigenda.htm.*

© OECD 2015

You can copy, download or print OECD content for your own use, and you can include excerpts from OECD publications, databases and multimedia products in your own documents, presentations, blogs, websites and teaching materials, provided that suitable acknowledgment of the source and copyright owner is given. All requests for public or commercial use and translation rights should be submitted to *rights@oecd.org*. Requests for permission to photocopy portions of this material for public or commercial use shall be addressed directly to the Copyright Clearance Center (CCC) at *info@copyright.com* or the Centre français d'exploitation du droit de copie (CFC) at *contact@cfcopies.com*.

Foreword

From their first commercialisation in the mid-1990s, genetically engineered crops (also known as transgenic crops) have been increasingly approved for cultivation, and for entering in the composition of foods or feeds, by a number of countries. To date, genetically engineered varieties of over 25 different plant species (including agricultural crops, flowers and trees) have received regulatory approvals in OECD and non-OECD countries from all regions of the world. Up to now, the large majority of plantings remain for soybean, maize, cotton and rapeseed (canola), as outlined in the OECD's *The Bioeconomy to 2030: Designing a Policy Agenda*. Over the 19-year period from 1996 to 2014, the surface area grown with transgenic crops worldwide has constantly raised, resulting in a significant increase of their harvested commodities used in foods and feeds (often designated as "novel" foods and feeds). This is highlighted in analyses and statistics from several sources which, despite some differences in total estimates, all concur in underlining the general increasing trend in volumes produced, number of countries involved and growth potential.

For instance, James reports in the *Global Status of Commercialized Biotech/GM Crops: 2014* a record 181.5 million hectares of genetically engineered plants grown, representing an annual growth rate of more than 3.5% from 2013. According to this study, the five main producers in 2014 were the United States, followed by Brazil, Argentina, India and Canada, covering together almost 90% of the total area. Interestingly, developing countries grew more of global transgenic crops (53%) than industrial countries, at 47%. Among the 28 countries having planted transgenic crops in 2014, only 9 of them were OECD countries, listed by decreasing area as follows: the United States, Canada, Australia, Mexico, Spain, Chile, Portugal, the Czech Republic and the Slovak Republic. However, an additional group of countries does not produce transgenic crops but imports the produced commodities, for use in their feed industry in particular, as it is the case in several jurisdictions of Europe as well as some other economies worldwide.

Information of these transgenic crops which have been approved for commercial release in at least one country (for planting and/or for use in foods and feeds processing) can be found in the OECD *BioTrack Product Database* (http://www2.oecd.org/biotech/). Each transgenic product and its Unique Identifier are described, as well as information on approvals in countries.

In parallel to the expansion of genetically engineered crops developed for their resistance to pests and diseases, varieties are being developed by breeders for new types of traits: adaption to climate change, improved composition (biofortification), enhanced meat productivity, easier processing and many other applications. The range of biotechnology applications to agricultural plant breeding is widening, and it seems that the trend will continue. Consequently, the volume of novel foods and feeds available on the market and exchanged internationally is expected to increase in the coming years.

Consumers from all over the world are requiring a high level of safety and full confidence in the products they eat. This is particularly important for the products of modern biotechnology which are sometimes questioned and subject to diverse levels of acceptation among countries. The approvals of transgenic crops follow a science-based risk/safety assessment regarding their potential release in the environment (biosafety) and their use in foods or feeds (novel food and feed safety). The OECD has undertaken activities related to environmental safety aspects since the mid-1980s, while the development of scientific principles for food safety assessment was initiated in 1990. The OECD helps countries in their risk/safety assessment of transgenic organisms by offering national authorities a platform to exchange experience on these issues, identify emerging needs, collate solid information and data, and develop useful tools for risk assessors and evaluators.

To date, 26 consensus documents relating to the safety of novel foods and feeds have been published; two have been revised ten years later. Most of these publications address compositional considerations of crops subject to plant breeding improvement with modern biotechnologies. These consensus documents are focused on key food and feed nutrients, anti-nutrients, toxicants and other constituents as relevant. They provide solid information commonly recognised by experts and collate the reliable range of data available in the scientific literature at the time of the publication. They can be used in the comparative approach to safety assessment. In addition, documents of a broader nature aiming to facilitate harmonisation have been developed: animal feedstuffs derived from transgenic commodities (2003), designation of an OECD "Unique Identifier" for transgenic plants (2002, revised in 2006); and molecular characterisation of transgenic plants (2010).

Volumes I and II of this series compile the consensus documents of the OECD Series on Safety of Novel Foods and Feeds issued since 2012 (Volume I covers 2002-08, Volume II covers 2009-14). The presentation of the OECD work, originally published in 2006, was used as a basis for the introduction section that explains the purpose of the consensus documents, their relevance to risk/safety assessment, and their preparation by the relevant OECD task force. The present compendium offers ready access to those documents which have been published thus far. As such, it should be of value to applicants for uses of transgenic crop commodities in foods and feeds, regulators and risk/safety assessors in national authorities, as well as to the wider scientific community.

Each of the consensus documents may be updated in the future as new knowledge becomes available. Users of this book are therefore encouraged to provide information or an opinion regarding the contents of the consensus documents or any of the OECD's other harmonisation activities. Comments can be provided to: ehscont@oecd.org.

The published consensus documents are also available individually from the OECD's Biotrack website, at no cost: www.oecd.org/biotrack.

Acknowledgements

This book results from the common effort of the participants in the OECD's Task Force for the Safety of Novel Foods and Feeds. Each chapter is composed of a "consensus document" which was prepared under the leadership of a participating country or several countries, as listed at the end of this volume. During their successive drafting, valuable inputs and suggestions for the documents were provided by a number of delegates and experts in the Task Force, whether from OECD member countries, non-member economies or observer organisations.

Each consensus document was issued individually, as soon as it was finalised and agreed for declassification, by the OECD Environment, Health and Safety Division in the Series on Safety of Novel Foods and Feeds. The manuscripts of Volumes I and II of this publication, containing the 2002-14 consensus documents, were prepared by Elisabeth Huggard, Arely Badillo, Carolina Tronco-Valencia and Jennifer Allain. They were edited by Bertrand Dagallier, under the supervision of Peter Kearns, at the EHS Division of the OECD Environment Directorate.

Table of contents

Tables

Figures

Executive summary

This document constitutes the first volume of the OECD Series on Novel Food and Feed Safety. It is a compendium collating in a single issue the individual "consensus documents" published by the Task Force for the Safety of Novel Foods and Feeds from 2002 to 2008. The second volume of the series will cover the documents issued from 2009 to 2014.

Modern biotechnologies are applied to plants, and also trees, animals and microorganisms. The safety of the resulting products represents a challenging issue, and in particular as genetically engineered crops are increasingly cultivated and foods or feeds derived from them are marketed worldwide. Modern biotechnology products should be rigorously assessed by governments to ensure high safety standards for environment, human food and animal feed. Such assessments are considered to be essential for a healthy and sustainable agriculture, industry and trade.

The OECD Task Force for the Safety of Novel Foods and Feeds (the Task Force) was established in 1999. Its purpose is to assist countries in evaluating the potential risks of transgenic products, foster communication and mutual understanding of relevant regulations in countries, and facilitate harmonisation in risk/safety assessment of products from modern biotechnology. This is intended to encourage information sharing, promote harmonised practices and prevent duplication of efforts among countries. Therefore the Task Force's programme, while consolidating high food and feed safety standards, contributes to reducing costs and potential for non-tariff barriers to trade. Being focused on foods and feeds derived from genetically engineered organisms (also named "novel" foods and feeds), the Task Force's activities and outputs are directly complementary to those of the Working Group on Harmonisation of Regulatory Oversight in Biotechnology, which deals with environmental safety.

The Task Force is composed of delegates from OECD member countries, non-member economies, international bodies and observer organisations involved in these matters, from all regions of the world. National participants and experts are from those government ministries and agencies which have responsibility for the risk and safety assessment of novel foods and feeds in the respective countries. The Task Force provides a platform for delegates to exchange experience and information, identify new needs and develop practical tools for helping the food and feed safety assessment. The main outputs are the "consensus documents", which compile science-based information and data relevant to this task. The key composition elements (nutrients, anti-nutrients, toxicants and sometimes other constituents) that they contain can be used to compare novel foods and feeds with conventional ones. These documents are published after consensus is reached among countries.

Part I of this publication (Volume I) contains two documents of broad application aimed to contribute to harmonised assessments of food and feed safety: *i)* considerations for the safety assessment of animal feedstuffs derived from genetically modified plants; and *ii)* guidance for the designation of a Unique Identifier for transgenic plants, which is a system developed by the OECD for accessing and exchanging information stored in databases on these products, currently used worldwide.

Part II of the publication (Volume I) gathers the consensus documents prepared by the Task Force on compositional considerations for transgenic crops. Each chapter contains background information on the considered species: its production, process and uses of its products for foods and feeds, and for most of them a brief summary on appropriate comparators for testing new varieties and screening characteristics used by breeders. The core of the chapter is then constituted by detailed information on compositional elements: key nutrients and anti-nutrients, toxicants and allergens where applicable, and sometimes other constituents such as secondary metabolites. The final sections suggest key products and constituents for analysis of new varieties for food use and for feed use. Volume I covers the following crops, presented in the order of their initial publication by the Task Force between 2002 and 2008: sugar beet, potato, maize, wheat, rice, barley, alfalfa and other temperate forage legumes, cultivated mushroom, sunflower and tomato.

This set of science-based information and data, agreed by consensus and published by the OECD, constitute a solid reference recognised internationally. It is already widely used in comparative approach as part of the risk/safety assessment of transgenic products. As such, this publication should be of value to applicants for commercial uses of genetically engineered crops, to regulators and risk assessors in national authorities in charge of granting approvals to transgenic plant products for their use as foods or feeds, as well as to the wider scientific community.

Introduction

OECD activities on novel food and feed safety

The OECD Task Force for the Safety of Novel Foods and Feeds (Task Force) was established in 1999, with primary goals to promote international regulatory harmonisation in the risk and safety assessment of biotechnology products among member countries.

The terms "novel foods and feeds" relate usually to foods and feeds derived from transgenic organisms, i.e. partly fully composed of such ingredients. By extension, these terms could also be understood as foods and feeds containing products obtained from other modern biotechnology techniques. Regulatory harmonisation is the attempt to ensure that the information used in risk/safety assessments, as well as the methods used to collect such information, are as similar as possible. It could lead to countries recognising or even accepting information from one another's assessments. The benefits of harmonisation are clear: it increases mutual understanding among member countries, which avoids duplication, saves on scarce resources and increases the efficiency of the risk/safety assessment process. This, in turn, improves food and feed safety while reducing unnecessary barriers to trade (OECD, 2000).

The Task Force comprises delegates from the 34 member countries of the OECD and the European Commission. A number of observer delegations and invited experts also participate in its work, including Argentina and the Russian Federation, as well as the Food and Agriculture Organization of the United Nations (FAO), the World Health Organization (WHO), the Business and Industry Advisory Committee to the OECD (BIAC), and other organisms as relevant such as the United Nations Environment Programme, the World Bank, the Center for Environmental Risk Assessment of the ILSI Research Foundation (CERA) and the African Biosafety Network of Expertise. Since 2002, several other non-member countries (Bangladesh, Brazil, the People's Republic of China, Colombia, India, Indonesia, Kenya, Latvia, Moldova, Philippines, South Africa, Thailand and others) have participated in activities of the Task Force under the auspices of OECD Global Relations Secretariat and its Global Forum on Biotechnology.

Typically, delegates of the Task Force are from those government ministries and agencies which have responsibility for the food or feed safety assessment of products of modern biotechnology, including foods and feeds derived from transgenic organisms. In some OECD countries this is the Ministry of Health; in others it is the Ministry of Agriculture. Other countries have specialised agencies with this responsibility. Often, it is a shared responsibility among more than one ministry or agency. The expertise that these delegates have in common is related to their experience with food and/or feed safety assessment.

The emergence of the concept of consensus documents

By 1997, several OECD countries had gained experience with safety assessment of foods derived through modern biotechnology. An OECD Workshop in Aussois, France, examined the effectiveness of the application of substantial equivalence in safety assessment. It was concluded that the determination of substantial equivalence provides equal or increased assurance of the safety of foods derived from genetically modified plants, as compared with foods derived through conventional methods (OECD, 1997).

At this event, it was also recognised that a consistent approach to the establishment of substantial equivalence might be improved through consensus on the appropriate components (e.g. key nutrients, key toxicants and anti-nutritional compounds) on a crop-by-crop basis, which should be considered in the comparison. It is recognised that the components may differ from crop to crop.

Following the Aussois workshop, there was a detailed analysis of whether there was a need to undertake work on food/feed safety at the OECD, and if so, what that work would entail. This analysis was undertaken by an Ad Hoc Group on Food Safety (established by the Joint Meeting).[1] It took into account the results of national activities and those of previous OECD work, as well as the activities of the FAO and WHO.

As a result of the Ad Hoc Group on Food Safety's activities, the Joint Meeting established the Task Force, with major part of its programme of work being the development of consensus documents on compositional data. These data are used to identify similarities and differences following the comparative approach as part of a food and feed safety assessment. They should be useful to the development of guidelines, both national and international and to encourage information sharing among OECD countries as well as with non-members.

Participation from non OECD member economies is strongly encouraged by the Task Force, as transgenic crops are grown in several of these countries and economies, their commodities traded internationally and widely used for food and feeds. This exchange has increased over the years and now more actively involves their expertise. For example, the consensus documents on the composition of cassava, grain sorghum and papaya were developed in co-operation of non-member countries with leadership/co-leadership of South Africa for the two first and Thailand for the latter. Similarly, Brazil is co-ordinating the preparation of a future document on the common bean while the Philippines are actively involved in the revision of the rice composition document. This concrete enlargement to non-members' inputs and competence broadens the expertise available to the Task Force, while addressing a wider range of food and feed products that are of global interest.

Background and principles surrounding the use of consensus documents

The OECD "consensus documents" are a compilation of current information that is important in food and feed safety assessment. Agreed by consensus among the Task Force participants, they provide a technical tool for regulatory officials, industry and other interested parties, as a general guide and reference source. They complement those of the OECD Working Group on Harmonisation of Regulatory Oversight in Biotechnology which deal with the environmental safety aspects (biosafety) (OECD, 2006a; 2006b; 2010a; 2010b). They are mutually acceptable to, but not legally binding on, member countries and are used as key references by other economies beyond the OECD for their assessment of novel foods and feeds. They are not intended to be a comprehensive description of all the issues considered to be necessary for a safety assessment, but a base set for an individual product that supports the comparative approach. In assessing an individual product, consideration of additional components may be required depending on the specific case in question.

The work of the Task Force builds on previous OECD experience in biotechnology safety-related activities, dating back to the mid-1980s. Initially, much of the work concentrated on the environmental and agricultural implications of the use of transgenic crops. By the end of 1990, however, work had been established to develop scientific principles for food safety assessment of products of modern biotechnology. This work was often undertaken in parallel to complementary activities of the FAO and WHO.

In 1990, a joint consultation of the FAO and WHO established that the comparison of a final product with one having an acceptable standard of safety provides an important element of safety assessment (WHO, 1991).

In 1993, the OECD further elaborated this concept and advocated the approach to safety assessment based on substantial equivalence as being the most practical approach to addressing the safety of foods and food components derived through modern biotechnology (as well as other methods of modifying a host genome, including tissue culture methods and chemical- or radiation-induced mutation).

A Joint FAO/WHO Expert Consultation on Biotechnology and Food Safety (1996) elaborated on compositional comparison as an important element in the determination of substantial equivalence. A comparison of critical components can be carried out at the level of the food source (i.e. species) or the specific food product. Critical components are determined by identifying key nutrients and key toxicants and anti-nutrients for the food source in question. The comparison of critical components should be between the modified variety and non-modified comparators with an appropriate history of safe use. The data for the non-modified comparator can be the natural ranges published in the literature for commercial varieties or those measured levels in parental or other edible varieties of the species (FAO/WHO, 1996). The comparator used to detect unintended effects for all critical components should ideally be the near isogenic parental line grown under identical conditions. While the comparative approach is useful as part of the safety assessment of foods derived from plants developed using recombinant DNA technology, the approach could, in general, be applied to foods derived from new plant varieties that have been bred by other techniques.

The Joint FAO/WHO Expert Consultation on Foods Derived from Biotechnology in 2000 (FAO/WHO, 2000) concluded that the safety assessment of genetically modified foods requires an integrated and stepwise, case-by-case approach, which can be aided by a structured series of questions. A comparative approach focusing on the determination of similarities and differences between the genetically modified food and its conventional counterpart aids in the identification of potential safety and nutritional issues and is considered the most appropriate strategy for the safety and nutritional assessment of genetically modified foods. The concept of substantial equivalence was developed as a practical approach to the safety assessment of genetically modified foods. It should be seen as a key step in the safety assessment process, although it is not a safety assessment in itself; it does not characterise hazard, rather it is used to structure the safety assessment of a genetically modified food relative to a conventional counterpart. The consultation concluded that the application of the concept of substantial equivalence contributes to a robust safety assessment framework.

Between 2000 and 2003, the *ad hoc* Intergovernmental Task Force on Foods Derived from Biotechnology to the Codex Alimentarius Commission (Codex Task Force) undertook work to develop principles and guidelines for foods derived from genetically engineered plants. The full report of the Codex Task Force included:

- principles for the risk analysis of foods derived from modern biotechnology
- a guideline for the conduct of food safety assessment of foods derived from recombinant-DNA plants
- a guideline for the conduct of food safety assessment of foods produced using recombinant-DNA microorganisms (Codex Alimentarius Commission, 2003).

One notable feature of the principles is that they make reference to a safety assessment involving the comparative approach between the food derived from modern biotechnology and its conventional counterpart. Annex II (safety assessment of foods derived from recombinant-DNA plants modified for nutritional or health benefits) and Annex III (safety assessment in situation of low-level presence of recombinant-DNA plant material in food) were added to the guidelines in 2008.

The OECD Task Force is working closely with the Codex Task Force in order to strengthen their complementary activities.

The process through which consensus documents are prepared

The consensus documents are prepared by the Task Force on official proposals by countries. Typically, the focus is a food crop or vegetable for which modern biotechnology can be used in the plant-breeding process. New improved varieties of these species are being developed by researchers for future release in at least one country, or even exist already at commercial level for some of them.

The Task Force establishes *ad hoc* drafting groups, composed of officials and scientific experts of the species in interested countries. These drafting groups work with all this diversity of inputs, under the co-ordination of "lead countries". The successive revised drafts are reviewed by the full Task Force, with careful examination of the proposed information, data, tables and figures. The several revisions and completions can require a few years, leading to a consensus from all delegations obtained on all elements. Following an OECD internal process for final approval, the document is published and becomes available online for worldwide users.

The OECD Biotrack website provides publications and news from the Task Force, the Series on Novel Food and Feed Safety, contact details of national safety systems and other information. It links to the biosafety (environmental safety) publications, the Series on Harmonisation of Regulatory Oversight in Biotechnology. It also gives free access to the OECD *BioTrack Product Database*. It is available at: www.oecd.org/biotrack,

Current and future trends

With the growing development of products from modern biotechnology, the production of transgenic crops has increased drastically in the last 20 years. It might even be expanded in the future if new varieties adapted to new needs are adopted. Prospects encompass agriculture, industry and energy sectors.

Resistances to pests and diseases were introduced in plants from the early time of genetic engineering, and still constitute the essential feature of the varietal improvement for agriculture, horticulture and forestry. In parallel, breeders are also working on incorporating new traits in crops for gaining other types of beneficial effects. Some of these varieties are about to enter the market or start being grown. In recent years, drought-tolerant varieties (maize, and now sugarcane) are designed to contribute to climate change adaptation. "Innovation in plant breeding (including biotechnology) that aims to develop crop varieties that are more resilient to climate change impact (e.g. resistance to drought, soil salinity or temperature extremes) is part of a larger basket of possible adaptation options in agriculture" (Agrawala et al., 2012). Other innovative traits can have a direct beneficial impact on foods and feeds, and some are already promising: staple crops (rice, tubers, other species) offering nutritive improvements with increased content (biofortification) of elements such as pro-vitamins or micro-nutrients, feed plants (such as maize and alfalfa) modified for higher digestibility and meat productivity, and many other products under development. The range of biotechnology applications to plant breeding continues to widen, leading to an expected increase of derived foods and feeds used and exchanged internationally in the coming years.

A reliable risk/safety assessment of novel foods and feeds is therefore more than ever a necessity for many world economies, in the context of international trade of commodities. Release of such products should be based on solid information and appropriate tools for leading to national decision making. Harmonised regulations, common practices and easy access to solid science-based compiled information are sought. The tools developed by the OECD Task Force designed to promote international harmonisation in the field of food/feed safety assessment are recognised and appreciated, and they might play an increasing role for fulfilling these needs in the future.

The Task Force is continuing its work on a range of issues. New projects have begun recently on the composition of two new species, the common bean and apple. Further species might be subject to similar activity in the future. The main area of the 2013-16 programme of work remains the development of consensus documents on compositional considerations. Emerging topics are also considered for remaining reactive to key demand, e.g. other new biotechnology techniques, innovative feed ingredients, animal composition data, all of them to be considered regarding food and feed safety issues.

In parallel, the consensus documents are reviewed periodically and updated as necessary to ensure that scientific and technical developments are taken into account. Users of these documents have been invited to provide the OECD with new scientific and technical information, and to make proposals for additional areas to be considered. For example, the low erucic acid rapeseed (canola) and soybean documents, both published originally in 2001, were completed and revised by the Task Force, leading to updated issues in 2011-12. The rice document (2004) has initiated a revision process (a new version expected in 2015) and others might follow in the coming years.

Note

1. The Joint Meeting was the supervisory body of the Ad Hoc Group and, as a result of its findings, established the Task Force as a subsidiary body. Today, its full title is the Joint Meeting of the Chemicals Committee and the Working Party on Chemicals, Pesticides and Biotechnology.

References and additional reading

Agrawala, S. et al. (2012), "Adaptation and innovation: An analysis of crop biotechnology patent data", *OECD Environment Working Papers*, No. 40, OECD Publishing, Paris, http://dx/doi.org/10.1787/5k9csvvntt8p-en.

Codex Alimentarius Commission (2003; Annexes II and III adopted in 2008), *Guideline for the Conduct of Food Safety Assessment of Foods Derived from Recombinant DNA Plants*, CAC/GL 45/2003, at: www.codexalimentarius.net/download/standards/10021/CXG_045e.pdf (accessed 28 Jan. 2015).

FAO/WHO (2000), *Safety Aspects of Genetically Modified Foods of Plant Origin,* Report of a Joint FAO/WHO Expert Consultation on Foods Derived from Biotechnology Food and Agriculture Organization, Rome, www.fao.org/fileadmin/templates/agns/pdf/topics/ec_june2000_en.pdf (access. 28 Jan. 2015).

FAO/WHO (1996), "Joint consultation report, Biotechnology and food safety", *FAO Food and Nutrition Paper No. 61*, Food and Agriculture Organization, Rome, at: www.fao.org/ag/agn/food/pdf/biotechnology.pdf (accessed 28 Jan. 2015).

James, C. (2015), "Global status of commercialized biotech/GM crops: 2014", *ISAAA Brief No. 49*, International Service for the Acquisition of Agri-Biotech Applications: ISAAA, Ithaca, New York, at: www.isaaa.org/resources/publications/briefs/49 (executive summary accessed 28 Jan. 2015).

OECD (2010a), *Safety Assessment of Transgenic Organisms: OECD Consensus Documents: Volume 4*, OECD Publishing, Paris, http://dx/doi.org/10.1787/9789264096158-en.

OECD (2010b), *Safety Assessment of Transgenic Organisms: OECD Consensus Documents: Volume 3*, OECD Publishing, Paris, http://dx/doi.org/10.1787/9789264095434-en.

OECD (2009), *The Bioeconomy to 2030: Designing a Policy Agenda*, OECD Publishing, Paris, http://dx.doi.org/10.1787/9789264056886-en.

OECD (2006a), *Safety Assessment of Transgenic Organisms: OECD Consensus Documents: Volume 2*, OECD Publishing, Paris, http://dx.doi.org/10.1787/9789264095403-en.

OECD (2006b), *Safety Assessment of Transgenic Organisms: OECD Consensus Documents: Volume 1*, OECD Publishing, Paris, http://dx.doi.org/10.1787/9789264095380-en.

OECD (2000), "Report of the Task Force for the Safety of Novel Foods and Feeds", prepared for the G8 Summit held in Okinawa, Japan on 21-23 July 2000, C(2000)86/ADD1, OECD, Paris, at: www.oecd.org/chemicalsafety/biotrack/REPORT-OF-THE-TASK-FORCE-FOR-THE-SAFETY-OF-NOVEL.pdf (accessed 28 Jan. 2015).

OECD (1997), "Report of the OECD Workshop on the Toxicological and Nutritional Testing of Novel Foods", held in Aussois, France, 5-8 March 1997, final text issued in February 2002 as SG/ICGB(98)1/FINAL, OECD, Paris, at: www.oecd.org/env/ehs/biotrack/SG-ICGB(98)1-FINAL-ENG%20-Novel-Foods-Aussois-report.pdf (accessed 28 Jan. 2015).

WHO (1991), *Strategies for Assessing the Safety of Foods Produced by Biotechnology*, Report of a Joint FAO/WHO Consultation, World Health Organization of the United Nations, Geneva, out of print.

Part I

Towards harmonised assessments for food and feed saftey

Chapter 1

Animal feedstuffs derived from genetically modified plants

This chapter was prepared by the OECD Task Force for the Safety of Novel Foods and Feeds with Canada and the United Kingdom as lead countries. It addresses considerations in the safety assessment of genetically modified feeds derived from crops, based on the scientific issues involved. Sections include information on genetically modified plants used as animal feed, assessment of genetically modified feedstuffs, the fate of DNA and protein in animal feeding, animal feeding studies as part of a safety assessment, post-market surveillance/monitoring, by-products of industrial crops, and future genetically modified feedstuffs with the question of agronomic versus quality traits.

Summary

Animal feed represents an important point of entry of plant products into the food chain. Consequently, it is important that novel feedstuffs be as carefully assessed for safety as those products used directly as human food. This chapter is intended to provide considerations in the safety assessment of genetically modified (GM) feeds derived from plants, based on the scientific issues involved.

The safety assessment of GM foods and feeds share many common elements, notably the molecular characterisation of the introduced genetic elements, the expression of the novel traits and the impact of these in the newly modified plant. These have been extensively considered elsewhere. This chapter focuses on those aspects of particular importance to the safety assessment of GM feed, in particular, the wholesomeness of the feed for livestock and the safety for consumers of products (e.g. meat, milk, eggs) obtained from livestock whose diet includes GM feedstuffs.

Establishing the degree of equivalence to other (conventional) varieties is a useful starting point for a safety assessment, and is as relevant to feed issues as to those of foods. Consideration should be given to the differential expression of introduced traits in the plant in the selection of material for comparison, particularly when plant parts not used for food purposes are included in animal feed (e.g. maize stover, cottonseed meal). Studies intended to demonstrate the safety of the isolated product of any introduced gene should take account of the maximum concentration found in any plant part or by-product consumed as feed and the consequent exposure of the animal.

The fate of DNA and novel proteins in the digestive tract of both humans and animals has been raised as an issue of concern. Intact DNA and protein can be detected in minimally processed feedstuffs such as hay and silage, but may be degraded by typical feed manufacturing processes. Both DNA and protein are usually extensively digested when consumed by the animal. However, evidence of the degradation of protein during feed preparation should not automatically be assumed to confer safety. Any introduced and expressed protein should be separately examined for its toxic potential regardless of its susceptibility to breakdown.

Fragments of non-transgenic plant DNA have been detected in animal tissues including milk. However, there is no basis to suppose that transgenic DNA poses hazards any different to other sources of DNA and the possibility of incorporation of functionally intact DNA (or protein) into animal products is extremely remote. Consequently, unless there is reason for specific concern, the routine testing of animal products for newly introduced DNA or any expressed products is not considered necessary.

Many new varieties of plants used as feedstuffs are introduced onto the market based on agronomic and compositional data alone. Feeding trials to confirm safety and/or nutritional value are generally unnecessary. To date, all approved GM plants with modified input traits (e.g. herbicide tolerance) have been shown to be compositionally equivalent to their conventional counterparts. Feeding studies with feeds derived from the approved GM plants have shown equivalent animal performance to that observed with the non-GM feed. Thus the evidence to date is that for GM varieties shown to be compositionally equivalent to conventional varieties, feeding studies made with target livestock species will add little to a safety assessment and generally are not warranted.

For plants engineered with the intention of significantly changing their composition/nutrient bioavailability and thus their nutritional characteristics, suitable comparators may not be available for a nutritional assessment based solely on compositional analysis. In such cases, feeding trials with one or more target species may be useful to demonstrate wholesomeness for the animal. Under these circumstances, the duration of feeding studies should be for the production cycle of the animal. Such feeding studies may be usefully supplemented with shorter term balance studies to confirm that the modification produces the intended nutritional benefit (e.g. higher metabolisable energy value, improved nitrogen retention).

In general, the use of a comparative growth study for the screening for any unintended effects of the genetic modification with adverse consequences for the host animal and consumers of animal products not detected by chemical analysis is not warranted for GM varieties any more than for conventionally derived varieties. However, if concerns remain regarding unintended effects of a particular modification, broiler chicks are useful for comparative growth studies. Because of their rapid weight gain, broiler chicks are particularly sensitive to any change in nutrient supply or the presence of toxic elements in their feed and are particularly useful for this purpose. The young of other livestock tend not to show such rapid growth rates, but may, on occasions form a more appropriate model. For feedstuffs intended for aquaculture, a fish species such as the catfish may substitute. Milk production is better used in place of growth rate for feed primarily intended for lactating ruminants. It is important to note that standardised and internationally recognised conditions for such tests have not been established.

In time, it may be possible to detect unintended effects by non-targeted profiling techniques based on the measurement of the transcriptome or proteome. Alternatively, measures to detect unintended effects may be rendered unnecessary by improved molecular characterisation and understanding of the implications of the molecular events for the metabolism of the plant.

Post-market surveillance may be easier to undertake in relation to feed than food. Control of diets, the ability to monitor animal health and the various assurance schemes used in some areas to track animal products to the point of sale would aid the evaluation of long-term effects of GM feeds. However, given that there is no basis to indicate any difference in long-term effects of GM feeds from those of conventionally derived feeds, it is not clear under what circumstances such post-marketing surveillance would be warranted. In addition, given the lack of a theoretical basis for the general transfer of functional protein or DNA to animal products, and in the absence of any documented adverse response to products of animals fed GM feedstuffs, post-market surveillance of consumers appears to be of very limited value. Post-market surveillance is likely to be useful only when designed to answer a specific question.

Recombinant technology has greatly expanded the opportunities to use plants for the production of a multitude of non-food/feed products. Some of these products raise serious issues of feed security. If a GM plant used for the production of industrial products has a conventional counterpart traditionally used for feed purposes, there is a risk of unauthorised material entering the food chain. For this reason a safety assessment for all parts of the industrial GM plant that might enter the feed chain should be conducted. This would inform the risk manager and allow actions proportionate to the risk to be taken. Alternatively, if measures have been implemented to prevent entry of plants producing unauthorised products into the feed supply, a safety assessment of that plant may not be necessary. The processes needed to ensure that the material does not become a component of feed should be proportional to the associated risk.

For the present, agronomic (input) traits will continue to dominate new introductions. However, transgenic plants designed specifically to address issues of feed quality, to resist stress and to grow in more marginal areas have reached the stage of field trials.

The safety assessment of the "next generation" feeds will involve a case-by-case approach with the probable need to introduce specific elements appropriate to each trait. Sound experimental designs for the assessment of these "value-added" novel products should be required.

The approach to safety assessment of novel feeds by national authorities range from use of existing food, feed or environmental legislation to the creation of legislation specific to novel foods and/or feeds. A number of OECD member countries are in the process of developing new legislation on assessment and labelling of GM feeds. Approaches to legislation include the establishment of new GM-specific food and feed laws, updating or establishing feed laws to include GM feeds, or simply considering that the definition of food includes feed.

Whatever the regulatory approach taken by national authorities, it is recognised that an assessment of the safety of GM products used as animal feed that is universally accepted and applied and which can be shown to be rigorous in its approach, is fundamental to retaining consumer confidence in animal products.

Scope and purpose

Animal feed represents an important point of entry of plant products into the food chain. Consequently, it is important that novel feedstuffs are as carefully assessed for safety for target animals and for consumers of animal products as those products used directly as human food. Novelty can be introduced into the feed chain in a number of ways: by the novel use of existing resources or by the introduction of novel traits to existing feedstuffs. This chapter focuses on the latter and considers only those feed ingredients derived from genetically modified (GM) plants. This chapter is intended to provide considerations in the safety assessment of GM feedstuffs, based on the scientific issues involved.

Relationship to food safety assessment

Many feeds for animals make use of the same plants (or by-products of the same plants) used for human food. Consequently, many elements of a safety assessment are common to both. Both require a precise characterisation of the introduced genetic elements, information on substances present as a result of the modification and evidence whether detectable unintended effects occurred because of the insertion(s). These issues have been extensively considered in the context of GM foods. With the exception of genes and gene products unique to feeds and the detection of unintended effects, these common issues will not be further considered here. The approach taken to these common issues is fully described in the reports of the Codex Ad Hoc Intergovernmental Task Force on Foods derived from Biotechnology and the various expert consultations and the references therein (FAO/WHO, 1996; 2000; 2001).

Feed use does introduce different concerns and parameters to the safety evaluation than that of human food use. In particular, the assessment of animal feeds must take into account any risk to the animals consuming the feed and any indirect risk to the consumer of animal products (meat, milk and eggs). There is also a potential for a greater exposure of animals to a GM plant or plant by-product than humans as it may comprise a large percentage of the diet fed on a daily basis, often for the complete life span of the animal. A broiler chick, for example, has a daily consumption of approximately 60 g maize kernel/kg body weight and a growing pig about 45 g/kg body weight compared to an adult human who consumes about 0.2 g/kg body weight each day. In addition, livestock are fed plants and components of plants that humans do not consume and exposure to novel gene products can differ from that of humans. However, feed use does allow the GM plant material to be fed directly to target species, providing tests of wholesomeness and introducing the possibility of using the delivery of nutrients as part of the safety assessment.

Genetically modified plants used as animal feed

Established patterns of use of plants for which approved genetically modified varieties exist

Currently, maize is the only cereal that has GM varieties in commercial production. It is also the cereal of choice for animal feed in most parts of the world, being replaced by other cereals, particularly wheat or barley, only when local availability and price favour substitution. Use of maize as animal feed accounts for approximately 75% of a total world production of 600 million tonnes annually. Consequently, there is a well-established global market for this commodity. Approximately 79 million tonnes were exported in 2001 (FAOSTAT, 2003). Maize may be fed as the whole grain or as various by-products of the corn-milling industry or as whole crop silage (OECD, 2002). Corn-soybean diets are extensively used for poultry and for growing pigs and maize silage for dairy cattle. In each case, the maize provides the major energy source. The protein-enriched maize gluten feed and gluten meal are valuable by-products of starch extraction, the former used for pigs and ruminants and the latter for poultry.

Soybean dominates the oilseed market with a global annual production in excess of 150 million tonnes. Feed use accounts for 97% of total production, mostly for domestic animal production, but some 40 million tonnes are traded annually, to areas generally not suited for soybean production, such as the European Union (EU). It is the preferred source of protein for most pig and poultry production diets (78% of total soybean production) with the remainder being used for ruminants, companion animals and in aquaculture (OECD, 2001). Because of the presence of anti-nutritional factors, there is very limited consumption of the unprocessed bean. Most is fed as the protein-enriched seed meal left after extraction of soybean oil. However, treated whole beans, hulls and the vegetative parts of the plant in fresh or conserved state are also used to a limited extent, primarily for cattle. Such use is domestic and rarely found outside the producer countries.

Seed meals left after oil extraction from other oilseed crops can also provide valuable sources of protein and energy for animals, and animal feeding represents the most cost-effective means of disposal of these by-products. Approximately 20 million tonnes of low erucic acid rapeseed meal are used annually in the rations of all classes of livestock, but the high-fibre content relative to soybean limits inclusion levels. Rapeseed and rapeseed oil are also occasionally used in small amounts to boost the energy content of some non-ruminant diets. Cottonseed meal (12 million tonnes a year) is fed predominantly to ruminants, which are protected from the toxic effects of gossypol by the presence of a rumen microbial flora capable of its degradation.

Other commodities or their by-products have a lesser and often localised role in animal feeding. Of relevance to this chapter are fodder beet (a sub-species of *Beta vulgaris*) and potato, both of which have approved GM varieties. Fodder beet, roots and tops (leaves) are used exclusively for ruminant feed, traditionally in areas where climatic conditions are less suitable for cereal production. The extent of use of potatoes as a feed ingredient varies considerably depending on locality. When used, either as whole tubers or trimmings, they are generally fed raw to ruminants but are heat-treated before feeding to pigs. Potatoes used for starch production also generate by-product streams that have found an outlet in animal feeding. These include the fibre-rich pulp remaining after starch extraction and protein-enriched liquid feed used primarily, but not exclusively, by the pig industry.

Traits introduced into plants used in animal feeding by recombinant DNA technology

Virtually all of the GM plants currently grown for commercial production have been modified to improve their agronomic properties. Traits have been introduced to confer resistance to common pests (European corn borer, Colorado beetle), to viral pathogens or to introduce tolerance to selected herbicides for better weed control (Table 1.1). At present, most varieties carry a single introduced trait, but there is growing trend towards "stacked-gene varieties" carrying two or more traits, either introduced simultaneously or obtained by crossing single-trait varieties.

To date, only a few GM varieties with modified composition have been approved for commercial production. The first to be accepted in the United States in 1995 was an oilseed rape modified to produce high concentrations of lauric acid in the oil for use as food and in the detergent industry. Both this construct and the high oleic acid soybean approved for release in the United States in 1997 and in Canada in 2000 have yet to be grown in commercial quantities.

The global market: Production, use and export of genetically modified plants used in feedstuffs

The global area devoted to transgenic plants in 2001 was estimated as 52.6 million hectares, an approximate 19% increase on the previous year's plantings. This area increased by a further 11.6% in 2002 reaching 58.7 million hectares. Four countries (listed in decreasing area): the United States, Canada, Argentina and the People's Republic of China grew 99% of the global crop with a further 12 countries accounting for the remaining 1%. Of these, only Australia and South Africa grew more than 100 000 hectares (James, 2002).

The United States, which produced 66% of total world plantings of transgenic crops in 2002, showed a net gain of 3.3 million hectares of crops compared to 2001 as a result of increased plantings of transgenic soybean, maize and cotton. Figures derived from US seed sales in 2002 indicate that transgenic cotton now represents 71% of all cotton seed sales, transgenic maize varieties 32% of the total maize seed sales and transgenic soybean 74% of the total sales (NASS, 2002). The second largest producer of GM varieties, Argentina, showed an overall gain of 1.7 million hectares in 2002 compared to the previous year, which resulted from significant increases in the area of transgenic soybean and cotton and, to a lesser extent, maize. Herbicide-resistant soybean now represents greater than 95% of all soybeans produced in Argentina and, for the first time, more than half (51%) of the 72 million hectares of soybeans grown worldwide were

GM varieties. In China, the area of transgenic cotton showed a substantial increase, rising from 1.5 million hectares in 2001 to 2.1 million hectares in 2002. Because of a high initial uptake of transgenic canola varieties, subsequent use of transgenic crops in Canada has grown relatively slowly compared to other countries. The total area devoted to transgenic varieties in Canada was 3.0 million hectares in 2000, rising to 3.2 million hectares in 2001 and to 3.5 million hectares in 2002.

Table 1.1. **Genetically modified plants used as feedstuffs that have obtained regulatory approval in at least one country grouped by introduced property**

Introduced genetic material	Introduced property	Recipient crops
Insect resistance		
Genes encoding a truncated endotoxin produced by strains of *Bacillus thuringiensis*:	Resistance to attack by:	
cry1A(b) and *cry1A(c)*	– Lepidoptera (including European corn borer)	Maize, cotton
*cry9C**	– Lepidoptera (including European corn borer)	Maize
Cry1F	– Lepidoptera (including European corn borer, corn earworm, fall army worm and black cutworm)	Maize
cry3A	Coleoptera (including Colorado beetle)	Potato
Virus resistance		
Gene encoding a viral coat protein	Resistance to attack by potato virus Y	Potato
Viral replicase gene	Potato leaf curl virus	Potato
Herbicide tolerance		
epsps (bacterial or engineered plant gene) (more rarely *gox* encoding an oxidoreductase)	Tolerance to glyphosate	Sugar and fodder beet, soybean, rape, cotton, maize
pat encoding PPT acetyl transferase	Tolerance to glufosinate ammonium	Maize, soybean, rice, sugar beet, rape
oxy encoding nitrilase	Tolerance to oxynil herbicides	Cotton, rape
csl-1 encoding an acetolactose synthase	Tolerance to sulphonylurea	Cotton, flax
Modified *als* genes encoding an acetolactose synthase	Tolerance to imidazolines	Maize, rape
Male sterility		
barnase encoding a ribonuclease	Male sterility (pollen)	Maize, rape
barstar encoding a ribonuclease inhibitor	Fertility restorer	Maize, rape
Modified composition		
Sense suppression of *gmFad2-1* encoding a δ-12 desaturase	Increased content of oleic acid	Soybean
Antisense suppression of *gbss* (granule-bound starch synthase)	High amylopectin starch	Potato
Bay *te* encoding 12:0 ACP thioesterase	Increased content of lauric and myristic acids	Rape

Notes: Some of the plants included in the table have yet to be released on the market.* Now removed from the market.

Source: Adapted from Aumaitre et al. (2002).

A major factor in determining future demand for non-GM/GM varieties will be the feed market to which the bulk of all maize and soybean is destined. Establishing an assessment of the safety of GM products used as animal feed that is universally accepted and applied and which can be shown to be rigourous in its approach is fundamental to retaining consumer confidence in animal products.

Assessment of genetically modified feedstuffs

Characterisation

The concept of substantial equivalence forms a useful conceptual basis for a safety assessment, and is as relevant to feed issues as to those of foods. However, the choice of comparators (see below), the key characteristics selected and interpretation of compositional data can differ among different authorities. The OECD consensus documents on common plants that are used for food and feed provide a valuable source of information and can be used to ensure a consistency of approach (www.oecd.org/biotrack). These documents delineate key nutrients, anti-nutrients and toxins contained in common food/feed plants and their products and by-products from common manufacturing processes that are used for food and feed purposes.

As indicated above, the characterisation of the host plant, the molecular characterisation of the donor genetic elements, the expression of the novel traits and the impact of these in the newly modified plant are well-established elements in the safety assessment of both GM food and GM feedstuffs. While it is not the intention to duplicate what has been extensively covered within the context of GM food safety assessment, animal feedstuffs make use of plants and plant parts not directly consumed by humans. Often these plant parts are not used for human food but are consumed by livestock. In addition, the nature of the genetic modification may be of relevance only to livestock feeding as in the case of forages. As a consequence, although the principles underpinning the safety assessment of food and feed may be similar, they may differ in detail and emphasis.

By-products and plant parts versus the whole plant

Feed manufacturers make considerable use of by-products from other industries using GM plants in which the introduced DNA/novel protein may be virtually absent or, as in the case of seed meals, considerably concentrated (Table 1.2). This has implications for levels of exposure, choice of comparators and for determining the concentration of novel protein used in acute/sub-chronic toxicity studies made with newly introduced and expressed proteins.

Table 1.2. **Typical protein, oil and cell wall contents (g/kg dry matter) of maize kernel and its by-products of processing fed to animals**

Fraction	Protein	Oil	Cell wall (NDF)[1]
Whole kernel	102	42	117
Germ meal	108	64	224
Gluten feed	220	51	383
Gluten meal	669	69	84
Fibre	147	42	538

Note: 1. Neutral detergent fibre.

Consideration also should be given to the differential (spatial and temporal) expression of introduced traits in the plant, particularly when plant parts not used for food purposes are included in animal feed (e.g. maize stover, cottonseed meal). Promoters may be chosen to preferentially express a trait in a given part of the plant prone to pathogen attack or, conversely, to reduce or avoid expression in those parts consumed as human food. For example, a GM plant may be constructed to express the Bt toxin only (or preferentially) in parts of the plant, such as the leaves, which are subject to first generation insect attack (Table 1.3). While such an approach may serve to reduce human exposure, the exposure of animals that consume most parts of a plant may be substantially increased. As Table 1.3 shows, dairy cattle consuming maize stover (aerial vegetation) have a substantially greater exposure to the *Cry1A(b)* protein than animals fed only maize kernel.

Table 1.3. **Concentration of *Cry1A(b)* protein (µg/g fresh weight tissue) in YieldGard™ (event MON810) hybrid maize**

Plant tissue	Parameter	1994 United States (6 sites)	1995 United States (5 sites)	1995 EU (4 sites)	1996 EU (3 sites)
Leaf[1]	Mean	9.35	8.95	8.60	12.15
	Standard deviation	1.03	2.17	0.74	3.86
	Range	7.93-10.34	5.21-10.61	7.59-9.39	7.77-15.06
Forage/whole plant[2]	Mean	4.15	3.34	4.80	4.88
	Standard deviation	0.71	1.09	0.75	0.52
	Range	3.65-4.65	2.31-4.48	4.11-5.56	4.32-5.34
Kernel	Mean	0.31	0.57	0.53	0.41
	Standard deviation	0.09	0.21	0.12	0.06
	Range	0.19-0.39	0.39-0.91	0.42-0.69	0.35-0.46

Notes: 1. The mean was calculated from the analyses of plant samples from each field site. 2. For the 1994 US trials, values represent the analysis of whole plants; for the remaining trials, values represent the analysis of forage tissue. Whole plants were collected two weeks after pollination; forage samples were collected at the soft dough or early dent stage. Means were determined from the analysis of plant samples from one site in the United States and all sites in the European Union. A plant sample was a pool of two individual plants.

Source: Sanders et al. (1998) with additional data on standard deviation and range provided by the authors.

Studies intended to demonstrate the safety of the product of any introduced gene should take account of the maximum level found in any plant part consumed by animals or in any by-product used as a feed ingredient. A margin of safety then should be established based on this value. This should be done regardless of the frequency of use of the plant part or by-product, or the potential for disruption of protein during any extraction process.

Whilst it is desirable to include unprocessed plant material in studies intended to demonstrate tolerance to, or the wholesomeness of, the GM plant, this may not always be possible. Some seeds, such as soybean, are processed before use because of the presence of known anti-nutritional factors (see OECD, 2001). In such cases, the processed seed should be substituted to avoid the possibility of any adverse effects being masked by the effects of the known anti-nutrients or toxicants. Extrapolation between plant parts is possible, but studies should reflect the botanical nature of the feed and, if necessary, separate studies should be made with seeds and vegetative material. Consideration should also be given to by-product streams in which protein is concentrated or in which lipophilic or hydrophilic metabolites could accumulate. Where a variety of such by-products are produced, it may be necessary to include in studies only those at the extreme (e.g. maize gluten meal in preference to gluten feed).

Fate of DNA and protein in animal feeding

The vast majority of proteins in feeds are not known to present any safety hazards to animals and only when DNA *per se* is consumed in high concentrations are the breakdown products of nucleic acid hazardous to humans (Simmonds, 1990). However, the introduction of GM plants into the food chain has rekindled interest in the fate of DNA in the digestive tract. The use of sensitive molecular biological techniques not previously available has demonstrated that DNA can survive in polymeric form to a far greater extent than was previously recognised and that DNA fragments can be taken up both by host tissues (Schubbert et al., 1994, Hohlweg and Doerfler, 2001) and the resident microflora (Mercer et al., 2001). The particular issue of the possible transfer of functional DNA to microorganisms has become associated with recombinant technology, largely because of the use of genes coding for antibiotic resistance as a means of selection.

Survival of DNA/protein during the harvest and storage of feedstuffs

Grain harvested at maturity generally has a relatively low moisture content and can be stored without further treatment until required for use. Little degradation of either protein or DNA occurs under ordinary storage conditions. Vegetative parts of the plant and whole plants harvested before grain maturity have higher moisture content and require further treatment to ensure their stability unless used immediately for grazing animals. Stabilisation may simply rely on air drying (more rarely artificial drying) over a period of days to produce hay or haylage. Ensiling is the preferred method of conservation for plants with a high moisture content and/or high soluble carbohydrate content and the silage so produced is typically used for the feeding of ruminants. Microbial fermentation of soluble sugars and protein rapidly reduces the pH to a point when all further microbial growth is inhibited. Enzymes, microorganisms, organic acids or added molasses may be added to encourage a rapid development of an acid-producing flora.

Autolysis of protein begins immediately after cutting and is retarded by rapid wilting and enhanced by slow wilting. Protein losses continue during ensiling to an extent highly dependent on the ensiled material, the microbial flora that develops and the rate of pH reduction (Fairburn et al., 1988). *Cry1A(b)* protein could not be detected in maize silage prepared from *Cry1A(b)* expressing plants (Fearing et al., 1997). DNA appears less affected than protein and intact DNA with no evidence of lower molecular weight degradation products has been extracted from wide variety of harvested vegetative materials and from ryegrass and maize silages (Chiter et al., 2000; Table 1.4). Single gene studies reflect the studies made on total DNA stability. The *Cry1A(b)* gene was detected in maize seven months after ensiling by amplification of a 211 bp sequence (Hupfer et al., 1999). Similarly *rubisco SS*, a plant plastid gene, could be readily amplified from maize silage (Chiter et al., 2000).

Intact DNA and protein in crops conserved by air drying or ensiling or in pulps obtained by low-temperature aqueous extraction (e.g. sugar beet) can be detected throughout the normal duration of storage (Chiter et al., 2000). Unless subject to some other form of processing, which is unlikely in the case of silage or hay, DNA and protein from such sources are consumed by the animal largely in an intact form.

Survival of DNA/protein during feed manufacture

Manufactured feeds are subject to shear forces and heat treatments of varying severity (pelleting, extrusion, expansion, etc.). However, manufacturing processes generally are optimised to protect the nutritional value of the protein in the feed, and to avoid breakdown products and the formation of adducts. Some enzyme additives can partially survive pelleting at 90°C in an active form and are little affected by lower temperatures

(Samarasinghe et al., 2000). This general protection of proteins against damage during processing is usually ascribed to the other organic fractions of the feed. Consistent with this view is the detection using enzyme-linked immunosorbent assay (ELISA) of the resistant version of the excitatory postsynaptic potential (EPSP) protein, which confers herbicide resistance, in extracted GM soybean meal (Ash et al., 2000). Since the detection method was antibody-based, this implies the considerable retention of structural integrity.

Grinding and dry milling have little effect on DNA structure unless they are accompanied by localised heating. More severe commercial treatments, particularly those involving heat and/or chemical extraction, invariably lead to disruption of DNA structure (Smith et al., 2003). Only highly fragmented DNA could be recovered from oilseed meals following chemical and mechanical extraction of the oil (Chiter et al., 2000). Similarly, no intact DNA could be found in the extensively processed by-products of the maize wet-milling industry (Table 1.4). This was confirmed in a separate study, where although specific nuclear and plasmid genes could be detected by polymerase chain reaction (PCR) in the wet gluten and germ fractions, after heat drying the DNA fragments were further degraded and could no longer be detected (Gawienowski et al., 1999).

Table 1.4. **Degree of damage to DNA recovered from commercially sourced feed ingredients**

Commodity	Number of samples examined	Degree of damage
Linseed	5	Intact
Linseed – expelled	2	Degraded
Soybean	8	Intact
Soybean – extracted	7	Degraded
Maize kernel	2	Intact
Forage maize	2	Intact
Maize silage	4	Intact
Maize gluten meal	2	Degraded
Rapeseed	3	Intact
Rapeseed – extracted	3	Degraded
Rapeseed – expelled	3	Degraded

Source: Forbes et al. (2000).

Survival of DNA/protein in the digestive tract

Most ingested proteins are rapidly degraded by proteases, primarily of host origin in the case of non-ruminants and of microbial and host origin in ruminants. Tests made with simulated gastric and intestinal conditions have confirmed that, with a single exception, protein products of the genes introduced into current commercial crops (see Table 1.1) are as rapidly degraded as other dietary proteins (Noteborn et al., 1994; Harrison et al., 1996; Wehrmann et al., 1996). The exception is the product of *cry9C*, a bacterial lectin, which, in common with a sub-group of other plant and microbial lectins and protease inhibitors, is highly resistant to proteolysis (EPA, 1998).

Evidence of the degradation of protein during feed preparation should not automatically be assumed to confer safety. However, the degree of protein degradation occurring during feed processing, if applicable, can add to an established margin of safety. Digestion by livestock of any introduced protein also should be considered in the safety assessment with regard to its impact on the animal and any consequences for consumers of livestock products. While *in vitro* methods to mimic the conditions found in the digestive tract of non-ruminant species are well established, they are less suitable for ruminants. To determine whether a protein will be degraded in ruminants, different methods exist. Rumen fluid may be obtained from fistulated ruminants; however, the properties of this fluid are diet-dependent (Tilley and Terry, 1963; Goering and Van Soest, 1970). Alternatively, to simulate the proteolytic activity of ruminal microorganisms, a protease extract from *Streptomyces griseus* may be used (Mathis et al., 2001). Regardless of susceptibility to breakdown during processing, all introduced and expressed protein should be separately examined for its toxic potential. This could involve a search for any structural resemblance to known toxins and/or animal studies.

Naked DNA/RNA released from the food matrix is readily degraded in most compartments of the gastro-intestinal tract. The longest periods of survival have been observed in the presence of saliva and in the oral cavity, where DNA has been detected after several hours (Mercer et al., 1999, 2001; Duggan et al., 2000). Elsewhere, while fragments capable of amplification may be detected for up to 30 minutes, any biological activity is extremely short-lived (Duggan et al., 2000). Most experiments have been carried out under artificial conditions and with naked DNA with the intention of demonstrating a capacity for transfection. While survival on release also may be very short *in vivo*, there is likely to be a constant leaching of DNA into the gut lumen as the feed matrix is disrupted. The *rubisco* gene, or at least an amplifiable fragment of the gene, could be recovered from the intestinal content of mice up to 49 hours after feeding and for a further 70 hours from the caecum (Hohlweg and Doerfler, 2001). Similarly, a 1914 bp DNA fragment containing the entire coding region for *cryIA(b)* was still amplifiable from the rumen fluid of sheep five hours after consumption of GM maize grain, although not from sheep fed silage prepared from the same maize line (Duggan et al., 2003).

Uptake of DNA by the microflora of the gastro-intenstinal tract

Transformation represents the most general and likely mechanism for the acquisition by gut bacteria of DNA released from feed. However, in one of the few attempts to demonstrate this transfer *in vivo*, a β-lactamase introduced into maize and conferring resistance to ampicillin, could only be detected in association with plant material and not with other intestinal contents or in faeces when fed to chicken (Chambers et al., 2002). The survival in the gut of the antibiotic marker gene mirrored that of other plant DNA targets and could be detected only in the crop and stomach.

Detection of transgenic DNA and protein in animal products

Following the work of Schubbert and colleagues (Schubbert et al., 1994, 1997, 1998; Hohlweg and Doerfler, 2001) the expectation is that DNA fragments of plant origin will be found in the tissues of farm animals, particularly in peripheral lymphocytes and the liver. While fragments of plastid encoded genes are far more likely to be detected than nuclear genes because of their copy number, the same principle(s) determining survival and uptake appear to apply. Whether any particular gene (including a transgene) is detected in tissues thus will be largely a product of the sensitivity of the detection method.

As expected, amplifiable fragments of plant DNA have been detected in animal tissues (Klotz and Einspanier 1998; Hohlweg et al., 2000; Einspanier et al., 2001) and in a variety of animal products including milk, although not in eggs. Fragments of transgenic DNA have yet to be detected in the major animal products (Table 1.5). In addition to the data introduced into the public arena, a number of other similar experiments have been completed by plant-breeding companies. No transgene (or its expressed product) has been detected in any animal product examined to date.

Table 1.5. Examination of animal products for the presence of transgenic DNA or protein

Host animal	Genetically modified plant	Tissues examined	Outcome
Ruminants			
Dairy cows[1]	Herbicide tolerant soybean	Blood, milk	Transgene not detected
Dairy cows[2]	Bt maize	Blood, milk, digesta, faeces	Transgene detected only in digesta
Beef steer[2]	Bt maize	Blood, muscle, liver, spleen	Transgene not detected
Dairy cows[3]	Bt maize (whole plant)	Milk	Transgene and *Cry1A(b)* protein not detected
Dairy cows[4]	Herbicide tolerant soybean	Milk	Transgene (*epsps*) not detected
Dairy cows[5]	Bt maize kernel	Milk	Transgene not detected
Poultry			
Chickens[2]	Bt maize	Muscle, liver, spleen, kidney, eggs	Transgene not detected
Laying hens[6]	Herbicide tolerant soybean	Eggs, liver, faeces	Transgene not detected
Broiler chickens[7]	Herbicide tolerant soybean	Muscle, skin, liver	Transgene not detected
Broiler chickens[3]	Bt maize	Breast meat	Transgene and *Cry1A(b)* protein not detected
Laying hens[3]	Bt maize	Eggs, liver, white and dark meat	*Cry1A(b)* protein not detected
Broiler chickens and laying hens[8]	Bt maize	Digesta, meat and eggs	Transgene detected only in feed, maize DNA detected in digesta and liver
Broiler chickens[9]	Bt maize	Blood, liver and muscle	Transgene (*Cry 9c*) not detected
Pigs			
Grower-finisher pigs[10]	Bt maize	Loin meat	Transgene and *Cry1A(b)* protein not detected
Grower-finisher pigs[11]	Bt maize	Blood, muscle, liver, spleen, lymph nodes	Transgene not detected
Grower-finisher pigs[12]	Bt maize	Blood, muscle, liver, spleen, lymph nodes, ovaries	Transgene not detected

References: 1. Klotz and Einspanier (1998); 2. Einspanier et al. (2001); 3. Faust (2000); 4. Phipps et al. (2002); 5. Phipps et al. (2001); 6. Ash et al. (2000); 7. Khumnirdpetch et al. (2001); 8. Aeschbacher et al. (2001); 9. Japan MAFF (2001); 10. Weber and Richert (2001); 11. Klotz et al. (2002); 12. Reuter and Aulrich (2003).

Existing knowledge of the metabolic processes involved in the digestion, absorption and utilisation of amino acid and peptides by livestock species does not wholly preclude the incorporation of intact proteins into animal products although it suggests it to be unlikely. Generally, proteins are synthesised *de novo* from an amino acid pool. Studies made of the supply of amino acids to the mammary gland, for example, have shown that the bulk of milk proteins are synthesised *in situ* from single amino acids and some small peptides. However, immunologic (IgG) proteins are taken up from the blood supply (Whitney et al., 1976). Uptake is receptor mediated making it unlikely that any ingested protein surviving to be detected in serum would have the physical characteristics necessary for absorption. Egg proteins are generally synthesised in the liver and transported as specifically tagged lipoproteins. Thus, it would be exceptionally unlikely for an expressed protein of any plant gene to be found intact in meat, milk or eggs and none have been detected to date (Table 1.5).

Daily exposure by mammals to fragments of food plant and microbial DNA that results in their random incorporation into the nucleus of somatic cell populations has no recognised long-term consequences. There is no basis to suppose that transgenic DNA poses hazards any different to other sources of DNA.

Animal feeding studies as part of a safety assessment

Unlike foods specifically for human use, GM plants can be fed directly, often in raw or minimally processed form, to the target species and the growth, health and welfare of the animal monitored, or the absorption and tissue distribution of particular metabolites measured. Limits to the amount of any one component that can be incorporated into an animal ration usually prevent chronic toxicity studies from being made at a relevant dose level with the target species, as is the case with whole-food testing on the human side. The "wholesomeness" of the product based on nutritional efficacy, however, can be directly demonstrated.

Most conventional varieties of maize, soybean and other feedstuffs are introduced to the market primarily on the basis of their composition. Experience has shown that nutritional value can be predicted with sufficient accuracy from compositional data to make a feeding trial unnecessary.

The traits introduced into most existing commercial GM varieties (see Table 1.1) are agronomic in character and have little or no effect on feed composition or the bioavailability of nutrients. Consequently, the gross composition of such GM varieties also falls within the range normally associated with conventional varieties of the same feedstuff and they would be expected to behave as any other variety. Many feeding studies have been made to test this assumption (Table 1.6). There is no evidence to date from such studies to suggest that the performance of animals fed the GM feed ingredient differed in any respect from those fed the non-GM counterpart or from the performance predicted by the composition of the feed (Faust, 2002). This suggests that for those GM plants with modified input traits, provided that compositional analyses demonstrate no meaningful differences from the comparator(s) nutritional equivalence can be assumed. For such GM varieties, routine feeding studies made with target species will add little to a safety assessment and generally are not warranted (see below).

Value of feeding trials with nutritionally modified feeds

Plants modified with the intention of significantly changing their composition and thus their nutritional characteristics may present added issues for a safety assessment. Provided the introduced changes affect only composition and not the bioavailability of individual nutrients (e.g. increased water-soluble carbohydrate in forages), or if the by-products fed are essentially free of the modified component (e.g. seedmeals left after modified oil extraction), valid nutritional comparisons with conventional varieties can still be made. This may involve augmenting the comparator diet to match the composition of the GM variety.

If the modification is expected to substantially change the bioavailability of a component (e.g. amylopectin-rich starch) then a suitable comparator for feeding studies cannot be designed on the basis of composition alone. In such cases it may be possible only to conduct feeding trials with one or more of the major target species to demonstrate wholesomeness for the animal. Under these circumstances, the duration of feeding studies should be for the production cycle of the animal.

Traditionally, other types of studies (e.g. determination of metabolisable energy [ME] value, balance study to demonstrate improved nitrogen retention) have been utilised to evaluate non-GM feedstuffs and could confirm that the intended modification produces the expected outcome.

Table 1.6. **Summary of reported studies made with livestock fed genetically modified feed in comparison to conventional feed**

Animal species	Genetically modified plant	Outcome
Ruminants		
Dairy cows[1]	Herbicide-tolerant soybean	No significant differences in milk production or composition but fat corrected milk (FCM) greater (P<0.05) in genetically modified groups.
Dairy cows[2]	Bt maize (chopped plants)	No differences in milk production or composition.
Sheep, beef cattle[3]	Bt maize silage	No significant differences in digestibility, weight gain or dry matter intake (DMI).
Dairy cows[4]	Bt maize silage	No significant differences in DMI, milk performance or milk composition.
Dairy cows, sheep[5]	Bt maize and maize silage	No significant difference in performance.
Sheep[6]	Herbicide tolerant sugar beet and leaf silage	No significant differences in the digestibility of nutrients observed.
Dairy cows[7]	Herbicide tolerant maize and maize silage	No significant differences in DMI, milk production or milk composition.
Dairy cows[8]	Bt maize and maize silage	No significant difference in performance.
Dairy cows[9]	Bt maize + Bt maize silage	No significant differences in DMI, milk performance or milk composition.
Beef cattle[10]	Bt maize, maize residues and maize silage	Silage: differences (P<0.05) seen between hybrids not consistently related to Bt. Residue: no significant differences.
Beef cattle[11]	Bt maize silage and maize residues	Silage: no significant difference in average daily gain (ADG) and DMI, feed:gain ratio greater in Bt (P<0.05). Residues: no significant differences.
Dairy cows[12]	Herbicide-tolerant cottonseed, Bt cottonseed	No significant differences in body condition, milk yield or milk composition.
Beef cattle[13]	Herbicide-tolerant maize	No significant differences in growth performance, carcass characteristics or meat composition.
Beef cattle[14]	Herbicide-tolerant maize (two events)	No significant differences in growth performance or carcass characteristics.
Dairy cows[15]	Herbicide-tolerant maize/ maize silage	No significant differences in milk production or composition.
Dairy cows[16]	Herbicide-tolerant maize/ maize silage	Intake of genetically modified maize silage and milk production significantly reduced (P<0.05). Ascribed to differences in silage quality.
Sheep[17,18]	Herbicide tolerant sugar/ fodder beet	No significant effects on feeding value.
Sheep[19]	Herbicide tolerant canola	No significant effects on digestibility or growth performance.
Pigs		
Pigs[20]	Herbicide tolerant maize	No significant effects on nutrient digestibility.
Pigs[21]	Bt maize	No significant differences in DMI or weight gain.
Pigs[22]	Bt maize	No significant difference in nutrient digestibility.
Pigs[23,24]	Bt maize	No significant differences in nutrient digestibility, DMI or weight gain.
Piglets[25]	Bt maize	No significant effects on feed:gain ratio, but ADG significantly increased in Bt-maize fed piglets.*
Growing pigs[26]	Herbicide tolerant maize, Bt maize	No significant differences in digestible energy (DE) compared to near isogenic parent lines.
Growing pigs[27]	Herbicide tolerant soybean	No significant differences in performance parameters or carcass measurements. Sensory scores not significantly influenced by diet.
Pigs[6]	Herbicide tolerant sugar beet	No significant effects on nutrient digestibility.
Pigs[28]	Bt maize	No significant differences in growth compared to near isogenic control.
Growing-finishing pigs[29]	Herbicide-tolerant maize	No significant differences in growth performance or carcass measurements.

Table 1.6. **Summary of reported studies made with livestock fed genetically modified feed in comparison to conventional feed** (*cont.*)

Animal species	Genetically modified plant	Outcome
Other species		
Catfish[1]	Herbicide-tolerant soybean	No significant differences in survival or feed:gain ratio. Weight gain/final weight greater ($P<0.05$) in one of two GM groups.
Rabbits[38]	Rapeseed	No significant differences observed.
Rabbits[39]	Bt maize	No significant differences observed.

Notes: * Suggested by authors that improved performance in Bt maize fed groups due to a lower fumonisin B1 content.

References: 1. Hammond and Padgette (1996); 2. Faust and Miller (1997); 3. Daenicke et al. (1999); 4. Rutzmoser et al. (1999); 5. Barriere et al. (2001); 6. Böhme et al. (2001); 7. Donkin et al. (2000); 8. Faust (2000); 9. Folmer et al. (2000a); 10. Folmer et al. (2000b); 11. Hendrix et al. (2000); 12. Castillo et al. (2001); 13. Simon et al. (2002); 14. Berger et al. (2002); 15. Ipharraguerre et al. (2002); 16. Grant et al. (2002); 17. Hvelplund and Weisbjerg (2001); 18. Weisbjerg et al. (2001); 19. Stanford et al. (2002); 20. Böhme and Aulrich (1999); 21. Weber et al. (2000); 22. Aulrich et al. (2001); 23. Reuter et al. (2001); 24. Reuter et al. (2002); 25. Piva et al. (2001); 26. Gaines et al. (2001a); 27. Cromwell et al. (2002); 28. Weber and Richert (2001); 29. Fischer et al. (2002); 30. Aulrich et al. (1998); 31. Brake and Vlachos (1998); 32. Halle et al. (1998); 33. Mireles et al. (2000); 34. Sidhu et al. (2000); 35. Kan et al. (2001); 36. Aeschbacher et al. (2001); 37. Gaines et al. (2001b); 38. Maertens et al. (1996); 39. Chrastinová et al. (2002).

Source: Modified from Flachowsky and Aulrich (2001).

Detection of unintended effects of transformation

The degree of equivalence to existing varieties is established on the basis of a comparison of compositional and agronomic data. However, even when this is considered "substantial", there remains a remote possibility of unintended effects of transformation in the plant not detected by a targeted chemical analysis or as a change in growth characteristics of the plant.

Unintended effects occurring during the development of new plant varieties are not unique to those produced using recombinant DNA technology but have the potential to occur in all forms of plant breeding. There is, at present, no reason to suppose that the incidence of unintended effects is significantly greater when recombinant DNA methods are used. Transgene integration occurs in plants through the same illegitimate recombination mechanisms that allow the chromosomal recombinations that are the basis for conventional plant breeding (Gelvin, 2000). Since no sequence homology is required, no sequences in the genome appear specifically favoured for integration. Consequently, it is not presently possible to predict the site of integration of transgenes into the host genome. However, both chromosomal recombination and transgene integration appear to occur more frequently in gene-rich regions, increasing the likelihood of mutations caused by disruption of gene function (Barakat et al., 2000).

If other studies (e.g. compositional, agronomic) indicate that unintended effects may have occurred, consideration could be given to the use of comparative growth studies as a means of investigating such effects. Fast-growing species such as the broiler chick increase their body weight approximately 45-fold during the approximately 40 days they take to reach market weight. Because of this rapid weight gain, broilers are particularly sensitive to any change in nutrient supply or the presence of toxic elements in their feed. Consequently, growth rate studies made with broilers can be used to examine GM products for unintended changes provided that they can be nutritionally matched to a parental line or other suitable control and that they are suitable for inclusion in broiler diets. Broilers have advantages over many other species used in commercial production as they tend to provide a genetically homogeneous population and can be used in relatively large numbers to increase the statistical power of the experiment.

The young of other livestock tend not to show such rapid growth rates, but will, on occasions, form a more appropriate model. For feedstuffs intended for aquaculture, extrapolating results to other fish species from a comparative growth study made with a fish species such as the catfish, is preferable to an extrapolation from results obtained with broilers. Also, the presence of a known toxicant may restrict feeding studies only to those animals known to tolerate the compound. Gossypol limits the use of cottonseed meal in animals other than ruminants. In this case, milk production might substitute for growth rate and could be used to screen feedstuffs intended for use with ruminants.

Before such studies could become routine, standardised and internationally recognised designs for such tests would need to be established. In particular, the number of animals and the statistical degree of confidence necessary to conclude unintended effects presently need to be specified. Large numbers of animals may be required to achieve appropriate statistical power. It must be recognised that these studies cannot provide definitive evidence of safety but may contribute to the safety assessment as a whole.

In time, it may be possible to detect unintended effects by using non-targeted profiling techniques based on the measurement of the transcriptome or proteome (see below). Alternatively, measures to detect unintended effects may be rendered unnecessary by improved molecular characterisation and understanding of the implications of the molecular events for the metabolism of the plant.

Non-targeted profiling

Non-targeted methods intended to profile gene expression, a significant proportion of the proteome (proteomics) or metabolite production (metabolomics) are being considered as supplements to the targeted methods currently used in establishing the degree of equivalence with parental lines/existing varieties (Fiehn et al., 2000). The rapid technical developments in the measurement of the transcriptome or proteome may not immediately allow fully validated methods to be established.

Post-market surveillance/monitoring

Post-market surveillance may be a more practical proposition when undertaken in relation to animal feed than to food since intake is accurately known and recorded and individual animals can be routinely monitored for health. The various "quality assured schemes" also can allow accurate tracing of animal products to their point of sale. However, given the lack of a theoretical basis for the general transfer of functional protein or DNA to animal products and in the absence of any documented adverse response to products of animals fed GM feedstuffs, post-market surveillance of consumers appears to be of very limited value.

Surveillance of animals, particularly the longer lived species, may be useful where the long-term clinical effects of an introduced trait can only be surmised from short-term biological studies. Similarly, it may be advantageous to augment toxicological studies in laboratory animals when livestock are exposed to a novel gene product with multiple biological activities. However, as with human studies, post-market monitoring does not provide a gold standard in safety assurance, nor does it substitute for other components of the assessment. Such large-scale studies may be confounded by many factors, such as environment and management (Byers, 1998), although these can be partially offset by the inclusion of a carefully selected control population. Post-market surveillance should be seen as a supplement to the assessment scheme. The specific purpose for post-market surveillance should be clearly stipulated before the surveillance is initiated.

By-products of industrial crops

Recombinant technology has greatly expanded the opportunities for the use of plants for the production of high-value products, particularly peptide- and protein-based therapeutics (Fischer and Emans, 2000; Walmsley and Arntzen, 2000; Daniell et al., 2001). Expression is generally targeted to the seed and to the major seed-producing commodity crops, but there are examples of expression in chloroplasts engineered by plastid transformation (Staub et al., 2000). Although many constructs have been produced, generally, expression levels are considered too low for commercial exploitation. To date, only a few have progressed to field and clinical trials although diagnostic kits based on plant-expressed products are on the market.

Intermediate value products such as enzymes have also been introduced into plants used for feed purposes and with some varieties there is the option of extracting the enzyme protein for food use or using the product intact as a feed ingredient (Jensen et al., 1996; Denbow et al., 1998). Bulk chemical production intended to provide low-cost feedstock has focused on modifying oil production in oilseeds, particularly rapeseed. One of the first GM plants to obtain release in the United States was an oilseed rape modified to produce high concentrations of lauric acid in the oil for food use, with the meal being used for feed. Subsequently, other modified rape varieties have undergone field trials (Murphy, 1996; Napier and Michaelson, 2001), although none are in commercial production.

If the industrial use of crops modified to express high-value pharmaceutical agents or bulk feedstock for the chemical industry becomes commonplace, then this raises serious issues of feed security. Usually the cost-effective mechanism for the disposal of residues left after extraction of seeds or vegetative material is into the animal feed chain. While high-value, low-volume products could absorb the costs of alternative forms of disposal, this is less likely to be the case for lower value bulk products.

Although disposal of hazardous by-products can be seen primarily as a problem of risk management, there is a persuasive argument for completing a safety assessment for all parts of industrial GM plants that could, and if from a conventional plant would, enter the food chain. A full safety assessment made on all plants used for bulk production and their by-products would inform the risk manager and allow actions proportionate to the risk to be taken. This might range from allowing some by-products left after extractions of the modified oil or other primary product to be used as a feed ingredient to a refusal to allow feed use due to the risk that it would contain residual material of a highly toxic nature.

Alternatively, if measures have been implemented to prevent the entry of plants producing pharmaceuticals products into the feed supply, a safety assessment of that plant may not be necessary. The processes needed to ensure that the material does not become a component of feed should be proportional to the associated risk.

Agronomic versus quality traits: Future genetically modified feedstuffs

Transformation events described in the scientific literature address a wide variety of issues including the nutritional, organoleptic and shelf-life of food plants; the expression of plant compounds impacting on public health; and the ability of crop plants to resist stress and to grow in more marginal areas. Taking applications for field trials as an indication of varieties likely to be at the forefront of those seeking regulatory approval, it is evident that agronomic (input) traits will continue to dominate new introductions for some time to come.

An important input target applied to a wide range of plant species including those used as feeds, has been the development of alternatives to Bt toxins able to offer protection against a broader range of insect pests. Like the Bt endotoxins, most transgenics of this type have introduced genes coding for proteins targeting some aspect of insect gut function. These include a variety of plant-derived lectins and digestive enzyme inhibitors (Schuler et al., 1998). Of these, the snowdrop lectin (*Galanthus nivalis* agglutinin) has received the greatest attention and has been successfully expressed in many different crops including cereals (Rao et al., 1998). Feed assessments of plants that have had these compounds incorporated should take into account that protease and amylase inhibitors and some lectins are recognised as anti-nutritional factors and processing of feeds is often required to ensure their removal from the finished feed or feed ingredient. Plant enzyme inhibitors belonging to a general class of defence protein, some of which (e.g. soybean Kunitz-type trypsin inhibitor) are recognised as cross-reacting allergens (Mena et al., 1992) and are highly resistant to digestion in the gut.

Major cereals, other than maize, have proved recalcitrant to transformation, delaying the introduction of transgenic varieties. However, considerable progress has been made during the last decade and *Agrobacterium*-mediated gene transfer has been added to the biolistic, electroporation and polyethylene glycol-induced methods developed previously (Ingram et al., 2001). Wheat and barley have particular implications for feed use. Any new transgenic varieties are first likely to parallel the work done with maize and soybean and focus on herbicide tolerance. Only thereafter are pest resistance and feed quality issues likely to be addressed commercially.

Transgenic plants addressing quality issues of importance to animal feeding are assumed likely to be included amongst the next "generation" of transgenic varieties. If so, they will be one of two types: those involving modifications to the composition of plants important in manufactured compound feed (essentially seeds), and those involving forages (essentially vegetation).

Seed proteins of both legumes and cereals are considered, from a nutritional standpoint, to have a less than ideal amino acid composition; legumes being considered deficient in sulfur amino acids and cereal grains deficient in lysine and threonine. Introduction of novel seed proteins that have more desirable amino acid profiles (Saalbach et al., 1994; Molvig et al., 1997) or down-regulating one or more proteins with less desirable characteristics (Kohnomurase et al., 1995) have resulted in beneficial changes in amino acid profiles. In an alternative approach, circumventing the normal feedback regulation in the biosynthetic pathway for selected amino acids has been shown to increase the concentration of the free acid with evidence of increased incorporation into storage proteins (Galili et al., 1994, 2000; Falco et al., 1995).

Forage quality is not high on the agenda of most plant-breeding companies. Where relevant work has been undertaken, it is in areas such as the modification of lignin biogenesis (Vogel and Jung, 2001) where results have implications for other industries and where forage crops provide valuable, fast-growing models. In comparison, relatively limited work to date has been undertaken addressing other important issues relating to dry matter digestibility (see Herbers and Sonnewald, 1996), protein quality (Bellucci et al., 1997) and nitrogen capture.

The brief forecast above of traits likely to be encountered, which covers both agronomic traits (disease resistance) and an increased emphasis on quality issues, highlights the need for a case-by-case approach to safety evaluation and the probable need to develop specific assessments appropriate to each event. The OECD Workshop on the Nutritional Assessment of Novel Foods and Feeds (February 2001) concluded that the concept of substantial equivalence as a starting point in the assessment process remained a valid tool for assessing novel foods and feeds with nutritionally modified characteristics. It is to be expected that added value plants will require identity preservation to distinguish them from other varieties.

Current legislative process applied to genetically modified feed

The present approach to safety assessment of novel feeds by national authorities ranges from use of existing food legislation (in the case of the United States where feed is considered in legislation to be food), to feed-specific legislation (as in Canada, the Czech Republic and Hungary, among others). Other OECD member countries are in the process of developing new legislation specific to GM foods and/or feeds and to the labelling of GM feeds. The European Union has recently proposed legislation to ensure that dual-use plants (food and feed applications) are simultaneously assessed for safety for both applications.

References

Aeschbacher, K. et al. (2001), "Genetically modified maize in diets for chickens and laying hens: Influence on performance and product quality", *Proceedings: International Symposium on Genetically Modified Crops and Co-products as Feeds for Livestock*, pp. 41-42, September, Nitra, Slovak Republic.

Ash, J.A. et al. (2000), "The fate of genetically modified protein from Roundup Ready® soybeans in the laying hen", *Poultry Science*, Vol. 79, Supplement 1, p. 26.

Aulrich, K. et al. (2001), "Genetically modified feeds in animal nutrition. 1st Communication: *Bacillus thuringiensis* (Bt) corn in poultry, pig and ruminant nutrition", *Archives of Animal Nutrition*, Vol. 54, No. 3, pp. 183-195.

Aulrich, K. et al. (1998), Inhaltsstoffe und Verdaulichkeit von Maiskörnern der Sorte Cesar und der gentechnish veränderten Bt-Hybride bei Legehennen, VDLUFA Kongressband 1998, 110. VDLUFA-Kongress, Giessen, pp. 465-468.

Aumaitre, A. et al. (2002), "New feeds from genetically modified plants: Substantial equivalence, nutritional equivalence, digestibility, and safety for animals and the food chain", *Livestock Production Science*, Vol. 74, No. 3, April, pp. 223-238, http://dx.doi.org/10.1016/S0301-6226(02)00016-7.

Barakat, A. et al. (2000), "The distribution of T-DNA in the genomes of transgenic *Arabidopsis* and rice", *FEBS Letters*, Vol. 471, Issues 2-3, April, pp. 161-164, http://dx.doi.org/10.1016/S0014-5793(00)01393-4.

Barriere, Y. et al. (2001), "Feeding value of corn silage estimated with sheep and dairy cows is not affected by genetic incorporation of the Bt 176 resistance to *Ostrinia nubilalis*", *Journal of Dairy Science*, Vol. 84, Issue 8, August, pp. 1 863-1 871, http://dx.doi.org/10.3168/jds.S0022-0302(01)74627-9.

Bellucci, R. et al. (1997), "Differential expression of a gamma-zein gene in *Medicago sativa*, *Lotus corniculatus* and *Nicotiana tabacum*", *Plant Science*, Vol. 127, No. 2, pp. 161-169, http://dx.doi.org/10.1016/S0168-9452(97)00092-7.

Berger, L.L. et al. (2002), "Effect of feeding diets containing corn grain with Roundup (event GA21 or NK603), control or conventional varieties on steer feedlot performance and carcass characteristics", *Journal of Animal Science*, Vol. 80, Supplement 1, p. 270.

Böhme, H. and K. Aulrich (1999), Inhaltsstoffe und Verdaulichkeit von transgenen zuckerrüben bzw. Körnermais im vergleich zu den isogenen Sorten beim Schwein, VDLUFA Kongressband 1999, 111, VDLUFA – Kongress, Halle/Saale, pp. 289-292.

Böhme, H. et al. (2001), "Genetically modified feeds in animal nutrition. 2nd Communication: Glufosinate tolerant sugar beets (roots and silage) and maize grains for ruminants and pigs", *Archives of Animal Nutrition*, Vol. 54, No. 3, pp. 197-207.

Brake, J. and P. Vlachos (1998), "Evaluation of transgenic Event 176 'Bt' corn in broiler chickens", *Poultry Science*, Vol. 77, No. 5, May, pp. 648-653.

Byers, T. (1998), "Nutrition monitoring and surveillance", in Willett, W. (ed.), *Nutritional Epidemiology, 2nd Edition*, Oxford University Press, New York.

Castillo, A.R. et al. (2001), "Effect of feeding dairy cows with either BollGard, BollGard II, Roundup Ready or control cottonseeds on feed intake, milk yield and milk composition", *Journal of Animal Science*, Vol. 79, Supplement 1, *Journal of Dairy Science*, Vol. 84, Supplement 10, p. 413.

Chambers, P.A. et al. (2002), "The fate of antibiotic resistance marker genes in transgenic plant material fed to chickens", *Journal of Antimicrobial Chemotherapy*, Vol. 49, No. 1, January, pp. 161-164.

Chrastinová, L. et al. (2002), "Genetically modified maize in diets for rabbits", *Proceedings of the Society of Nutrition Physiology*, Vol. 11, p. 195.

Chiter, A. et al. (2000), "DNA stability in plant tissues: Implications for the possible transfer of genes from genetically modified food", *FEBS Letters*, Vol. 481, No. 2, September, pp. 164-168.

Cromwell, G.L. et al. (2002), "Soybean meal from Roundup Ready or conventional soybeans in diets for growing-finishing pigs", *Journal of Animal Science*, Vol. 80, No. 3, March, pp. 708-715.

Daenicke, R., K. Aulrich and G. Flachowsky (1999), "GVO in der Fütterung", *Mais*, Vol. 27, pp. 135-137.

Daniell, H. et al. (2001), "Medical molecular farming: Production of antibodies, biopharmaceuticals and edible vaccines in plants", *Trends in Plant Science,* Vol. 6, No. 5, May, pp. 219-226.

Denbow, D.M. et al. (1998), "Soybeans transformed with a fungal phytase gene improves phosphate availability in broilers", *Poultry Science*, Vol. 77, No. 6, June, pp. 878-881.

Donkin, S.S. et al. (2000), "Effect of feeding Roundup Ready corn silage and grain on feed intake, milk production and milk composition in lactating dairy cattle", *Journal of Animal Science*, Vol. 78, Supplement 1, p. 273.

Duggan, P.S. et al. (2003), "Fate of genetically modified maize DNA in the oral cavity and rumen of sheep", *British Journal of Nutrition*, Vol. 89, No. 2, February, pp. 159-166.

Duggan, P.S. et al. (2000), "Survival of free DNA encoding resistance from transgenic maize and the transformation activity of DNA in ovine saliva, ovine rumen fluid and silage effluent", *FEMS Microbiology Letters*, Vol. 191, No. 1, October, pp. 71-77.

Einspanier, R. et al. (2001), "The fate of forage plant DNA in farm animals: A collaborative case-study investigating cattle and chicken fed recombinant plant material", *European Food Research and Technology*, Vol. 212, pp. 129-134.

EPA (1998), "Data evaluation report", MRID 44258108, Office of Pesticide Programs, Environmental Protection Agency, Washington, DC.

Fairburn, R. et al. (1988), "Proteolysis associated with ensiling of chopped alfalfa", *Journal of Dairy Science*, Vol. 71, Issue 1, pp. 152-158, http://dx.doi.org/10.3168/jds.S0022-0302(88)79536-3.

Falco, S.C. et al. (1995), "Transgenic canola and soybean seeds with increased lysine", *Bio-technology*, Vol. 13, No. 6, June, pp. 577-582.

FAOSTAT (2003), FAO Statistics on-line website, http://faostat.fao.org/, Food and Agriculture Organization, Rome, (accessed in 2003 on the previous website).

FAO/WHO (2001), "Evaluation of allergenicity of genetically modified foods", Report of a Joint FAO/WHO Expert Consultation on Allergenicity of Foods Derived from Biotechnology, 22-25 January, Food and Agriculture Organization, Rome, available at: www.fao.org/3/a-y0820e.pdf.

FAO/WHO (2000), "Safety aspects of genetically modified foods of plant origin", Report of a Joint FAO/WHO Expert Consultation on Foods Derived from Biotechnology, Food and Agriculture Organization, Rome, available at: www.fao.org/fileadmin/templates/agns/pdf/topics/ec_june2000_en.pdf.

FAO/WHO (1996), "Joint consultation report, Biotechnology and food safety", *FAO Food and Nutrition Paper No. 61*, FAO, Rome, available at: www.fao.org/ag/agn/food/pdf/biotechnology.pdf (accessed 28 January 2015).

Faust, M.A. (2002), "New feeds from genetically modified plants: The US approach to safety for animals and the food chain", *Livestock Production Science*, Vol. 74, Issue 3, April, pp. 239-254, http://dx.doi.org/10.1016/S0301-6226(02)00017-9.

Faust, M.A. (2000), "Livestock products: Composition and detection of transgenic DNA/proteins", Proceedings of a Symposium on Agriculture, Biotechnology, Market, ADAS-ASAS ed Baltimore, Maryland.

Faust, M.A. and L. Miller (1997), "Study finds no Bt in milk", *Iowa State University Integrated Crop Management Newsletter*, IC-478, Autumn.

Fearing, P.L. et al. (1997), "Quantitative analysis of CRYIA(b) expression in Bt maize plants, tissues and silage and stability of expression over successive generations", *Molecular Breeding*, Vol. 3, Issue 3, June, pp. 169-176.

Fiehn, O. et al. (2000), "Metabolite profiling for plant functional genomics", *Nature Biotechnology*, Vol. 18, No. 11, November, pp. 1 157-1 161.

Fischer, R. and N. Emans (2000), "Molecular farming of pharmaceutical proteins", *Transgenic Research*, Vol. 9, Issue 4-5, pp. 279-299.

Fischer, R.L. et al. (2002), "Comparison of swine performance when fed diets containing Roundup Ready corn (event NK603), control or conventional corn grown during 2000 in Nebraska", *Journal of Animal Science*, Vol. 80, Supplement 1, p. 224.

Flachowsky, G. and K. Aulrich (2001), "Nutritional assessment of feeds from genetically modified organisms", *Journal of Animal and Feed Sciences*, Vol. 10, Supplement 1, pp. 181-194.

Folmer, J.D. et al. (2000a), "Effect of Bt corn silage on short-term lactational performance and ruminal fermentation in dairy cows", *Journal of Dairy Science*, Vol. 83, p. 1 182.

Folmer, J.D. et al. (2000b), "Utilisation of Bt corn residue and corn silage for growing beef steers", *Journal of Animal Science*, Vol. 78, Supplement 2, p. 85.

Forbes, J.M. et al. (2000), "Effect of feed processing conditions on DNA fragmentation", UK MAFF Report CS0116, London.

Gaines, A.M. et al. (2001a), "Swine digestible energy evaluations of Bt (MON810) and Roundup Ready corn compared to commercial varieties", *Journal of Animal Science*, Vol. 79, Supplement 1, *Journal of Dairy Science*, Vol. 84, Supplement 10, p. 109.

Gaines, A.M. et al. (2001b), "Nutritional evaluation of Bt (MON810) and Roundup Ready corn compared to commercial hybrids in broilers", *Journal of Animal Science*, Vol. 79, Supplement 1, *Journal of Dairy Science*, Vol. 84, Supplement 10, p. 51.

Galili, S. et al. (2000), "Enhanced levels of free and protein-bound threonine in transgenic alfalfa (*Medicago sativa* L.) expressing a bacterial feedback-insensitive aspartate kinase gene", *Transgenic Research*, Vol. 9, No. 2, April, pp. 137-144.

Galili, G. et al. (1994), "Production of transgenic plants containing elevated levels of lysine and threonine", *Biochemical Society Transactions*, Vol. 22, No. 4, November, pp. 921-925.

Gawienowski, M.C. et al. (1999), "Fate of maize DNA during steeping, wet-milling and processing", *Cereal Chemistry*, Vol. 76, No. 3, May, pp. 371-374, http://dx.doi.org/10.1094/CCHEM.1999.76.3.371.

Gelvin, S.B. (2000), "*Agrobacterium* and plant genes involved in T-DNA transfer and integration", *Annual Review of Plant Physiology and Plant Molecular Biology*, Vol. 51, June, pp. 223-256, http://dx.doi.org/10.1146/annurev.arplant.51.1.223.

Goering, H.K. and P.J. Van Soest (1970), "Forage fiber analysis. Apparatus, reagents, procedures, and some applications", *Agriculture Handbook*, No. 379, Agricultural Research Service, United States Department of Agriculture, Washington, DC.

Grant, R.J. et al. (2002), "Influence of glyphosate tolerant (trait NK603) corn silage and grain on feed consumption and milk production in Holstein dairy cattle", *Journal of Animal Science*, Vol. 80, Supplement 1, p. 384.

Halle, I. et al. (1998), Einsatz von Maiskörnern der Sorte Cesar und des gentechnisch veränderten Bt-hybriden in der Broilermast, Proc.5, Tagung, Schweine und Geflügelernährung, Wittenberg, Tagungsband, pp. 265-271.

Hammond, B.G. and S.R. Padgette (1996), "The feeding value of soybeans fed to rats, chickens, catfish and dairy cattle is not altered by genetic incorporation of glyphosate tolerance", *Journal of Nutrition*, Vol. 126, No. 3, March, pp. 717-727.

Harrison, L.A. et al. (1996), "The expressed protein in glyphosate-tolerant soybean, 5-Enolypyruvylshikimate-3 phosphate synthase from *Agrobacterium sp.* strain CP4, is rapidly digested *in vitro* and is not toxic to acutely gavaged mice", *Journal of Nutrition*, Vol. 126, pp. 728-740, http://jn.nutrition.org/content/126/3/728.full.pdf.

Hendrix, K.S. et al. (2000), "Feeding value of whole plant silage and crop residues from Bt or normal corns", *Journal of Animal Science*, Vol. 78, Supplement 1, p. 273.

Herbers, K. and U. Sonnewald (1996), "Manipulating metabolic partitioning in transgenic plants", *TIBTECH*, Vol. 14, pp. 198-205.

Hohlweg, U. and W. Doerfler (2001), "On the fate of plant or other foreign genes upon the uptake in food or after intramuscular injection in mice", *Molecular Genetics and Genomics*, Vol. 265, No. 2, April, pp. 224-233.

Hohlweg, U. et al. (2000), "Fremde DNA überwindet die Darm-Blut-und Plazentaschranke", *Bioforum*, Vol. 3/2000, pp. 120-122.

Hupfer, C. et al. (1999), "The effect of ensiling on PCR-based defection of genetically modified Bt maize", *European Food Research and Technol*ogy, Vol. 209, pp. 301-304.

Hvelplund, T. and M.R. Weisbjerg (2001), "Comparison of nutrient digestibility between Roundup Ready® beets and pulp derived from Roundup Ready® beets and conventional beets and pulps", *Journal of Animal Science*, Vol. 79, Supplement 1, p. 417.

Ingram, H.M. et al. (2001), "Genetic transformation of wheat: Progress during the 1990s into the millennium", *Acta Physiologiae Plantarum*, Vol. 23, Issue 2, pp. 221-239.

Ipharraguerre, R. et al. (2002), "Performance of lactating dairy cows fed glyphosate-tolerant corn (event NK603)", *Journal of Animal Science*, Vol. 80, Supplement 1, p. 358.

James, C. (2002), "Global review of commercialised transgenic crops: 2002", *ISAAA Briefs*, No. 27, ISAAA, Ithaca, New York.

Japan MAFF (2001), "No traces of modified DNA in poultry fed on GM corn", *Nature*, Vol. 409, p. 657.

Jensen, L.G. et al. (1996), "Transgenic barley expressing a protein-engineered thermostable (1,3-1,4)-β-glucanase during germination", *Proceedings of the National Academy of Sciences of the United States of America*, Vol. 93, No. 8, April, pp. 3 487-3 491.

Kan, C.A. et al. (2001), "Comparison of broiler performance when fed B.t., patental/isogenic control or commercial varieties of dehulled soybean meal", *Proceedings: International Symposium on Genetically Modified Crops and Co-products as Feeds for Livestock*, pp. 19-22, September, Nitra, Slovak Republic.

Khumnirdpetch, V. et al. (2001), "Detection of GMOs in the broilers that utilized genetically modified soybean meals as a feed ingredient", Plant & Animal Genome IX Conference, January, San Diego, California (Poster 585).

Klotz, A. and R. Einspanier (1998), "Nachweis von 'novel-feed' im Tier? Beeinträchtigung des Verbrauchers von Fleisch oder Milch ist nicht zu erwarten", *Mais*, Vol. 3, pp. 109-111.

Klotz, A. et al. (2002), "Degradation and possible carry over of feed DNA monitored in pigs and poultry", *European Food Research and Technology*, Vol. 214, Issue 4, April, pp. 271-275.

Kohnomurase, J. et al. (1995), "Improvement in the quality of seed storage protein by transformation of *Brassica napus* with an antisense gene for cruciferin", *Theoretical and Applied Genetics*, Vol. 91, No. 4, September, pp. 627-631, http://dx.doi.org/10.1007/BF00223289.

Maertens, L. et al. (1996), "Digestibility of non-transgenic and transgenic oilseed rape in rabbits", in: *Proceedings of the 6th World Rabbit Congress*, Vol. 1, pp. 231-235, Toulouse, France.

Mathis, CP. et al. (2001), "A collaborative study comparing an *in situ* protocol with single time-point enzyme assays for estimating ruminal protein degradability of different forages", *Animal Feed Science and Technology*, Vol. 93, Issues 1-2, September, pp. 31-42, http://dx.doi.org/10.1016/S0377-8401(01)00273-5.

Mena, M. et al. (1992), "A major barley allergen associated with bakers asthma disease is a glycosylated monomeric inhibitor of insect alpha-amylase – cDNA cloning and chromosomal location of the gene", *Plant Molecular Biology*, Vol. 20, , No. 3, November, pp. 451-458.

Mercer, D.K. et al. (2001), "Transformation of an oral bacterium via chromosomal integration of free DNA in the presence of human saliva", *FEMS Microbiology Letters*, Vol. 200, No. 2, June, pp. 163-167.

Mercer, D.K. et al. (1999), "Fate of free DNA and transformation of the oral bacterium *Streptococcus gordonii* DL1 by plasmid DNA in human saliva", *Applied Environmental Micribiology*, Vol. 65, No. 1, January, pp. 6-10.

Mireles, A. et al. (2000), "GMO (Bt) corn is similar in composition and nutrient availability to broilers as non-GMO corn", *Poultry Science*, Vol. 79, Supplement 1, p. 65.

Molvig, L. et al. (1997), "Enhanced methionine levels and increased nutritive value of seeds of transgenic lupins (*Lupinus angustifolius* L) expressing a sunflower seed albumin gene", *Proceedings of the National Academy of Sciences of the United States of America*, Vol. 94, pp. 8 393-8 398.

Murphy, J.M. (1996), "Engineering oil production in rapeseed and other oil crops", *TIBTECH*, Vol. 14, pp. 206-213.

Napier, J.A. and L.V. Michaelson (2001), "Towards the production of pharmaceutical fatty acids in transgenic plants", *Journal of the Science of Food and Agriculture*, Vol. 81, Issue 9, July, pp. 883-888, http://dx.doi.org/10.1002/jsfa.892.

NASS (2002), "Prospective plantings", National Agricultural Statistics Service, United States Department of Agriculture, available at: http://usda.mannlib.cornell.edu/usda/nass/ProsPlan//2000s/2002/ProsPlan-03-28-2002.pdf (updated annually).

Noteborn, H.P.J.M. et al. (1994), "Safety assessment of the *Bacillus thuringiensis* insecticidal crystal protein CRY1A(b) expressed in transgenic tomatoes", *Proceedings of the ACS Meeting*, Chapter XX, Washington, DC.

OECD (2002), "Consensus document on compositional considerations for new varieties of maize (*Zea mays*): Key food and feed nutrients, anti-nutrients and secondary plant metabolites", OECD, Paris, available at http://www.oecd.org/env/ehs/biotrack/46815196.pdf

OECD (2001), "Consensus document on compositional considerations for new varieties of soybean: Key food and feed nutrients and anti-nutrients", OECD, Paris, cancelled and replaced with revised version in August 2012, available at http://www.oecd.org/officialdocuments/publicdisplaydocumentpdf/?cote=ENV/JM/MONO(2012)24&doclanguage=en

OECD (2000), "Report of the Task Force for the Safety of Novel Foods and Feeds for the G8 Summit (Okinawa)", C(2000)86/ADD1, OECD, Paris, available at http://www.oecd.org/chemicalsafety/biotrack/REPORT-OF-THE-TASK-FORCE-FOR-THE-SAFETY-OF-NOVEL.pdf

OECD (1997), "Report of the OECD workshop on toxicological and nutritional testing of novel foods", SG/ICGB(1998)1, OECD, Paris, final version available at: www.oecd.org/officialdocuments/publicdisplaydocumentpdf/?doclanguage=en&cote=sg/icgb(98)1/final.

OECD (1993), *Safety Evaluation of Foods Derived by Modern Biotechnology, Concepts and Principles*, OECD, Paris, available at: http://www.oecd.org/env/ehs/biotrack/41036698.pdf.

Phipps, R.H. et al. (2002), "Detection of transonic DNA in milk from cows receiving herbicide tolerant (CP4EPSPS) soyabean meal", *Livestock Production Science*, Vol. 74, pp. 269-273.

Phipps, R.H. et al. (2001), "Detection of transgenic DNA in bovine milk: Results for cows receiving a TMR containing Bt insect protected maize grain (MON810)", *Journal of Animal Science*, Vol. 79, Supplement 1, *Journal of Dairy Science*, Vol. 84, Supplement 10, p. 114.

Piva, G. et al. (2001), "Performance of broilers and piglets fed Bt corn", *Proceedings: International Symposium on Genetically Modified Crops and Co-products as Feeds for Livestock*, pp. 27-30, September, Nitra, Slovak Republic.

Rao, K.V. et al. (1998), "Expression of snowdrop lectin (GNA) in transgenic rice plants confers resistance to rice brown planthopper", *Plant Journal*, Vol. 15, No. 4, August, pp. 469-477.

Reuter, T. and K. Aulrich (2003), "Investigations on genetically modified maize (Bt-maize) in pig nutrition: Fate of feed ingested foreign DNA in pig bodies", *European Food Research and Technology*, Vol. 216, Issue 3, March, pp. 185-192.

Reuter, T. et al. (2002), "Investigations on genetically modified maize (Bt-maize) in pig nutrition: Fattening performance and slaughtering results", *Archives of Animal Nutrition*, Vol. 56, No. 5, October, pp. 319-326.

Reuter, T. et al. (2001), "Influence of Bt-corn on growing and slaughtering results of pigs", *Proceedings of the Society of Nutrition Physiology*, Vol. 10, p. 150.

Rutzmoser, K., J. Meyer and A. Obermaier (1999), "Verfütterung von Silomais der Sorten "Pactol" und "Pactol CB" (gentechnisch veränderte Bt-Hybride) an Milchkühe. Bodenkultur und Pflanzenbau, Schriftenreihe der Bayer", *Landesanstalt für Bodenkultur und Pflanzenbau*, Vol. 3, pp. 25-34.

Saalbach, I. et al. (1994), "A chimeric gene encoding the methionine-rich 2S albumin of the Brazil nut (*Bertholletia excelsa* HBK) is stably expressed and inherited in transgenic grain legumes", *Molecular Genetics and Genomics*, Vol. 242, No. 2, January, pp. 226-236.

Samarasinghe, K. et al. (2000), "Activity of supplemental enzymes and their effect on nutrient utilisation and growth performance of growing chickens as affected by pelleting temperature", *Archives of Animal Nutrittion*, Vol. 53, No. 1, pp. 45-58.

Sanders, P.R. et al. (1998), "Safety assessment of the insect-protected corn", in Thomas, J.A. (ed.), *Biotechnology and Safety Assessment*, 2nd edition, pp. 241-256, Taylor and Francis, Basingstoke, United Kingdom.

Schubbert, R. et al. (1998), "On the fate of orally ingested foreign DNA in mice: Chromosomal association and placental transmission to the fetus", *Molecular Genetics and Genomics*, Vol. 259, No. 6, October, pp. 569-576.

Schubbert, R. et al. (1997), "Foreign (M13) DNA ingested by mice reaches peripheral leucocytes, spleen and liver via the intestinal wall mucosa and can be covalently linked to mouse DNA", *Proceedings of the National Academy of Sciences*, Vol. 94, pp. 961-966.

Schubbert, R. et al. (1994), "Ingested foreign (phage M13) DNA survives transiently in the gastrointestinal tract and enters the bloodstream of mice", *Molecular Genetics and Genomics*, Vol. 242, No. 5, March, pp. 495-504.

Schuler, T.H. et al. (1998), " Insect resistant transgenic plants", *TIBTECH*, Vol. 16, pp. 168-175.

Sidhu, R.S. et al. (2000), "Glyphosate-tolerant corn: The composition and feeding value of grain from glyphosate-tolerant corn is equivalent to that of conventional corn (*Zea mays* L.)", *Journal of Agricultural and Food Chemistry*, Vol. 48, No. 6, June, pp. 2 305-2 312.

Simmonds, H.A. (1990), "Nutrition and gout", *BNF Nutrition Bulletin*, Vol. 15, pp. 169-179.

Simon, J.J. et al. (2002), "Effect of Roundup Ready corn (event NK603) on performance in beef feedlot diets", *Journal of Animal Science*, Vol. 80, Supplement 1, p. 46.

Smith, N. et al. (2003), "The effect of processing on DNA fragmentation in animal feeds", *Animal Feed and Science Technology*.

Stanford, K. et al. (2002), "Effects of feeding Glyphosate-tolerant canola meal on lamb growth, meat quality and apparent feed digestibility", *Journal of Animal Science*, Vol. 80, Supplement 1, p. 71.

Staub, J.M. et al. (2000), "High-yield production of a human therapeutic protein in tobacco chloroplasts", *Nature Biotechnology*, Vol. 18, No. 3, March, p. 333.

Tilley, J.M.A. and R.A. Terry (1963), "A two-stage technique for the *in vitro* digestion of forage crops", *Journal of the British Grassland Society*, Vol. 18, Issue 2, June, pp. 104-111, http://dx.doi.org/10.1111/j.1365-2494.1963.tb00335.x.

Vogel, P.K. and H.-J.G. Jung (2001), "Genetic modification of herbaceous plants for feed and fuel", *Critical Reviews in Plant Science*, Vol. 20, Issue 1, pp. 15-49, http://dx.doi.org/10.1080/20013591099173.

Walmsley, A.M. and C.J. Arntzen (2000), "Plants for delivery of edible vaccines", *Current Opinion in Biotechnology*, Vol. 11, No. 2, April, pp. 126-129.

Weber, T.E. and B.T. Richert (2001), "Grower-finisher growth performance and caracass characteristics including attempts to detect transgenic plant DNA and protein in muscle from pigs fed genetically modified Bt corn", *Journal of Animal Science*, Vol. 79, Supplement 2, p. 67.

Weber, T.E. et al. (2000), "Grower-finisher performance and carcass characteristics of pigs fed genetically modified 'Bt' corn", *Purdue 2000 Swine Day Report*.

Wehrmann, A. et al. (1996), "The similarities of *bar* and *pat* gene products make them equally applicable for plant engineers", *Nature Biotechnology*, Vol. 14, No. 10, October, pp. 1 274-1 278.

Weisbjerg, M.R. et al. (2001), *Undersøgelser af genmodificerede foderroer til malkekøer*. DJF Rapport 25, Husdrybrug, Ministeriet for Fødevarer, Landbrug og Fiskeri, Denmark.

Whitney, R. McL. et al. (1976), "Nomenclature of the proteins of cow's milk: Fourth revision", *Journal of Dairy Science*, Vol. 59, No. 5, May, pp. 795-815.

WHO (1991), *Strategies for Assessing the Safety of Foods Produced by Biotechnology*, Report of a Joint FAO/WHO Consultation, World Health Organization of the United Nations, Geneva, out of print.

Chapter 2

Unique Identifier for transgenic plants

This chapter contains guidance for the designation of a Unique Identifier for transgenic plants developed by the OECD Working Group on Harmonisation of Regulatory Oversight in Biotechnology, published in 2002 and further revised 2006 to cover plants with several "stacked" traits. The Unique Identifier is a nine-digit alphanumeric code based on the transformation event, with a single digit for verification. Assigned by the developer of the genetically engineered product, the Unique Identifier also contains the applicant identification. OECD unique identifiers are used as "keys" to access information on genetically engineered organisms approved for commercial application contained in the OECD BioTrack Product Database *and interoperable systems such as the CBD Biosafety Clearing House, the FAO GM Foods Platform and other databases from public and private sectors worldwide.*

OECD guidance on the Unique Identifier for transgenic plants – 2006 revised version

This guidance for a Unique Identifier for transgenic plants was developed by OECD's Working Group on Harmonisation of Regulatory Oversight in Biotechnology. The purpose of the Unique Identifier is for use as a "key" to unlock or access information in the OECD's *BioTrack Product Database*,[1] a database of products of modern biotechnology which have been approved for commercial application (cropping, and/or use in foods and feeds), as well as interoperable systems.

The OECD Workshop on Unique Identification Systems for Transgenic Plants, hosted by Switzerland in Charmey in October 2000, constituted a major step in the development of this guidance. A number of options for developing a Unique Identifier were under consideration. Subsequently, these options and related issues were discussed in detail within the Working Group on Harmonisation of Regulatory Oversight in Biotechnology.

In 2002, the delegations agreed on a guidance document in three parts, i.e. an introduction, a section on how to develop and generate unique identifiers, and a section on future developments.

The OECD's Business and Industry Advisory Committee (BIAC) has played an important part during all stages of the discussion through their Expert Group on Biotechnology. This was essential because, according to the guidance, it is the developers of transgenic products (mostly from the industry sector) who generate the unique identifiers.

Since the initial publication of the guidance document in 2002, the OECD's unique identification system for transgenic plants has been utilised without difficulty as providing relevant "keys" to access information of each transgenic product in the *BioTrack Product Database*. In addition, it has been recognised as an appropriate identification system of products included in the Biosafety Clearing-house of the Cartagena Protocol on Biosafety developed under the auspices of the Convention on Biological Diversity (CBD).

With the recent increases of commercialisation of plant products having one or more traits obtained through the use of recombinant DNA techniques and stacked by conventional crosses in the backdrop, it was proposed to standardise the way to designate a Unique Identifier for such plant products. The original guidance document allowed two different options for such a product. At the 18th meeting of the Working Group (in June 2006), the text modifying item 8 was agreed accordingly. The revised document was published in 2006.

Introduction

The purpose of the Unique Identifier is for its use as a key to accessing information on the genetically engineered products which have been approved for commercial application. The Unique Identifier system was developed from plant products in the OECD *BioTrack Product Database*, and it is directly applicable to plant products entered into other databases. While the concepts and elements were developed for plants, they may be considered for potential applicability to other products.

The OECD has been working on a "Unique Identifier for transgenic plants" since 2000. The guidance was initially published in 2002 and revised in 2006.

The consensus was on the need for a Unique Identifier defined as a simple alphanumeric code based on the transformation event (rather than other options such as a new variety), with a single digit for verification. The Unique Identifier should be a "key" to unlocking more detailed information in the *BioTrack Product Database* and interoperable systems (for examples, the CBD Biosafety Clearing-House and the FAO GM Foods Platform). As such, it should be kept short, simple and user-friendly.

It should also be built in a flexible way and might potentially serve as a core Unique Identifier for future developments. It should also take into account experience with, and be applicable to, existing products.

Each applicant has their own internal mechanism to avoid applying the same designation of the "transformation event" to different products. Consequently, incorporating the applicant information into the Unique Identifier is the only way to enable applicants to generate the Unique Identifier for their product while at the same time ensuring its uniqueness from those generated by other applicants. Furthermore, this provides applicants with the flexibility to generate the Unique Identifier when they believe it to be appropriate or necessary.

Development and designation of the Unique Identifier

Item 1

The purpose of the Unique Identifier is for its use as a key to accessing information in the OECD *BioTrack Product Database* and interoperable systems for the products of modern biotechnology which have been approved for commercial application. This guidance addresses the development of a Unique Identifier for use in the *BioTrack Product Database*. It was developed from plant products in the database and its use is directly applicable to plant products entered into the database.

While the concepts and principal components were developed for plants they may be considered for their potential applicability to other products.

Item 2

Applicants should designate a Unique Identifier for their product to the national authority, at the latest at the time of application for the first commercial approval.

Item 3

The national authority should, at the time of the first approval for commercialisation, notify the OECD of the designated Unique Identifier, in order to enable access to the relevant information in the BioTrack Product Database for all subsequent applications for commercialisation in other countries.

Item 4

The Unique Identifier is a code of a fixed length of nine alphanumeric digits for a transformation event derived from modern biotechnology.[2] It should be unique to that transformation event.

Item 5

The Unique Identifier is composed of three elements that must be separated by dashes (-). The total length is nine digits, the last of which is a verification digit. The transformation event and the applicant designation should total eight alphanumeric digits.

- two or three alphanumerical digits to designate the applicant
- five or six alphanumerical digits to designate the "transformation event"[3]
- one numerical digit as a verification, as foreseen in item 7.

For example,

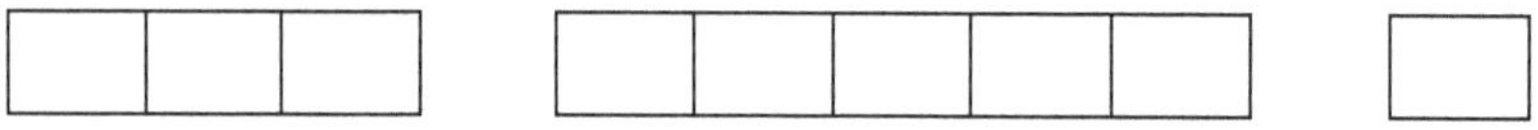

or

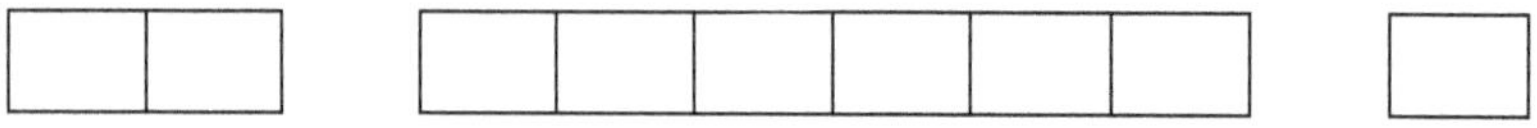

Item 6

The Unique Identifier should include the "applicant information" of two or three alphanumeric digits (for example, the first two or three digits of the applicant organisation name), followed by a dash. Any new applicant that is not identified within the database shall not be permitted to use the existing codes listed in the applicant's code table within the database. The applicant shall inform the national authorities, who will update the *BioTrack Product Database*, by including a new code that will be designed to identify the new applicant in the code table.

Item 7

The Unique Identifier should include one verification digit, which shall be separated from the rest of the Unique Identifier digits by a dash. The verification digit is intended to reduce errors by ensuring the integrity of the alphanumeric code, entered by the users of the database.

The rule for calculating the verification digit is as follows. The verification digit is made up of a single numerical digit. It is calculated by adding together the numerical values of each of the alphanumerical digits in the Unique Identifier. The numerical value of each of the digits ranges from ∅ to 9 for the numerical digits (∅ to 9) and 1 to 26 for the alphabetical digits (A to Z) (see Annex 2A.1). The total sum, if made up of several numerical digits, will be further calculated by adding the remaining digits together using the same rule, in an iterative process, until the final sum is a single numerical digit.

For example, the verification digit for the code CED-AB891 is calculated as follows:

- Step one : 3+5+4+1+2+8+9+1 = 33
- Step two: 3+3 = 6; therefore the verification digit is 6
- This Unique Identifier then becomes CED-AB891-6.

Item 8

The above guidance is sufficient to generate unique identifiers for the majority of existing plant products. Regarding plant products having two or more traits obtained through the use of recombinant DNA techniques and stacked by conventional crosses, the Unique Identifier should consist of the unique identifiers from each parental transgenic plant (e.g. MON-15985-7 x MON-∅1445-2).

Future development

It was recognised that it may be necessary to revisit in the future the potential use of prefixes or suffixes if there is a need to incorporate further information fields. The use of prefixes or suffixes, on an *ad hoc* or voluntary basis, to incorporate further information fields for use in the *BioTrack Product Database*, as appropriate or requested by a country, will continue to be discussed and should be made public by national authorities.

This guidance for the development and designation of the Unique Identifier may be reassessed in the light of experience gained.

Notes

1. Available at:. http://www2.oecd.org/biotech/
2. Zero should be reflected by the symbol ∅ to avoid confusion with the letter O.
3. When the transformation event of an existing plant product, prior to the adoption of this guidance, is shorter or longer than five or six digits, the applicant should select five or six digits within the transformation event in order to fit it into this limit.

Annex 2A.1
Digits and alphabetic characters to be used in the Unique Identifier

Form of digits to be used in the Unique Identifier

∅
1
2
3
4
5
6
7
8
9

Form of alphabetic characters to be used, plus numerical equivalents, for calculating verification digit

A=1	N=14
B=2	O=15
C=3	P=16
D=4	Q=17
E=5	R=18
F=6	S=19
G=7	T=2∅
H=8	U=21
I=9	V=22
J=1∅	W=23
K=11	X=24
L=12	Y=25
M=13	Z=26

Part II

Compositional considerations for transgenic crops

Chapter 3

Sugar beet (*Beta vulgaris*)

*This chapter, prepared by the OECD Task Force for the Safety of Novel Foods and Feeds with Germany as the lead country, deals with the composition of sugar beet (*Beta vulgaris*). It contains elements that can be used in a comparative approach as part of a safety assessment of foods and feeds derived from new varieties. Background is given on growing, processing and use of sugar beet and derived products. Then key nutrients and anti-nutrients are detailed for sugar beet roots, sugar, pulp and molasses. Relevant nutrients of sugar beet for animal feed use are suggested, followed by considerations for the assessment of new varieties.*

Background

Growing of sugar beet

Sugar beet (*Beta vulgaris* L.ssp. *vulgaris* var. *Altissima Doell*) is cultivated worldwide, but primarily in warm and temperate climates with sufficient precipitation. Today's sugar beets have a sucrose content of approximately 15-20% depending on climate, soil type, variety and cultivation methods.

The worldwide growing area for sugar beet is about 8.2 million hectares (OECD, 1993b) and the annual production of sugar beets is about 250 million t (FAOSTAT, 2000). The leading producing countries in 1999 were France, Germany, the United States, Turkey, Poland, Ukraine, Italy and the United Kingdom (Langendorf et al., 1999/2000; FAOSTAT, 2000; CEDUS, 1999). By 1998/99, 28.4% of all sugar produced was from sugar beets (Langendorf et al., 1999/2000; CEDUS, 1999).

Processing of sugar beet

In order to guarantee a continuous beet supply for processing, beets are usually stored in field clamps and/or at the factory yard. Maximum storage and thus the possible processing period depend on climate conditions, from a few weeks (Mediterranean) up to several months (Scandinavia). Generally, the harvested beet metabolises some of the stored sugar so that sugar losses are unavoidable and these losses increase with temperature (to up to 300 grammes sucrose/tonne of beets/day). Frost damage also results in an increase in components undesirable for sugar processing. In northern regions, the clamps are covered to avoid the irreversible effects of frost damage.

Figure 3.1 shows the typical processing line from beet to sugar including the fate of the by-products. For processing, beets are first washed with water to remove dirt and other large debris, then they are sliced into cossettes. The cossettes are extracted with water at temperatures around 70°C for about 100 minutes. The raw juice obtained is purified by a treatment with milk of lime and carbon dioxide. The material precipitated thereby, the carbonation sludge, is removed by filtration and pressed as carbonation lime. The resulting juice is called "thin juice", which is concentrated by evaporation to "thick juice". The evaporation is carried out in multi-stage evaporators working at a temperature range of 98-130°C at different pressures. The resulting "thick juice" is further concentrated to crystal magma from which crystalline sugar is recovered by centrifugation. During the centrifugation process, the crystals are separated from the syrup. The crystals are dried, cooled and stored for further use. The remaining syrup, the so-called molasses, is mainly used as animal feed or as fermentation substrate. The recovery of residual sucrose from molasses is applied in some regions, but to a minor extent. The material remaining from the treated cossettes is referred to as wet pulp. This pulp is pressed and dried to remove water and is commonly pelleted with added molasses. Carbonation lime is used as fertiliser (for further information see also Van der Poel et al., 1998; Schiweck et al., 1993).

Uses of sugar beet and derived products

The main purpose[3] of sugar beet processing is sugar (sucrose) recovery. The worldwide production of sugar from sugar beet is close to 40 million tonnes per year; world sugar consumption is about 120 million onnes per year; and the supply per capita varies between 10 and 50 kg/year (FAO, 1999; Langendorf et al., 1998/99). Sugar is mainly used as a food ingredient.

The sugar beet crop provides a number of by-products after harvest and processing which are valuable feed stuffs (Figure 3.1). Feed products from sugar beet are high in fibre and energy. Therefore, they are primarily used in feeding ruminants (dairy cows, beef cattle, sheep), but also non-ruminants. To meet the animals' requirement, feed rations containing sugar beets or their by-products are usually combined with other feed products.

Figure 3.1. **Principle steps of sugar beet processing and common product uses**

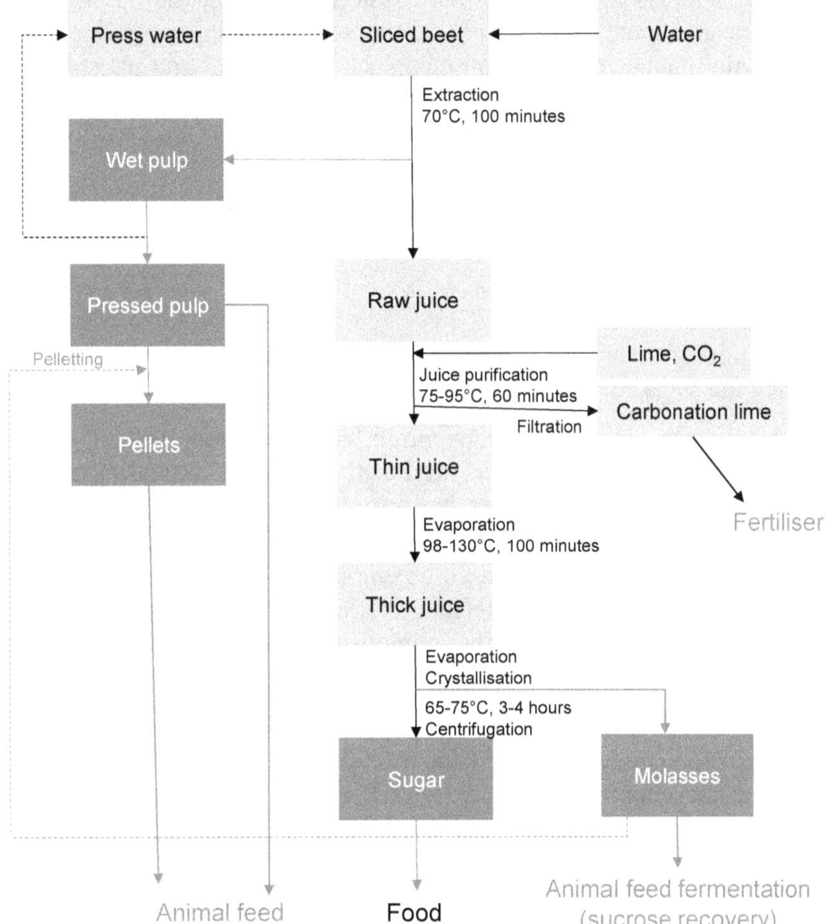

Sugar beet tops are usually ploughed under. In rare cases tops are ensiled or directly used in ruminant feeding.

Wet pulp is typically pressed (22-30% dry matter) and dried (85-90% dry matter). To increase the ease of handling and storage, dried pulp is usually (95%) pelleted with added molasses. Pressed or dried pulp is also directly used for feeding purposes. In some regions, mixtures of pulp and molasses pulps as such are used for animal feed.

Molasses is mainly used in animal feeding (about 60% of total molasses) as feed ingredient, pelleting aid or ensiling agent. Another major use (15%) is as raw material in fermentation (yeast, citric acid, alcohol, etc.) (Langendorf et al., 1999/2000). Special applications of molasses, e.g. as a source for single substances (e.g. betaine) are of minor economic importance. The recovery of remaining sucrose from molasses through ion exchange or other technologies is at present rarely applied, with the exception in regions of the United States where the Steffen process is applied (i.e. removal of sucrose from molasses as calcium saccharate precipitate). To a minor extent molasses is used for various industrial purposes, such as fuels, rubber, printing, chemical and construction industries.

Vinasses results from fermentation of molasses and is used as soil conditioner or animal feed.

Another by-product of sugar production is carbonation lime (produced during beet juice purification). Lime is used in agriculture after mechanical conditioning as a fertiliser providing calcium and increasing the pH of the soil and thus improving its structure. It contains a certain amount of plant nutrients such as nitrogen and phosphorus and can therefore also be used as a fertiliser for agricultural application, as well as an ingredient in potting soils used in mushroom production, and as a binder for briquetting and/or pelleting dry materials.

Nutrients and anti-nutrients in sugar beets

Sugar beet roots

The composition of sugar beet roots is mainly rated in view of their technical quality relevance during the sugar recovery process and their agronomic properties.

The term "technical quality" is a convention based on compositional parameters by which sugar technologists and breeders assess the relative suitability of sugar beets for processing. It is mainly determined by the sucrose and non-sucrose components such as potassium, sodium and α-amino-N. In different countries the definition of the technical quality may vary as it is based on empirical factors. The respective technical quality is used by breeders as one selection criterion for developing new varieties. Therefore, during the past decades the composition of cultivated beet has largely improved with regard to the technical quality (Märländer and Ladewig, 1997). The technical quality is assessed in field trials prior to market approval.

As sugar beet roots are seldom used for food or feed as such, a distinction between nutrients and anti-nutrients in a toxicological sense is not made. Data arise either from animal feeding experiments applying analytical methods established in animal nutrition (Table 3.1) or from the technical quality determined basically by sucrose, cations and amino nitrogen content (Table 3.2). The beet composition depends considerably on the growing conditions of the plants, such as location, climate and agronomical factors, mainly fertilising, variety and population density (Rother, 1998).

Table 3.1. **Chemical composition of sugar beet roots (23.0-24.6% dry matter)**

	Ranges* (% of dry matter)
Crude ash	5.08.1
Crude protein	4.7-6.8
Ether extract	0.3-0.6
Crude fibre	4.9-6.3
Sucrose	64.7-70.0

Note: * Reported as means of different feeding tables.

Source: NOVUS International Inc. (1996).

Table 3.2. **Major minerals and α-amino-nitrogen in sugar beet roots (23.0-24.5% dry matter)**

	Ranges (% of dry matter)
Sodium (Na)	0.4-0.8
Potassium (K)	5.6-7.2
Phosphorus (P)	1.4-2.2
α-amino-N	0.7-1.1

Source: Adapted from IfZ-Institut für Zuckerrübenforschung (1999) which included Überregionaler Sortenvergleich 1997, 1998 and 1999.

Sucrose is the main constituent of the sugar beet root dry matter.

The non-sucrose substances in sugar beet roots include other soluble saccharides, cell wall components, saponins, proteins, free amino acids, betaine, as well as organic and inorganic ions and other nitrogen-free acids. Inorganic anions include phosphates, chlorides, sulfates and nitrates of ubiquitous cations mainly potassium, sodium, calcium, magnesium and ammonium.

Anti-nutritional or other adverse effects to human or animal health due to beet components have not been found during their long history of safe use.

Sugar

White sugar is defined as "purified and crystallised sucrose with a polarisation of not less than 99.5°Z" (degree sugar) (FAO, 1999). The remainder consists of water, ash, invert sugar (i.e. glucose and fructose) and some colouring organic compounds.

Pulp

Protein and lipid contents of beet pulp products are usually low. In addition, beet protein consists mainly of non-essential amino acids (Tables 3.3-3.5).

Table 3.3. **Chemical composition of dried sugar beet pulp (84.0-91.0% dry matter)**

	Ranges* (% of dry matter)
Crude ash	3.8-6.7
Crude protein	6.6-9.7
Ether extract	0.5-1.6
Crude fibre	15.0-21.3
Sucrose	4.7-10.0

Note: * Reported as means of different feeding tables.

Source: NOVUS International Inc. (1996).

Table 3.4. **Major minerals in dried sugar beet pulp (84.0-91.0% dry matter)**

	Ranges* (% of dry matter)
Calcium (Ca)	0.6-1.1
Phosphate (P)	0.1-0.2
Magnesium (Mg)	0.1-0.3
Sodium (Na)	0.1-0.5
Potassium (K)	0.2-1.6

Note: * Reported as means of different feeding tables.

Source: NOVUS International Inc. (1996).

Table 3.5. **Amino acids in dried sugar beet pulp (84.0-91.0% dry matter)**

	Ranges* (% of dry matter)
Lysine	0.33-0.6
Methionine	0.01-0.15
Methionine + cystine	0.02-0.26
Threonine	0.25-0.47
Tryptophan	0.05-0.10
Isoleucine	0.23-0.36
Leucine	0.36-0.60
Valine	0.36-0.57
Histidine	0.19-0.29
Arginine	0.24-0.41
Phenylalanine	0.22-0.34

Note: * Reported as means of different feeding tables.

Source: NOVUS International Inc. (1996).

Molasses

The total sugar content in molasses is approximately 50% (Table 3.6). Minor carbohydrates are glucose, fructose, raffinose and some other oligo- or polysaccharides. Their concentration is below 1% and depends to a significant extent on the manufacturing process.

Table 3.6. **Chemical composition of sugar beet molasses (73-79% dry matter)**

	Ranges* (% of dry matter)
Crude ash	6.6-10.0
Crude protein	6.6-11.1
Ether extract	0.0-0.3
Crude fibre	0.0-0.3
Sucrose	43.0-50.5

Note: * Reported as means of different feeding tables.

Source: NOVUS International Inc. (1996).

Major cations are potassium followed by sodium, calcium and magnesium. Their content depends mainly on soil type and water availability. Additionally, the calcium and sodium content is influenced by processing practices (Table 3.7).

Table 3.7. **Major minerals in sugar beet molasses (73-79% dry matter)**

	Ranges* (% of dry matter)
Calcium (Ca)	0.1-0.5
Phosphate (P)	0.02-0.06
Magnesium (Mg)	0.01-0.3
Sodium (Na)	0.6-1.9
Potassium (K)	3.2-4.7

Note: * Reported as means of different feeding tables.

Source: NOVUS International Inc. (1996).

About 20% of the total mass consists of non-sucrose organic matter, in particular of non-protein nitrogen (NPN) containing substances, such as betaine (Table 3.9). In addition, molasses contains free and bound amino acids (Table 3.8) and pryrrolidone carboxylic acid (a conversion product of glutamine) (Table 3.9). In the manufacturing process, most of the amino acids undergo changes so that less than the amounts expected from beet roots are found in molasses (Reinefeld et al., 1982a, 1982b; Schiweck et al., 1993).

Table 3.8. **Amino acids in sugar beet molasses (73-79% dry matter)**

	Ranges* (% of dry matter)
Lysine	0.04
Methionine	0.04-0.01
Methionine + cystine	0.1-0.11
Threonine	0.1-0.11
Tryptophan	0.1-0.24
Isoleucine	0.1-0.27
Leucine	0.12-0.26
Valine	0.17-0.20
Histidine	0-0.02
Arginine	0.02
Phenylalanine	0.04-0.06
Glutamic acid	3-4

Note: * Reported as means of different feeding tables.

Source: NOVUS International Inc. (1996).

Table 3.9. **Contents of nitrogen-containing organic compounds in beet molasses (73-79% dry matter)**

	Ranges* (% of dry matter)
Total N-containing compounds	11-16
Betaine	4-5
Amino acids, pyrrolidone carboxylic acid, peptides, nucleic acid components	3-4
Amino acid sugar complexes	1-2

Note: * Reported as means of different feeding tables.

Source: Van der Poel et al. (1998).

Molasses contains up to 4% of organic acids predominantly lactic acid from the degradation of invert sugar (up to 1.7%) followed by malic, citric, fumaric and oxalic acid.

Molasses contains only low levels of trace elements except for iron. The main inorganic anions are chloride, sulfate, nitrate and traces of phosphate and nitrite (Table 3.10).

Identification of key sugar beet products consumed by animals

Several whole and processed fractions of the sugar beet plant may contribute to the animal diet. Sugar beets can contain oxalate up to 55g/kg dry matter, which is present primarily in the leaves (Thacker and Kirkwood, 1990). The sparing soluble calcium oxalate is known to have a reduced availability to animals. This is to be taken into account when rations are formulated.

Table 3.10. **Contents of major anions in beet molasses (73-79% dry matter)**

	Ranges* (% of dry matter)
Chloride	1.0-3.0
Sulfate	0.6-2.0
Phosphate	0.1-0.5
Nitrate	0.3-0.8
Nitrite	3.0-170 mg/kg

Note: * Reported as means of different feeding tables.

Source: Van der Poel et al. (1998).

Sugar beet roots as such are seldom used in livestock feeding. However, the tops are fed fresh or as silage, primarily to cattle. Fresh roots are fed to dairy cattle on a limited basis because sugar feeding involves the risk of acidosis. On the other hand, feeding restricted amounts of sugar beets may have a favourable effect on feed consumption and rumen fermentation and on crude fibre digestibility (Kluge, 1986; Flachowsky et al., 1988/89).

Sugar beet roots are suitable feedstuffs for pigs. Due to their high digestibility and energy concentration they have the potential for high growth performance rates when they are incorporated in the diet at a level up to 35% (Jeroch et al., 1993). Because of the low protein content of sugar beets, sugar beet-based diets for monogastrics would need adequate protein or amino acid supplementation.

By-products from sugar beet processing, for example beet pulp and molasses, are the main sugar beet products fed to animals.

Sugar beet pulp is more effectively used in ruminant than in pig feeding due to its high fibre content (up to 25% in the dry matter). It has the potential to replace high portions of cereals in concentrate mixtures for dairy cattle. Incorporation rates of 30% in the dry matter of diets for dairy cows and 50% for growing cattle are possible.

Molasses can be used in feeding ruminants, but only to a limited extent. It has to be homogeneously distributed over the total diet. Maximal incorporation rates are reported not to exceed 15% of dry matter intake. In pig feeding, the maximum possible inclusion rate increases with age. For growing pigs, the maximum level is reported to be 20% of the dry matter.

The limiting factors of the by-products from sugar processing are the low protein content and the high content of fibre, which are known to have a low efficiency of energy utilisation in monogastrics. Additionally, the high concentration of highly fermentable substances (sugars) might negatively affect rumen fermentation. Effects of specific ingredients (undesired substances and anti-nutrients) on animal health or on meat and milk quality are not known.

In assessing the nutrient quality of all sugar beet products which could be fed to animals, crude nutrients in roots, pulp, molasses and tops appear to be suitable indicators. The relevant nutrients in sugar beet matrices for animal feed use are shown in Table 3.11.

Table 3.11. **Relevant nutrients of sugar beet for animal feed use**

	Roots	Pulp	Molasses	Tops
Crude nutrients (crude ash, crude protein, ether extract, crude fibre)	X	X	X	X
Sucrose	X	X	X	
Pectins	X	X		

Consideration for the assessment of new sugar beet varieties

Agronomic characteristics are important to consider since unspecific or unpredicted phenotypic traits or changes in phenotypic traits may be indicative of unintended effects of potential safety concern that would require further investigation. Parameters that are analysed for variety registration include yield, content of sucrose, potassium, sodium and α-amino-N. In addition, field emergence, bolting resistance and certain disease tolerances of varieties are tested in variety trials.

The comparison of the chemical composition of a modified variety and a non-modified comparator should include the key nutrients crude ash, crude protein, crude fibre, sucrose and phosphorus as listed in Tables 3.1 and 3.2 for sugar beet roots. If the analyses of these parameters indicate that a novel variety is within the ranges given in the literature, apart from the intentional modifications resulting in recombinant DNA and new proteins, it can be considered equivalent with respect to its overall composition. Knowledge of the sugar recovery process permits the conclusion that sugar, as well as the intermediate products and molasses, contain neither DNA nor protein. The safety assessment would focus on the recombinant DNA and newly expressed proteins in pulp in view of animal feed use, i.e. its specific behaviour, if any, in the gastrointestinal tract of the animals.

If, apart from the intentionally modified DNA and resulting new proteins, the genetic modification results in a qualitative change rather than a quantitative shift of the beet constituents outside the naturally occurring ranges, the safety assessment would focus on those differences, possibly requiring nutritional and/or toxicological studies.

For livestock feed, the comparison of the chemical composition of a modified variety and a non-modified comparator should include the nutrients listed in Table 3.11 for roots. A more thorough nutritional and toxicological evaluation has to be decided on a case-by-case basis. The required data would be a function of the nature and degree of the difference of the feed ingredient from an accepted source, target animal species and also the potential dietary exposure.

Notes

3. A regional speciality in Central Europe is the use of beet syrup for human consumption. Beets are pressed (not extracted) and the juice obtained is directly thickened to syrup.

References

CEDUS (1999), "Le sucre", mémo statistique, Centre d'études et de documentation du sucre, Paris.

FAO (2000), *Report of a Joint FAO/WHO Expert Consultation on Foods Derived from Biotechnology*, World Health Organization, Geneva, 29 May-2 June 2000, http://www.fao.org/fileadmin/templates/agns/pdf/topics/ec_june2000_en.pdf (accessed 11 February 2015).

FAO (1999), "Codex Alimentarius", *FAO/WHO Food Standards*, Volume 11: Sugar, Cocoa Products, Chocolate and Miscellaneous, Food and Agriculture Organization.

FAO (1996), *Report of a Joint FAO/WHO Expert Consultation on Biotechnology and Food Safety*, Food and Agriculture Organization, Rome, Italy, 20 September- 4 October 1996 http://www.fao.org/ag/agn/food/pdf/biotechnology.pdf (accessed 11 February 2015)

FAOSTAT (2000), FAO Statistics on-line website, http://faostat.fao.org/, Food and Agriculture Organization, Rome, (accessed in 2000 on the previous website).

Flachowsky, G. et al. (1988/89), "Getreideersatz durch Trockenschnitzel und Preßschnitzelsilage in der Mastrindfütterung. Tierernährung und Fütterung, Erfahrungen, Ergebnisse, Entwicklungen", *Dt. Landwirtschaftsverl Vol. 16*, p. 64-70.

Forbes, J.M. et al. (1998), "Effect of feed processing conditions on DNA fragmentation", Scientific Report No. 376 to the Ministry of Agriculture, Fisheries and Food, United Kingdom.

IfZ-Institut für Zuckerrübenforschung (1999), *Koordinierte Versuche Zuckerrüben 1999 in Deutschland*, Göttingen, Germany.

Jeroch, H. et al. (eds.) (1993), Futtermittelkunde, Gustav Fischer Verlag, Jena, Stuttgart, Germany.

Klein, J. et al. (1998), "Nucleic acid and protein elimination during the sugar manufacturing process of conventional and transgenic sugar beets", *Journal of Biotechnology*, Vol. 60, No. 3, pp. 145-153, February.

Kluge, H. (1986), "Untersuchungen zum Einfluss der Kohlenhydratzusammensetzung der Futterration auf Kenndaten des Panseninhalts und der Verdaulichkeit bei Milchkühen", Thesis A, MLU Halle.

Langendorf, D. et al. (1998/99), *Sugar Economy*, Edition 45 WVZ, Dr. Albert Bartens KG, Berlin.

Langendorf, D. et al. (1999/2000), *Sugar Economy*, Edition 45 WVZ, Dr. Albert Bartens KG, Berlin.

Lüdecke, H. and I. Feyerabend (1958), *Zucker*, Vol. 11, pp. 389-93.

Mahn, K. (2000), "Menge und Verteilung qualitätsbestimmender Inhaltsstoffe und deren Bedeutung für die technische Qualität von Zuckerrüben", Dissertation, Cuvillier Verlag Göttingen, Germany.

Märländer, B. and E. Ladewig (1997), "Entwicklung wertbestimmender Eigenschaften bei Zuckerrüben", *Vortr. Pflanzenzüchtg*, Vol. 39, pp. 126-140.

NOVUS International Inc. (1996), *Raw Material Compendium* (2nd ed.), Brussels.

OECD (1997), "Report of the OECD Workshop on Toxicological and Nutritional Testing of Novel Foods", SG/ICGB(1998)1, OECD, Paris, final version available at: www.oecd.org/officialdocuments/publicdisplaydocumentpdf/?doclanguage=en&cote=sg/icgb(98)1/final.

OECD (1993a), *Safety Evaluation of Foods Derived by Modern Biotechnology: Concepts and Principles*, OECD, Paris, available at: www.oecd.org/science/biotrack/41036698.pdf.

OECD (1993b, *Traditional Crop Breeding Practices: An Historical Review to Serve as a Baseline for Assessing the Role of Modern Biotechnology*, OECD, Paris.

Potter, R.L. et al. (1990), "Isolation of proteins from commercial beet sugar preparations", *Journal of Agricultural and Food Chemistry*, Vol. 1 890, No. 38, pp. 1 498-1 502.

Reinefeld, E. et al. (1982a), Beitrag zur Kenntnis des Verhaltens der Aminosäuren im Verlauf der Zuckergewinnung 1, Mitteilung: Gaschromatographische Untersuchungen an Zuckerfabrikprodukten auf freie und durch Säurehydrolyse freisetzbare Aminosäuren, *Zuckerind*, Vol. 107, pp. 283-292.

Reinefeld, E. et al. (1982b), Beitrag zur Kenntnis des Verhaltens der Aminosäuren im Verlauf der Zuckergewinnung 2, Mitteilung: Untersuchungen über die Auswirkung von Prozeßparametern der Saftreinigung", *Zuckerind*, Vol. 107, pp. 1 111-1 120.

Rother, B. (1998), "Die technische Qualität von Zuckerrüben unter dem Einfluss verschiedener Anbaufaktoren", Dissertation, Cuvillier Verlag, Göttingen, Germany.

Schiweck, H. et al. (1993), "Das Verhalten der stickstoffhaltigen Nichtzuckerstoffe von Rüben während des Fabrikationsprozesses", *Zuckerind*, Vol. 118, pp. 15-23.

Schiweck, H. and M. Clarke (1994), "Sugar", in: *Ullmann's Encyclopedia of Industrial Chemistry*, VCH Weinheim.

Thacker, P.A. and R.N. Kirkwood (ed.) (1990), *Nontraditional Feed Source for Use in Swine Production*, Butterworth Publisher, Stoneham.

Van der Poel, P.W. et al. (1998), *Sugar Technology*, Dr. Albert Bartens KG, Berlin.

WHO (1991), *Strategies for Assessing the Safety of Foods Produced by Biotechnology*, Report of a Joint FAO/WHO Consultation, World Health Organization of the United Nations, Geneva, out of print.

Chapter 4

Potato (*Solanum tuberosum* ssp. *tuberosum*)

*This chapter, prepared by the OECD Task Force for the Safety of Novel Foods and Feeds with Germany as the lead country, deals with the composition of potato (*Solanum tuberosum *ssp.* tuberosum*). It contains elements that can be used in a comparative approach as part of a safety assessment of foods and feeds derived from new varieties. Background is given on production, human and animal consumption, and industrial uses of potatoes. Key food and feed nutrients, toxins, allergens and anti-nutrients are then detailed, followed by considerations suggested for the assessment of new potato varieties.*

Background

This chapter discusses key components (nutrients, anti-nutrients and toxicants) of potato for which data have been collected on varieties developed through conventional breeding techniques and that may contribute to an assessment of substantial equivalence (Love, 2000; Rogan et al., 2000).

Production of potatoes[1]

The world production of potatoes (*Solanum tuberosum* ssp. *tuberosum*) amounted to almost 308 million tonnes in 2000 (FAOSTAT, 2001), and potatoes were grown in over 120 countries (Burton, 1989).

Yield and composition of tubers may vary in wide ranges due to variety and growing conditions.

Potatoes for human consumption

The average consumption of potatoes differs widely between countries. Relevant statistical data are given by the Food and Agriculture Organization (FAO) (Table 4.1).

Table 4.1. **Average potato consumption, 1998**

	Potato consumption (kg/cap/year)
World	30
Africa	11
Asia	19
Europe	94
European Union (15)	78
North America	63
South America	31
Developing countries	17
Developed countries	75

Source: FAOSTAT (2001).

Especially in industrialised countries, direct consumption of potatoes has declined dramatically, whereas consumption of potato products (e.g. chips, crisps) has increased. For example, in Germany consumption of fresh potatoes declined from 87 kg/cap/year in 1971 to 42 kg/cap/year in 1999, but during the same period consumption of potato products increased from 14 kg/cap/year to 29 kg/cap/year (basis: fresh potatoes).

Potatoes for direct consumption should be cooked before eating because of the indigestibility of non-gelatinised starch and the presence of anti-nutritional proteins. Different kinds of preparation are in use resulting in various amounts of nutrient losses (e.g. ascorbic acid: 13% loss during cooking of unpeeled potatoes vs. 41% loss of peeled potatoes [Weber and Putz, 1998]).

Due to consumer request, potatoes are increasingly supplied in processed form. A schematic description of different methods of potato processing is given in Figure 4.1.

Figure 4.1. **Schematic description of potato processing**

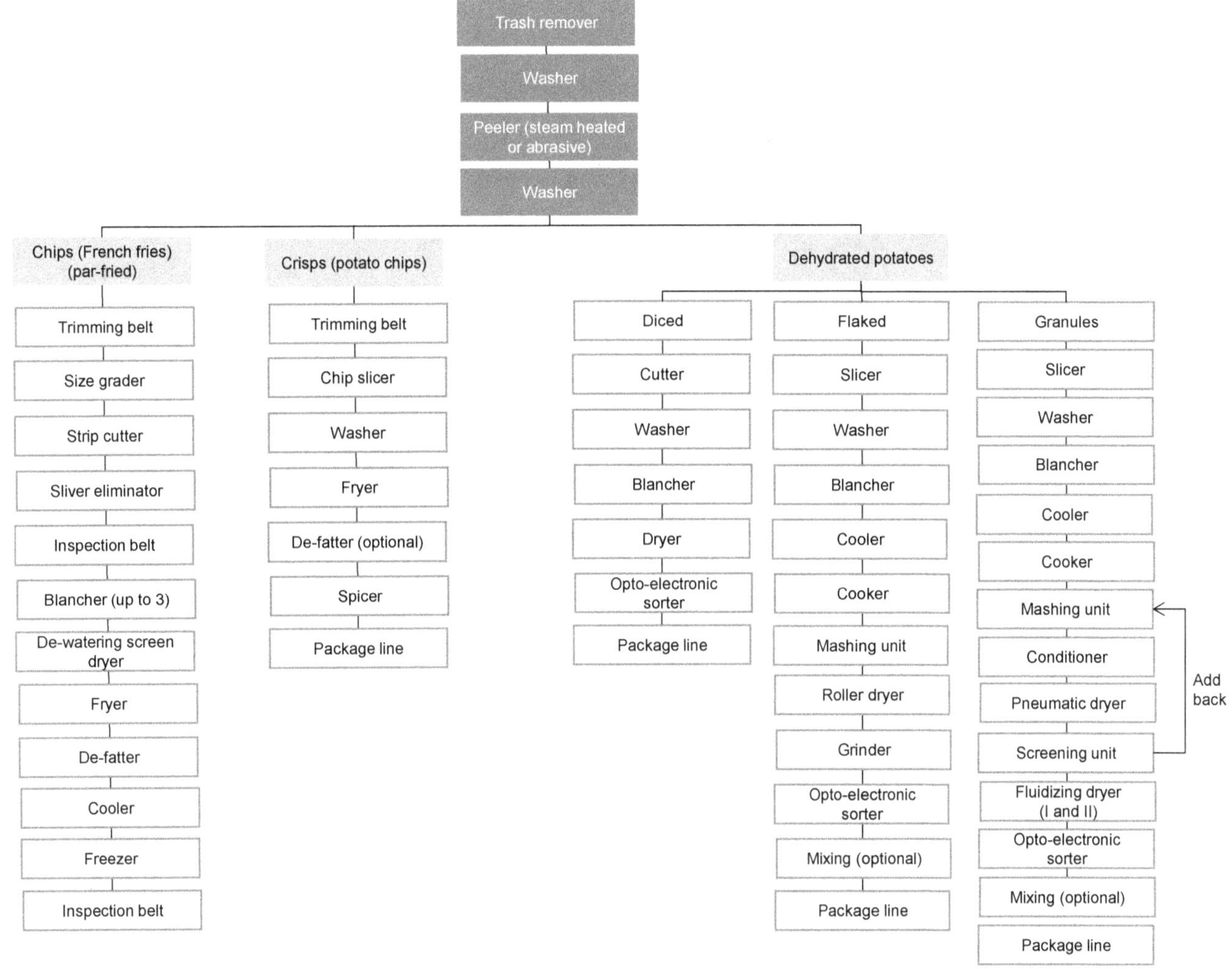

Source: Lisinska and Leszczynski (1989).

Industrial uses of potatoes

Especially in Europe, potatoes are used as raw material for starch manufacturing. Within the European Union, the annual potato starch production is about 1.9 million tonnes (Anonymous, 1995). Due to the high water content of the tubers and accompanying storage problems, separation of potato starch is carried out mainly in autumn and the beginning of winter because of their frost sensitivity.

Potato starch is easily separated from tubers because of its large grain size and the structure of the tubers. In addition to large factories with excellent equipment, very small and simple processing units exist (especially in developing countries). Figure 4.2 maps a typical potato starch processing line.

Potato starch, which is a mixture of amylose and amylopectin (75:25), shows specific properties different from starch of other sources. Therefore, several applications prefer potato starch, e.g. coating of papers, sizing of cotton, finishing in textile industry

(Treadway, 1975). Potato starch is also used in the food industry, particularly, in pre-gelatinised or modified form. Additional specific applications for potato starch can be foreseen with the development of potato varieties containing mainly one or the other starch component.

By-products (pulp, coagulated protein from fruit water) are normally used in animal feeding, but trends exist for food use too.

Figure 4.2. **Schematic description of potato starch processing**

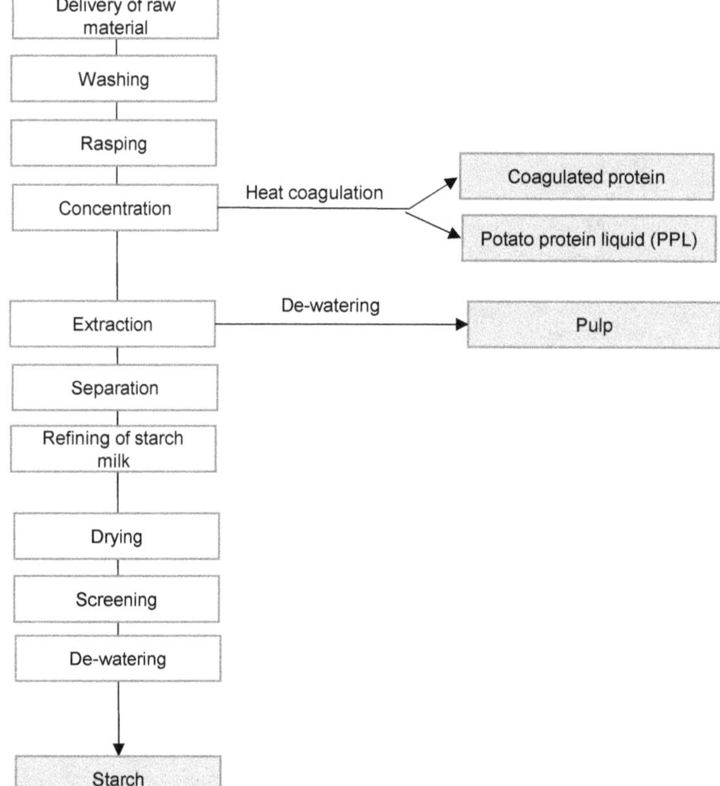

If coagulated protein is prepared from potatoes with a high glycoalkaloid content (particularly from unpeeled potatoes), the protein cannot be used in the food industry due to its high toxin content.

Potatoes are also used for industrial alcohol production. The basic method for alcohol production is to crush and cook the potatoes in water. The resulting gelatinised starch is hydrolysed to sugars (either by acids or by technical enzymes), and pumped to vats, where it is fermented by yeast addition. The fermentation is complete after two to three days and the alcohol is distilled off. The potential yield of alcohol from 1 tonne of potatoes varies between 60 and 140 litres.

The residues of the distillation process are used as feed stuff.

Potatoes as animal feed

The extent to which potatoes are used as animal feed varies considerably. It depends mostly on the price and availability of substitutes. Because of their low nutrient concentration, potatoes are an inefficient basic feed. On the other hand, the nutrient yield per hectare is higher than in any other crop. Therefore, home-produced feeding potatoes have an advantage over other crops (Burton, 1989). Normally, potatoes are fed in combination with other feedstuffs to meet the animal's requirements and to take advantage of complementary effects (Burton, 1989; Schindler, 1996). Balanced supplementation with amino acids, minerals and vitamins has to be considered.

In countries with a significant potato processing industry (for both human food and industrial use), the residues and by-products (peel, trimmings, rejected potatoes, separated pulp and proteins) are used as feedstuff (often after dehydration). In countries without a processing industry, potatoes which do not meet food standards are traditionally fed to stock (Burton, 1989).

Potatoes are normally fed raw to ruminants, but fed steamed to pigs. Practical feeding instructions for the various species are described in papers and textbooks on feeding stuffs and feeding (e.g. Church, 1984; Pond et al., 1995). The contribution rates to which potatoes are incorporated in diets for the various animal categories are as follows (Kling and Wöhlbier, 1983):

- swine: 2.4-7.8 kg/day according to live weight (30-110 kg); total consumption during the whole growing finishing period is about 700 kg of potatoes
- beef cattle: 5-15 kg per day
- dairy cattle: 5-10 kg per day.

Key food and feed nutrients

Due to its vegetative origin, the potato tuber is extremely sensitive to environmental impacts. Depending on variety, climate, soil type and farming practice, the composition of potato tubers may vary widely. The colour of potato tubers depends on the variety. Key nutrients of the potato tubers of safely consumed varieties are listed in Table 4.2. The cited ranges of values do not imply that values outside these ranges are necessarily unusual or harmful in any way.

Table 4.2. **Key nutrients of potato tubers (fresh weight basis)**

	Mean	Ranges
Dry matter (DM)	23.7%	13.1-36.8%
Starch	17.5%	8.0-29.4%
Protein	2.0%	0.69-4.63%
Fat	0.12%	0.02-0.2%
Dietary fibre[1]	1.7%	1-2%
Crude fibre	0.71%	0.17-3.48%
Minerals (crude ash)	1.1%	0.44-1.87%
Sugars	0.5%	0.05-8.0%
Ascorbic acid + Dehydroascorbic acid mg/kg	100-250	10-540

Sources: Adapted from Lisinska and Leszczynski (1989) and [1] Woolfe (1987).

Dry matter

The dry matter (solids; DM) content of tubers is composed of various substances, soluble or insoluble in water. The dry matter content is correlated with the specific gravity, ranging from 1.0485 to 1.151 g/cm^3 (Lisinska and Leszczynski, 1989). Specific gravity is a quality factor and is used for dry matter determination.

Potatoes high in dry matter content (18-24%) are suitable for the manufacture of dehydrated food products and animal feed. Potatoes for deep-fat frying (crisps and chips), in particular, require an optimum range of dry matter content (21-24%).

During storage, losses of dry matter may be up to 8% of fresh weight (FW) or 2% DM caused by tuber respiration. Respiration intensity depends on storage conditions.

Starch

Potato dry matter consists of 75-80% starch. Starch is the most important carbohydrate determining the quality of potato tubers used as food or feed. Tubers with a high starch content are more susceptible to mechanical damage (black spot susceptibility). The texture of cooked tubers tends toward mealiness if starch content is very high.

Potato starch plays an important role as both a food ingredient and as an industrial raw material (native as well as modified potato starch).

Protein

Potato protein is of high nutritional value despite protein denaturation during processing. It contains high levels of the essential amino acids lysine, methionine, threonine and tryptophan (Table 4.3).

Table 4.3. **Amino acid composition of potato tuber protein**

Amino acid	Ranges
Alanine	**4.62-5.32%**
Arginine	4.74-5.70%
Aspartic acid	11.9-13.9%
Cysteine	0.20-1.25%
Glutamic acid	10.2-11.8%
Glycine	4.30-6.05%
Histidine	2.10-2.50%
Isoleucine	3.73-5.80%
Leucine	9.70-10.3%
Lysine	6.70-10.1%
Methionine	1.20-2.15%
Phenylalanine	4.80-6.53%
Proline	4.70-4.83%
Serine	4.90-5.92%
Threonine	4.60-6.50%
Tryptophan	0.30-1.85%
Tyrosine	4.50-5.68%
Valine	4.88-7.40%

Source: Adapted from Lisinska and Leszczynski (1989).

The major proteins present in potato tubers are albumin, globulin, prolamine and glutenin. Another protein fraction is made up of glycoproteins (patatin, lectin), metaloprotein and phosphoproteins. Potato species and varieties can be discriminated by gel-electrophoresis of soluble tuber proteins.

Fat

The lipid content of potatoes is mainly composed of free fatty acids, fats and phospholipids. Linoleic acid comprises up to 40-50% of all fatty acids, linolenic acid 20-30%, oleic acid 1-5%, palmitic acid 20% and stearic acid 5%. Since the fat content of potato tubers is very low (0.02-0.2% FW), potatoes are not regarded as an important fat source.

Among phospholipids, the most important compounds are lecithins. Free carotenoids and their esters of fatty acids are present in potato tubers in very small amounts (0.1-0.4% of total lipid content).

The predominance of unsaturated fatty acids in the lipids confers easy oxidation. This is a critical factor in manufacture and storage, in particular, for dehydrated potato products.

Dietary fibre and crude fibre

Dietary fibre consists of insoluble and soluble polysaccharides, but also of lignin and of resistant starch. The definition of dietary fibre focuses on its "non-availability". In this view, dietary fibre is the sum of components which are not digested by enzymes of the human small intestine. Nevertheless, many of them are fermented by microorganisms in the large intestine. Processing of food, e.g. cooking or frying, may change some fibre properties (pectin breakdown) and the amount of resistant starch.

Crude fibre consists of cellulose, hemicellulose, pentosans and pectic substances. They are particularly concentrated within the cell wall. The composition of the cell wall is responsible for the textural characteristics of potato tubers. Cell wall breakdown during cooking, in combination with swollen and gelatinised starch granules, leads to cell rupture, whereas breakdown of the middle lamella allows cell separation (→ soft cooking tubers). Pectin release and pectin de-esterification accompany cell wall breakdown.

Sugars

The sugar content of potato tubers varies highly depending on the variety, maturity and physiological stage of the potatoes.

Sugar content changes during storage. Specific changes in the sucrose content are used as an indicator of the age of potato tubers.

A high sugar content (especially of the reducing sugars glucose and fructose) disqualifies potato tubers from their use as raw material for processing, especially for deep-fat fried and dehydrated products. Potatoes for the crisps industry should not exceed 0.15% of reducing sugars in fresh weight, whereas potatoes for the production of chips and dehydrated potatoes should contain less than 0.25% of reducing sugars.

Storage at +4°C inhibits sprouting; however, in most varieties the concentration of reducing sugars (resulting from starch hydrolysis) will increase at that temperature.

Vitamins

During preparation and processing of tubers, water soluble vitamins may be washed out. In addition, vitamins may be destroyed by heat and oxidation. Losses of 20-80% have been reported (Kolbe, 1997). The ascorbic acid content (10-540 mg/kg) may also be decreased during storage as it is used up as an antioxidant. Nevertheless, ascorbic acid from potatoes may contribute to the daily intake of humans, up to 40% of the recommended amount.

Minerals

Potassium is the major cation in potato tubers (0.22-0.94% FW). Its percentage of total mineral content is about 50% (Lisinska and Leszczynski, 1989) and its contribution to the human diet is up to 30% of the recommended daily potassium intake. Therefore, in low potassium diet regimes, this mineral should be watered out prior to further preparation. The potassium content is positively correlated with the content of organic acids. Sodium content of potato tubers is very low (3% of total mineral content).

Toxins and allergens

Glycoalkaloids

Potatoes naturally contain several types of alkaloids. The most important group of alkaloids in commercial potato varieties are the glycoalkaloids (GA), in which one or more sugar molecules (usually three) are linked to the steroidal alkaloid solanidine.

The total glycoalkaloid content (TGA) of potato tubers varies widely. Values between 2 and 410 mg/kg FW were found (Lisinska and Leszczynski, 1989), but in most cases the TGA concentration in whole tubers is between 10 and 150 mg/kg FW (Van Gelder, 1990). Ninety-five percent of the total glycoalkaloids in potato tubers consists of α-chaconine (solanidine-glucose-rhamnose-rhamnose) and α-solanine (solanidine-galactose-glucose-rhamnose).

Other combinations between the solanidine alkaloid and sugar molecules may be present in small amounts:

- β-chaconine (solanidine-glucose-rhamnose)
- γ-chaconine (solanidine-glucose)
- β1-solanine (solanidine-galactose-glucose)
- β2-solanine (solanidine-galactose-rhamnose)
- γ-solanine (solanidine-galactose).

Several other glycoalkaloids might be present in certain potato varieties, especially if these have been recently crossed with wild Solanum species. Glycoalkaloids are not evenly distributed within the tubers, but are present in higher concentrations at the periphery (reviewed by Smith et al., 1996). Therefore, tuber size is important for the GA level. Large and often unpredictable variations in GA levels can arise from differences in variety, locality, season, cultural practice and stress factors. Today, the widely accepted safety limit for the level of TGA in tubers is 200 mg/kg FW (Boemer and Mattis, 1924; Smith et al., 1996).

Glycoalkaloids are particularly concentrated in the outer region of the tuber. However, in green and sprouted tubers, the TGA concentration is also high in the internal part. In any case, peeling reduces the TGA content substantially. Glycoalkaloids are not destroyed during cooking and frying.

Glycoalkaloid poisoning causes several symptoms ranging from gastrointestinal disorders through confusion, hallucination and partial paralysis to convulsions, coma and death (Smith et al., 1996). Available information suggests that the susceptibility of humans to glycoalkaloid poisoning is high and very variable: oral doses in the range of 1-5 mg/kg body weight are marginally to severely toxic to humans (Hellenäs et al., 1992) whereas 3-6 mg/kg body weight can be lethal (Morris and Lee, 1984).

In pig feeding with a high potato portion (steamed, but unpeeled potatoes; see above) a TGA concentration of 150 mg/kg FW seems to be without any risks and does not result in growth depression. In cattle feeding, no risk is known when maximal portions are incorporated in the ration (see above), as long as the sprouts, which contain TGA concentrations of 2 000-5 000 mg/kg FW, are removed (Jeroch et al., 1993).

Recently, potato tubers have been shown to also contain small quantities of calystegines, which are nortropane alkaloids with glycosidase inhibitory activity. Calystegines are concentrated predominantly in potato eyes and sprouts (Keiner et al., 2000). The biological significance of this group of alkaloids for humans is not yet known.

Allergens

Until recently potatoes were not considered a source of allergens. However, potato contains multiple heat-labile proteins which can induce immediate hypersensitivity reactions when raw potatoes are consumed (Jeannet-Peter et al., 1999).

A study on patatin, the main storage protein in potatoes, reports induction of allergic reactions in sensitive children (Seppälä et al., 1999). The authors consider additional studies necessary, in order to confirm the allergenicity of patatin. In addition to patatin, concomitant IgE binding to several proteins belonging to the family of soybean trypsin inhibitors was observed (Seppälä et al., 2001).

Anti-nutrients

Protease inhibitors

Potato tubers contain several protease inhibitors that inhibit the activity of trypsin, chymotrypsin and other proteases, thus decreasing the digestibility and the biological value of the ingested protein. The concentration of trypsin inhibitors (TI) can be as high as 174 mg g^{-1} protein (Baker et al., 1982). Assuming a protein content of 2% FW in potato tubers (Table 4.2), this may result in a TI content of up to 3.5 g/kg potato tubers.

Protease inhibitors in potatoes are largely inactivated by boiling and other thermal processes. Serious anti-nutritional reactions could occur, however, if raw or inadequately cooked potatoes are consumed or fed.

Lectins

Lectins are (glyco)proteins which occur in virtually all living organisms and have the common property of binding to specific carbohydrate structures on cell surfaces, e.g. on intestinal or blood cells (Liener, 1989, Allen et al., 1996, Ciopraga et al., 2000). Some lectins found in beans are known to cause serious health effects when ingested by humans and animals. As lectins are inactivated during heating, only consumption and feeding of raw or inadequately cooked potatoes may cause adverse effects. Negative effects of lectins on animals' health and their performance are not yet known in detail (Kling and Wöhlbier, 1983; Smart et al., 1999).

Considerations for the assessment of new potato varieties

Agronomic characteristics of new potato varieties are important to consider since unspecific or unpredicted phenotypic traits or changes in phenotypic traits may be indicative of unintended effects of potential safety concerns that would require further investigations. In registration of new potato varieties, phenotypic traits and agronomic characteristics are tested, including yield, susceptibility and tolerance towards specific diseases. In addition, table potatoes are tested using sensory analysis, while processing potatoes are tested as chips, crisps and dehydrated potatoes.

The comparison of the chemical composition of tubers from a modified variety with tubers from the non-modified comparator, grown at the same time under the same conditions, should include the following components (according to Love, 2000):

- dry matter
- sugars, especially reducing sugars
- protein
- vitamin C
- glycoalkaloids.

If the analyses of these parameters indicate that a novel variety is within the ranges given in the literature, apart from the intentional modifications resulting in recombinant DNA and new proteins, it can be considered equivalent with respect to its overall composition. The safety assessment would then focus on the newly introduced (e.g. recombinant DNA and heterologous proteins) or intentionally altered constituents (e.g. starch components).

If, apart from the intentionally modified DNA and resulting new proteins, the genetic modification results in a qualitative change rather than a quantitative shift of the potato constituents outside the naturally occurring ranges, the safety assessment would focus on those differences, possibly requiring nutritional and/or toxicological studies.

Notes

1. For information on the environmental considerations for safety assessment of potato, see OECD (1997b).

References

Allen, A.K. et al. (1996), "Potato lectin: A three-domain glycoprotein with novel hydroxyproline-containing sequences and sequence similarities to wheat-germ agglutinin", *International Journal of Biochemistry & Cell Biology*, Vol. 28, No. 11, pp. 1 285-1 291.

Anonymus (1995), Stärkekontingente, Kartoffelwirtschaft, 02.08.

Baker, E.C. et al. (1982), "Development of a pilot-plant process for the preparation of a trypsin inhibitor-rich fraction from potatoes", *Industrial & Engineering Chemistry Product Research and Development*, Vol. 21, No. 1, pp. 80-82, http://dx.doi.org/10.1021/i300005a017.

Boemer, A. and H. Mattis (1924), Der Solaningehalt der Kartoffeln. Z. Unters. Nahr. Genussm. Gebrauchsgegenstaende, Vol. 47, pp. 97-127.

Burton, W.G. (1989), *The Potato*, 3rd edition, Longman Group UK Limited.

Church, D.C. (1984), *Livestock Feeds and Feeding*, 2nd edition, Prentice Hall, Inc., Englewood Cliffs, New Jersey.

Ciopraga, J. et al. (2000), "Isolectins from Solanum tuberosum with different detailed carbohydrate binding specificities: Unexpected recognition of lactosylceramide by N-acetyllactosamine-binding lectins", *Journal of Biochemistry*, Vol. 128, No. 5, November, pp. 855-867.

FAO (2000), *Report of a Joint FAO/WHO Expert Consultation on Foods Derived from Biotechnology*, World Health Organization, Geneva, 29 May-2 June 2000, http://www.fao.org/fileadmin/templates/agns/pdf/topics/ec_june2000_en.pdf (accessed 11 February 2015).

FAO (1996), *Report of a Joint FAO/WHO Expert Consultation on Biotechnology and Food Safety*, Food and Agriculture Organization, Rome, Italy, 20 September- 4 October 1996 http://www.fao.org/ag/agn/food/pdf/biotechnology.pdf (accessed 11 February 2015)

FAOSTAT (2001), FAO Statistics on-line website, http://faostat.fao.org/, Food and Agriculture Organization, Rome, (accessed in 2002).

Hellenäs, K.-E. et al. (1992), "Determination of potato glycoalkaloids and their aglycone in blood serum by high-performance liquid chromatography. Application to pharmacokinetic studies in humans", *Journal of Chromatography*, Vol. 573, No. 1, pp. 69-78.

Jeanett-Peter, N. et al. (1999), "Facial dermatitis, contact urticaria, rhinoconjunctivitis, and asthma induced by potato", *American Journal of Contact Dermatitis*, Vol. 10, No. 1, pp. 40-42.

Jeroch, H. et al. (1993), Futtermittelkunde Gustav Fischer Verlag Jena, Stuttgart, Germany.

Keiner, R. et al. (2000), "Accumulation and biosynthesis of calystegines in potato", *Journal of Applied Botany*, Vol. 74, No. 3-4, pp. 122-125.

Kling, M. and W. Wöhlbier (1983), Handelsfuttermittel, Eugen Ulmer Verlag, Stuttgart, Germany.

Kolbe, H. (1997), "Einflußfaktoren auf die Inhaltsstoffe der Kartoffel, Teil VII: Vitamine", *Kartoffelbau*, Vol. 48, pp. 34-39.

Liener, I.E. (1989), "The nutritional significance of lectins", in: Kinsella, J.E. and W.G. Soucie (eds.), *Food Proteins*, American Oil Chemists' Society, Champaigne, Illinois, pp. 329-353.

Lisinska, G. and W. Leszczynski (1989), *Potato Science and Technology*, Elsevier Applied Science, London.

Love, S.L. (2000), "When does similar mean the same: A case for relaxing standards of substantial equivalence in genetically modified food crops", *HortScience*, Vol. 35, No. 5, August, pp. 803-806, available at: http://hortsci.ashspublications.org/content/35/5/803.full.pdf.

Morris, S.C. and T.H. Lee (1984), "The toxicity and teratogenicity of Solanaceae glycoalkaloids, particularly those of the potato (Solanum tuberosum): A review", *Food Technology Australia*, Vol. 36, pp. 118-124.

OECD (1997), "Report of the OECD workshop on toxicological and nutritional testing of novel foods", SG/ICGB(1998)1, OECD, Paris, final version available at: www.oecd.org/officialdocuments/publicdisplaydocumentpdf/?doclanguage=en&cote=sg/icgb(98)1/final.

OECD (1997b), *OECD Consensus Document on the Biology of Solanum tuberosum subsp. tuberosum (Potato),* Series on Harmonization of Regulatory Oversight in Biotechnology, No. 8, OECD, Paris, available at: www.oecd.org/science/biotrack/46815598.pdf.

OECD (1993), *Safety Evaluation of Foods Derived by Modern Biotechnology. Concepts and Principles*, OECD, Paris, available at: www.oecd.org/science/biotrack/41036698.pdf. Pond, W.G. et al. (1995), *Basic Animal Nutrition and Feeding*, John Wiley & Sons.

Rogan, G.J. et al. (2000), "Compositional analysis of tubers from insect and virus resistant potato plants", *Journal of Agricultural Food and Chemistry*, Vol. 48, No. 12, December, pp. 5 936-5 945.

Schindler, M. (1996), "Kartoffeln in der Futterration", *Kartoffelbau*, Vol. 47, pp. 384-385.

Seppälä, U. et al. (2001), "Identification of four novel potato (Solanum tuberosum) allergens belonging to the family of soybean trypsin inhibitors", *Allergy*, Vol. 56, No. 7, pp. 619-626.

Seppälä, U. et al. (1999), "Identification of patatin as a novel allergen for children with positive skin prick test responses to raw potato", *Journal of Allergy and Clinical Immunology*, Vol. 103, No. 1, Part 1, January, pp. 165-171.

Smart, J.D. et al. (1999), "Lectins in drug delivery: A study of the acute local irritancy of the lectins from Solanum tuberosum and Helix pomatia", *European Journal of Pharmaceutical Sciences*, Vol. 9, No. 1, pp. 93-98.

Smith, D.B. et al. (1996), "Potato glycoalkaloids: Some unanswered questions", *Trends in Food Science & Technology*, Vol. 7, Issue 4, April, pp. 126-131, http://dx.doi.org/10.1016/0924-2244(96)10013-3.

Treadway, R.H. (1975), "Potato starch", Chapter 15 in: Talburt, W.F. and O. Smith (eds.), *Potato Processing*, 3rd ed., Avi Publishing Company, Westport, Connecticut.

Van Gelder, W.M.J. (1990), "Chemistry, toxicology, and occurrance of steroidal glycoalkaloids: Potential contaminants of the potato (Solanum tuberosum L.)", in: Rizk, A.-F.M. (ed.), *Poisonous Plant Contamination of Edible Plants*, CRC Press, pp. 117-156.

Weber, L. and B. Putz (1998), "Vitamin C in kartoffeln", *Kartoffelbau*, Vol. 49, pp. 278-281.

Woolfe, J.A. (1987), *The Potato in the Human Diet*, Cambridge Press, Cambridge, United Kingdom.

WHO (1991), *Strategies for Assessing the Safety of Foods Produced by Biotechnology*, Report of a Joint FAO/WHO Consultation, World Health Organization of the United Nations, Geneva, out of print.

Chapter 5

Maize (*Zea mays*)

*This chapter, prepared by the OECD Task Force for the Safety of Novel Foods and Feeds with the Netherlands and the United States as lead countries, deals with the composition of maize (*Zea mays*). It contains elements that can be used in a comparative approach as part of a safety assessment of foods and feeds derived from new varieties. Background is given on maize production and processing, followed by appropriate varietal comparators and characteristic screened by breeders. Then nutrients in maize and its products, anti-nutrients, allergens and secondary metabolites are detailed. The final sections suggest key products and constituents for analysis of new varieties for food use and for feed use.*

Background

Production of maize for food and feed

Maize is the world's third leading cereal crop, following wheat and rice. It is grown as a commercial crop in over 25 countries worldwide. Field maize has been grown for 8 000 years in Mexico and Central America and for 500 years in Europe.

Maize is naturally cross-pollinated and until about 1925 mainly open pollinated varieties were grown. Today mainly hybrids are grown. To produce hybrid seed, the tassels are removed from the plants prior to pollen shedding so that only one sort of pollen will be received by the silks. The hybrid plants grown from this seed give more vigorous growth and higher yields.

Sweet maize, derived from field maize by crossbreeding, introducing a sugar gene, has been grown in the United States since 1930 and in Europe since 1979. Maize for popcorn is a minor crop. The cultivation and use mainly takes place in the United States (Jugenheimer, 1976).

Worldwide production of maize is about 600 million tonnes a year (Corn Refiners Association, 2001; Pingali, 2001). In the European Union (EU), the annual total production of maize is 38.9 million tonnes. The major producers, the United States and the People's Republic of China, account for 43.2% and 17.9% of the field maize production respectively. In the EU, 6.6% of the total amount of field maize is grown. The United States accounts for 81% of the production of sweet maize, whereas in the EU, only 7% is grown.

Table 5.1. **World maize grain production, 2000/01**

	Production (million tonnes)	% of total
United States	253.2	43.2
China (People's Republic of)	105.0	17.9
European Union	38.9	6.6
Brazil	38.5	6.6
Mexico	18.5	3.2
Argentina	15.0	2.6
India	12.0	2.0
South Africa	8.0	1.4
Canada	6.8	1.2
Indonesia	6.2	1.1
Egypt	5.8	1.0
Hungary	4.5	0.8
Thailand	4.4	0.8
Philippines	4.3	0.7
Romania	4.0	0.7

Source: Based on local marketing years in thousands metric tonnes, adapted from USDA (2001) cited in Corn Refiners Association (2001).

In the EU, 2.9 million tonnes of field maize are used as food, and 21 million tonnes as feed (Eurostatistics, 1994). In 1995-97, 66% of all the maize produced worldwide was used for animal feed and 17% for human consumption. In developing countries, 30% of the maize produced was used for human consumption and 57% for animal feed, whereas in Western Europe, North America and other high-income countries, 4% was used for human consumption and 76% for animal feed during the same period (Pingali, 2001). The consumption of sweet maize was 79 thousand tonnes (frozen), 298 thousand tonnes (canned) and 45 thousand tonnes (fresh) in 1995 in Europe (AGPM, 1996).

Field maize and its products are used in food products (oil, grits, meal, flours, ethanol, syrup, starch) and feed (hulls, gluten, hominy). Sweet maize and its products are used in food (kernels, meal) and feed (hulls, 60-65% of volume). Popcorn maize kernels are used for popcorn and as a basis for confections.

Processing of maize

Wet milling

The maize kernel is composed of a hard outer layer (pericarp), the germ and endosperm (NCGA, 1999). The pericarp is a very hard fibrous coat of cellulose and hemicellulose that must be broken or removed in order for the kernel to be beneficial for consumption or for processing (Eckhoff and Paulsen, 1996). The tip of the pericarp that attaches the kernel to the cob is softer and easily broken providing an access into the kernel, particularly in the steeping process. The germ, the only living part of the corn kernel, contains about 50% oil on a dry weight basis, while the endosperm contains 70% starch (White and Pollak, 1995). Processes have been devised to separate these components of the maize kernel and, in the process, derive many food (67%) and feed (33%) products (Newcomb, 1995).

The wet milling process is the most important one; it employs modern technology as shown in Figure 5.1.

Generally, the type of corn used for wet milling is yellow dent. However, it is estimated that "waxy type" corn may make up as much as one-third of the corn processed, while a very small amount of high amylose corn is also processed (White and Pollak, 1995). Maize prepared for wet milling must be cleaned as thoroughly as possible. It is then steeped in hot water (49°C-54°C) and sulfur dioxide (0.1-0.2%) to soften the pericarp. Water-soluble nutrients adhering to the surface of the maize enter the steep water that is drawn off and evaporated, leaving solubles which are mixed with cleanings and screenings to make maize gluten feed. The softened kernel is cracked by machine and hydrocloned, a flotation process that separates the germ portion from the endosperm. The germ portion is pressed to separate the oil that is used for margarine, cooking oil and baking and frying fats for human use. The pressed germ is dried and added to maize gluten feed. The endosperm portion is finely milled and passed through screens to remove the fibrous hulls. Fibrous hulls are also added to maize gluten feed. The screened endosperm portion is centrifuged to separate the starch portion from the gluten. The gluten portion is dried and used as maize gluten meal. Maize starch, the primary product of wet milling, is obtained by washing and drying the starch portion.

About 40% of the starch is consumed directly as food or used for other industrial purposes, while about 60% is converted to various sweeteners (White and Pollak, 1995). The primary sweeteners (maize syrups) are regular, high fructose, dextrose and maltodextrins. The major use is for syrup containing approximately 55% fructose that is much sweeter than sucrose. Maltodextrins are not sweet, but contribute viscosity, mouthfeel and body to food products. Starch also serves as a major source of sugar for the fermentation of beverage alcohol. Dextrose that is enzymatically produced from starch has many food uses.

Figure 5.1. **Wet processing of maize**

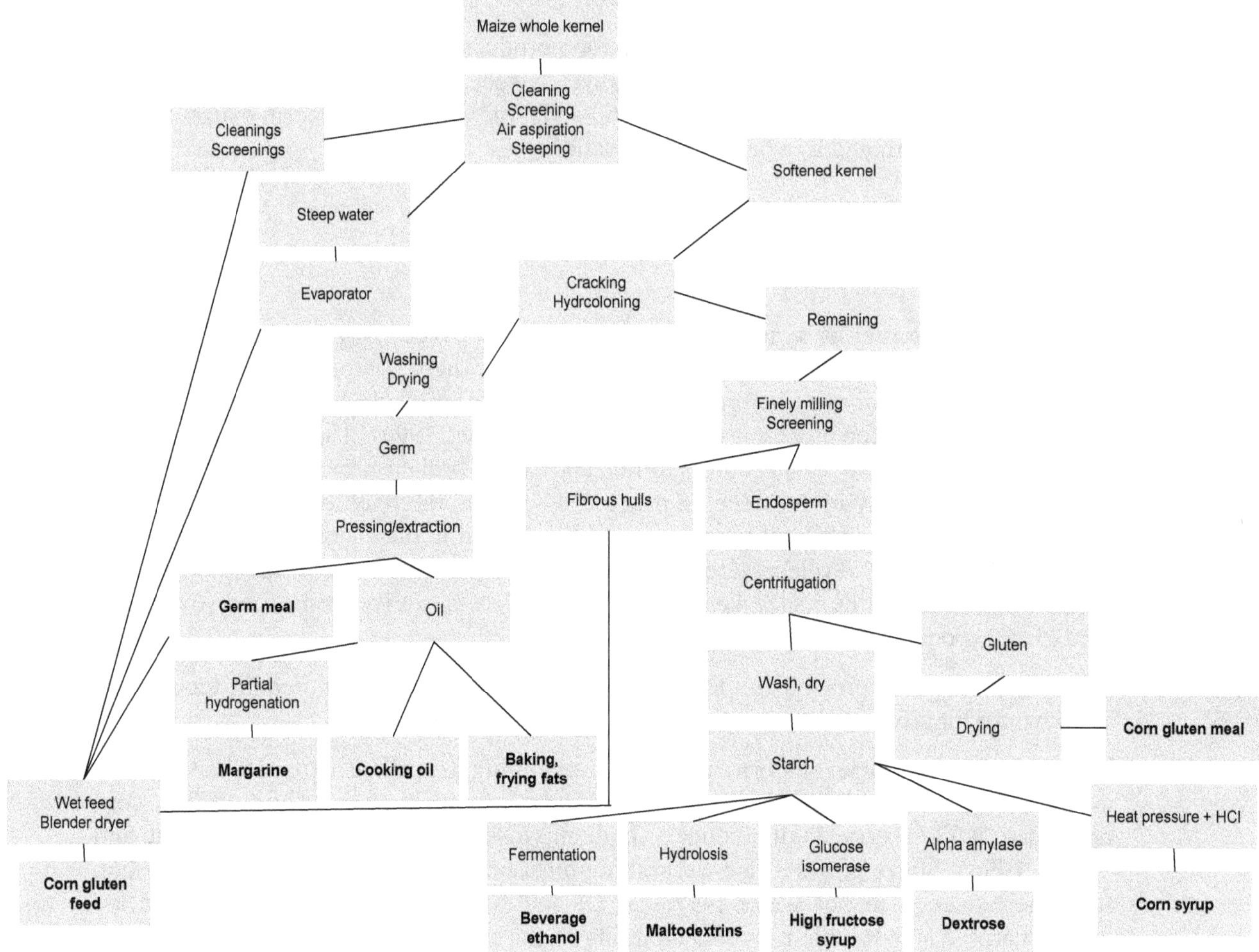

Dry milling

Dry milling is the oldest way of processing the corn kernel for human and animal food use. Dry milling is a term that usually refers to one of three different processes. The first process it stone grinding after screening and cleaning. Stone grinding is widely used in Africa, Latin America, Asia and by small mills in the United States and Canada (White and Pollak, 1995). Because of the oil content, the storage life and flavour stability of whole cornmeal is short. The industry has, therefore, devised processes that remove the oil, producing more refined products.

The second process is the dry-grind ethanol process for producing ethanol for commercial purposes (Eckhoff and Paulsen, 1996). Maize kernels are cleaned, ground, cooked, saccharified and put into a fermenter to convert starch to ethanol. The by-product, distillers dried solubles, is an important livestock feed.

The third process is called the tempering degerminating system (TD), and is the most widely used in the food processing industry. Maize kernels are cleaned and tempered by soaking them in water, strengthening the pericarp and the germ to protect them from shattering in subsequent mechanical separation procedures. Tempering is followed by degerminating, drying and mechanical separation. The preferred degerminating equipment is the Beall type degerminator, though several other types of machines are used, i.e. Entoleter, granulator, disc mill, roller mill or decorticator (Eckhoff and Paulsen, 1996). The tempering and degerminating steps are the most important because the clean separation of the germ is paramount to obtaining high-quality products in the downstream separation process.

The usual products and yields of the TD process are flaking grits (12%), coarse grits (15%), regular (fine) grits (23%), meal (6%), flour (4%), oil (1%) and hominy feed (35%). Corn bran is high in fibre, low in calories and readily absorbs water, making it a useful additive in human prepared foods. Flaking grits are used almost exclusively in the manufacture of corn flakes. Fine grits are frequently utilised by the snack, breakfast cereal and brewing industries. Cooked coarse grits are eaten as a breakfast food. Maize flour is used as an ingredient in muffins, breadings, batters, pancakes, doughnuts, breakfast foods and as binders in processed meats. Dried-milled maize products serve as a substrate for brewing beer. Corn grits and whole kernels are used to produce many distilled hard liquors. A minimum of 51% maize is used in the fermentation of the mash that is distilled into bourbon (White and Pollak, 1995).

Figure 5.2. **Dry milling of maize**

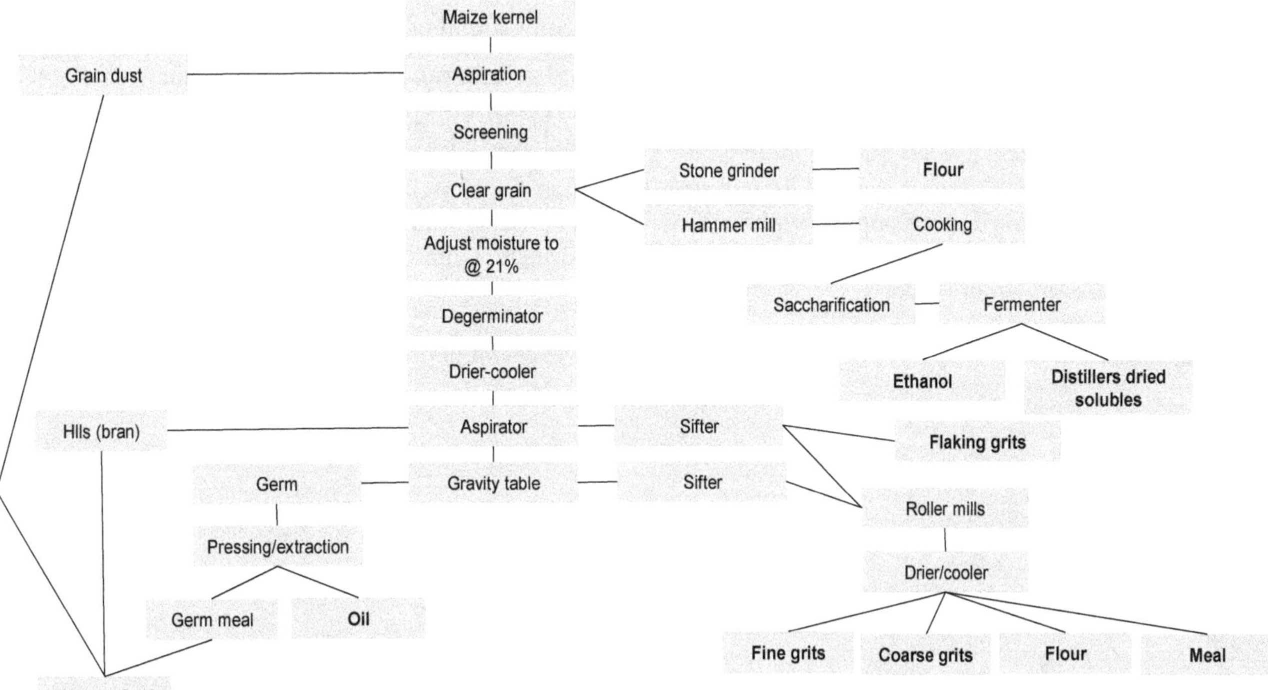

Masa production

Cooking (85-95°C) maize in the presence of alkali (lime) and fine grinding it produces a dough material called Masa. Masa is the starting material for tortillas, taco shells, tortilla chips and maize chips, which are widely consumed in the south-western United States, Mexico, Central and South America. Both white and yellow corn are used to make Masa. Totally hard endosperm dent corn is preferred because its superior cooking characteristics maximise the handling and mechanical qualities of the finished products. However, good quality Masa can be produced from soft endosperm corn by altering the cooking time (Eckhoff and Paulsen, 1996).

Feed processing

As described earlier, animal feed is produced as a by-product of milling. Alternatively, the whole corn plant may be used for animal (primarily ruminant) feed. It is harvested at various stages of growth, usually after the ear is formed, but is usually mechanically chopped prior to full maturity of the ear when the plant contains about 35-40% moisture. The material can be fed directly or preserved as silage in an upright sealed silo or in a trench or bunker, so as to limit the exposure to oxygen. The resulting silage is allowed to age under anaerobic conditions producing a palatable feed that retains up to 90% of its nutrients (Ensminger et al., 1990).

When the ear is allowed to go to maturity and the moisture content recedes to around 15%, it can be harvested by mechanically picking, or by mechanically picking and shelling in one operation. Stalks can be grazed in the field by ruminants, or harvested for roughage or animal bedding. Dry ear corn can be mechanically ground and fed to ruminants or it can be shelled. If the moisture content of the harvested maize grain is above about 13%, it is either dried or sometimes stored in an airtight silo and fed to ruminants or swine as high-moisture corn.

The cobs can be used in animal feed or for commercial uses.

Corn grain is the feed of choice. It can be fed as is to ruminants as whole grain, rolled (cracked), ground or steam flaked with there being little difference in digestible and net energy in diets containing less than 20% roughage (NRC, 1996).

Maize grain is usually ground or rolled when fed to swine and poultry, but pelleting is becoming more popular with poultry producers (Newcomb, 1995). For use in pet foods, maize is usually ground, cooked, pelleted or extruded. Common feed processing methods are shown in Figure 5.3.

Figure 5.3. **Feed processing of maize**

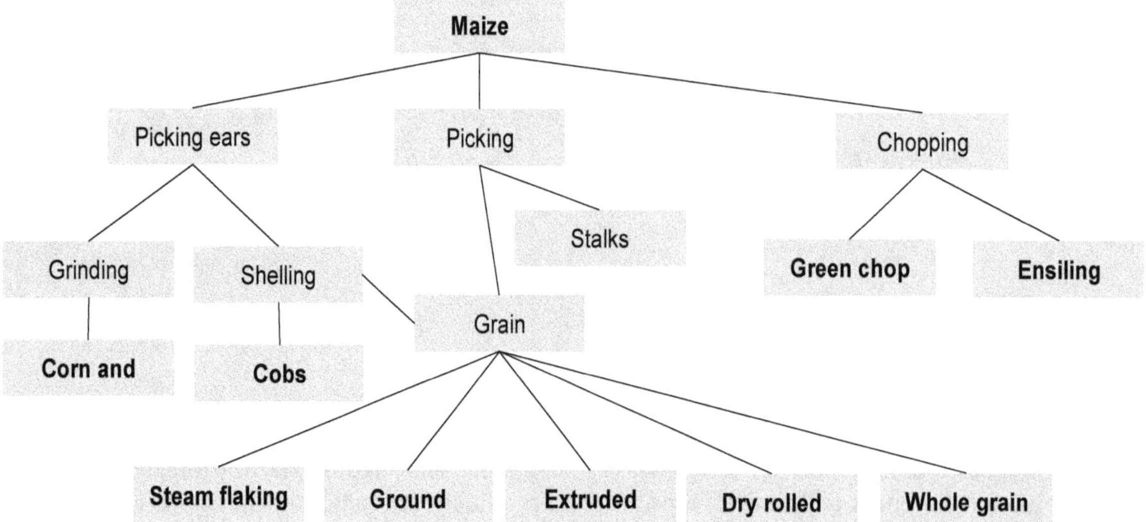

Appropriate comparators for testing new varieties

This chapter suggests parameters that maize developers should measure. Measurement data from the new variety should ideally be compared to those obtained from the near isogenic non-genetically modified organisms line grown under identical conditions. A developer can also compare values obtained from new varieties with the literature values of conventional counterparts present in this chapter. Critical components include key nutrients and key toxicants for the food source in question. Key nutrients are those components in a particular product which may have a substantial impact in the overall diet. These may be major constituents (fats, proteins and carbohydrates) or minor compounds (vitamins and minerals). Key toxicants are those toxicologically significant compounds known to be inherently present in the species, i.e. compounds whose toxic potency and level may impact on human and animal health. Similarly, the levels of known anti-nutrients and allergens should be considered. As part of the comparative approach, selected secondary plant metabolites, for which characteristic levels in the species are known, are analysed as further indicators of the absence of unintended effects of the genetic modification on the metabolism.

Traditional characteristics screened by maize developers

Phenotypic characteristics provide important information related to the suitability of new varieties for commercial distribution. Selecting new varieties is started based on parent data. Plant breeders developing new varieties of maize evaluate many parameters at different stages in the developmental process. In the early stages of growth, breeders evaluate stand count and seedling vigour. As the plant matures, disease data is evaluated, e.g. grey leaf spot, anthracnose, fusarium and head smut infestations. At near maturity or maturity, root lodging, stalk lodging, brittle snap, time to silk and time to shed are evaluated. The mature plant is measured for plant height, ear height, drop ear and husk cover. The harvested grain is measured for yield, moisture and test weight. In some cases, plants are modified for specific increases in certain components, and the plant breeder would be expected to analyse for such components (UPOV, 1994).

Nutrients in maize and maize products

Kernels

Dent field maize is harvested at maturity. The kernel goes through maturity stages denoted by “milk”, “dough” and “dent”. Maize kernels consist of endosperm (containing starch) and germ (containing oil). They are wrapped in the pericarp, a cellulose layer. At maturity of field maize, which usually occurs about 50-60 days after pollination, moisture content is 30% (White and Pollack, 1995). Sweet maize is harvested when the kernels are in the “milk” stage, when moisture content is about 75%. The moisture content of dried popcorn maize kernels is about 10%. It should be noted that values for some components (e.g. minerals) could vary considerably as result of differences in genetics and environmental and agronomic conditions (FAO, 1992).

In addition to these general types of maize, several maize variants have been developed with specific improvements in composition. Quality Protein Maize (QPM) variants have been developed with improved levels of lysine and tryptophan, the two limiting essential amino acids in maize protein. Other specialty types of maize are characterised by a higher oil content, higher amylose content or higher amylopectin content (waxy maize) (Jugenheimer, 1976). If the characteristic level for a specific component, which is altered in a specialty type of maize, is outside the general range of values found in the scientific literature, the comparison with the parent line will be decisive.

Table 5.2. **Proximate analysis of field maize kernels**

% of dry weight

Reference	Watson 82	Watson 87	USDA[1]	Souci et al.[1]	NRC[1,2]	Commercial range[3]	Range
Moisture		7-23	10.37	12.0-13.0	10-11.9	9.4-14.4	7-23
Protein	8.1-11.5	6-12	10.5[4]	9.37-12.1[5]	9.3-9.8	9.57-12.7	6-12.7
Total fat	3.9-5.8	3.1-5.7	5.29[4]	3.66-4.91[5]	4.1-4.4	3.6-5.3	3.1-5.8
Ash	1.27-1.52	1.1-3.9	1.34[4]	1.28-1.73[5]	1.5	1.28-1.5	1.1-3.9
Neutral detergent fibre (total fibre)[6]	8.3-11.9	8.3-11.9			9.5-10.8	10.1-11.7	8.3-11.9
Acid detergent fibre (cellulose)[6]	3.0-4.3	3.3-4.3			3.1-3.3	3.7	3.0-4.3
Total dietary fibre[6]				11.1[5]			11.1
Carbohydrates			82.85[4]			82.2-82.9	82.2-82.9

Notes: 1. Possibly including GMO-varieties. 2. Values taken from NRC (1994, 1998, 2000, 2001). Values from NRC (1994 and 1998) are calculated from given values on total weight basis, using reported moisture content of 11.00%. 3. Commercial range on non-GMO controls, compiled from data from AgrEvo (1998), Dow AgriSciences LLC (2000), Monsanto (1997, 2000) and Pioneer HyBrid International (1998). 4. Values calculated from given percentage of total weight, using the reported moisture content of 10.37%. 5. Values calculated from given percentage of total weight, using 12.50% as the average moisture content (reported values range from 12.0-13.0%). 6. Proximate analysis of maize usually includes acid detergent fibre (ADF) and neutral detergent fibre (NDF). The terms ADF and NDF are still commonly used in the feed industry and values for comparison are readily available. For food use, however, the concept of dietary fibre is preferred, although different definitions and methods of analysis are being used (see: Panel on the Definition of Dietary Fibre and the Standing Committee on the Scientific Evaluation of Dietary Reference Intakes, Food and Nutrition Board, 2001). The value for total dietary fibre from Souci et al. is obtained using a modification of the analytical method recommended by the Association of Official Analytical Chemists. Total dietary fibre determined this way includes lignin and non-starch polysaccharides (including cellulose, hemicellulose and pectin).

Sources: Watson (1982, 1987); USDA (2001); Souci et al. (2000); NRC (1994, 1998, 2000, 2001).

Table 5.3. **Proximate analysis of sweet maize and popcorn maize kernels**

Reference		Sweet maize				Popcorn maize	
		NEVO[1,2]	USDA[1]	Souci et al.[1]	Range	NEVO[1,3]	USDA[1,4]
Moisture	% of fw	84	75.96	74.70	74.70-84	10	4.10
Protein	% of dw	15.6[5]	13.4[5]	11.3-14.6[6]	11.3-15.6	12.2[5]	12.5[5]
Total fat	% of dw	8.75[5]	4.91[5]	4.86[6]	4.86-8.75	4.4[5]	4.38[5]
Ash	% of dw		2.58[5]	2.77-3.86[6]	2.58-3.86		1.88[5]
Total dietary fibre	% of dw	15.6[5]	11.2[5]		11.2-15.6	5.56[5]	15.7[5]
Carbohydrates	% of dw	72.5[5]	79.18[5]		72.5-79.18	79[5]	81.2[5]

Notes: 1. Possibly including GMO-varieties. 2. Values for boiled kernels. 3. Dried kernels. 4. Air-popped kernels. 5. Values calculated from a given percentage of total weight, using the indicated moisture content. 6. Values calculated from a given percentage of total weight, using average moisture content of 74.70% (reported values range from 73.90-75.60%).

Sources: USDA (2001); NEVO (2001); Souci et al. (2000).

Table 5.4. **Levels of minerals and vitamins in field maize kernels**

Reference		Watson 82	Watson 87	USDA[1,2]	Souci et al.[1,3]	NRC[1,4]	Commercial range[5]	Range
Sodium (Na)	mg/100 g	0-150	0-150	39	1.1-11	10-22		0-150
Potassium (K)	mg/100 g	320-720	320-720	320	340	340-440	360-370	320-720
Calcium (Ca)	mg/100 g	10-100	10-100	7.8	4.4-22	22-40	3-5	3-100
Phosphorus (P)	mg/100 g	260-750	260-750	234	190-290	300-320	290-320	234-750
Magnesium (Mg)	mg/100 g	90-1 000	90-1 000	142	82-140	120-130	120-130	82-1 000
Iron (Fe)	mg/100 g	0.1-10	0.1-10	3.02	1.7	3.3-5.5	2.3-2.5	0.1-10
Copper (Cu)	mg/100 g	0.09-1.0	0.09-1.0	0.35	0.27	0.25-0.34	0.19-0.21	0.09-1.0
Selenium (Se)	mg/100 g	0.0045	0.001-0.1	0.017	0.005-0.018	0.0034-0.014		0.001-0.1
Zinc (Zn)	mg/100 g	1.2-3.0	1.2-3.0	2.47	1.9	2.0-2.7	2.0-3.0	1.2-3.0
Vitamin A	mg/kg RE[6]	2.5 IU[6]/g	2.5 mg/kg	0.52	0.49-2.18			0.49-2.18
Vitamin B1 (thiamin)	mg/kg	3.0-8.6	3.0-8.6	4.3	2.3-6.9	3.9	3.5	2.3-8.6
Vitamin B2 (riboflavin)	mg/kg	0.25-5.6	0.25-5.6	2.2	1.1-2.7	1.1-1.3	5.6	0.25-5.6
Vitamin B6 (pyridoxine)	mg/kg	9.6	5.3	6.9	4.6	5.6-7.9		4.6-9.6
Vitamin C (ascorbic acid)	mg/kg			0	0			
Vitamin E	mg/kg	3.0-12.1	17-47 IU/kg	8.4 mg/kg ATE[6]	4.1-31.1mg vit.E act[6]	9.3-25 ATE		
Folate, total	mg/kg			0.21 (folic acid=0)	0.23-0.46 mg/kg folic acid	0.17-0.45		
Niacin (nicotinic acid)	mg/kg	9.3-70	9.3-70	40.5	11-23 mg/kg nicotinamide	27		9.3-70

Notes: All values are expressed on a dry weight basis. 1. Possibly including GMO-varieties. 2. Values calculated from given values on total weight basis, using reported moisture content of 10.37%. 3. Values calculated from given values on total weight basis, using average moisture content of 12.50% (reported values range from 12.0-13.0). 4. Values taken from NRC (1994, 1998, 2000, 2001). Values from NRC (1994 and 1998) calculated from given values on total weight basis, using the reported moisture content of 11.00%. 5. Commercial range on non-GMO controls, compiled from data from AgrEvo (1998), Dow AgriSciences LLC (2000), Monsanto (1997, 2000) and Pioneer Hybrid International (1998). 6. RE: retinol equivalents; IU: international units; ATE: alpha tocopherol equivalents = vit. E act.

Sources: Watson (1982, 1987); USDA (2001); Souci et al. (2000); NRC (1994, 1998, 2000, 2001).

Table 5.5. **Levels of minerals and vitamins in sweet maize and popcorn maize kernels**

Reference		Sweet maize				Popcorn maize	
		NEVO[1,2]	USDA[1]	Souci et al.[3]	Range	NEVO[1,4]	USDA[1,5]
Sodium (Na)	mg/100 g	6.3	62	0.59-1.98	0.59-62	5.6	4.2
Potassium (K)	mg/100 g	1 560	1 120	900-1 150	900-1 560	278	314
Calcium (Ca)	mg/100 g	69	8.3	8.6-13.7	8.3-69	22	10
Phosphorus (P)	mg/100 g	625	370	320-328	320-625	278	313
Magnesium (Mg)	mg/100 g	281	154	106-120	106-281		137
Iron (Fe)	mg/100 g	3.1	2.2	1.6-2.3	1.6-3.1	3.3	2.77
Copper (Cu)	mg/100 g	0.25	0.22	0.08-0.18	0.08-0.25		0.44
Selenium (Se)	mg/100 g	trace	0.025	0.0025-0.011	0.0025-0.025		0.10
Zinc (Zn)	mg/100 g	6.25	1.9	2.21-3.95	1.9-6.25		3.59
Vitamin A	mg/kg RE	0.44	1.16	0.40	0.40-1.16	0.89	0.21
Vitamin B1 (thiamin)	mg/kg	7.5	8.3	5.9	5.9-8.3	3.3	2.1
Vitamin B2 (riboflavin)	mg/kg	4.4	2.5	4.7	2.5-4.7	0.89	3.0
Vitamin B6 (pyridoxine)	mg/kg	6.3	2.3	8.7	2.3-8.7	2.4	4.7
Vitamin C (ascorbic acid)	mg/kg	0	283	470	283-470	0	0
Vitamin E	mg/kg	56	3.7 mg/kg ATE[7]	3.75 mg/kg vit. E act.[7]			1.25 mg/kg ATE
Folate, total	mg/kg	2.1	19.2 (folic acid=0)	1.7 mg/kg folic acid		0.12	2.4
Niacin (nicotinic acid)	mg/kg	106	70.8	67.2 mg/kg nicotinamde		11	20.3

Notes: All values are expressed on a dry weight basis. 1. Possibly including GMO-varieties. 2. Values calculated from given values on total weight basis, using moisture content indicated in Table 5.3. 3. Values calculated from given values on total weight basis, using average moisture content of 74.70% (reported values range from 73.90-75.60%). 4. Values for boiled kernels. 5. Dried kernels. 6. Air-popped kernels. 7. ATE: alpha tocopherol equivalents = vit. E act.

Sources: USDA (2001); NEVO (2001); Souci et al. (2000).

Table 5.6. **Amino acid composition of maize kernels in percentage of kernel dry weight**

Reference	Field maize							Sweet maize	Popcorn maize
	Watson[1]	White and Pollack[2]	USDA[3,4]	Souci et al.[3,5]	NRC[3,6]	Comm. range[7]	Range	USDA[3,4,5]	USDA[3,4,8]
Essential amino acids									
Methionine	0.10-0.21	0.16-0.25	0.22	0.10-0.46	0.19-0.20	0.17-0.28	0.10-0.46	0.28	0.26
Cysteine	0.12-0.16	0.20-0.27	0.19	0.08-0.32	0.20-0.21	0.17-0.26	0.08-0.32	0.11	0.23
Lysine	0.20-0.38	0.26-0.34	0.30	0.05-0.55	0.27-0.30	0.21-0.38	0.05-0.55	0.57	0.35
Tryptophan	0.05-0.12	0.04-0.06	0.07	0.05-0.13	0.07-0.07	0.05-0.08	0.04-0.13	0.10	0.09
Threonine	0.29-0.39	0.28-0.39	0.39	0.37-0.58	0.33-0.33	0.27-0.49	0.27-0.58	0.54	0.47
Isoleucine	0.26-0.40	0.27-0.38	0.38	0.40-0.71	0.31-0.33	0.22-0.50	0.22-0.71	0.54	0.45
Histidine	0.20-0.28	0.24-0.32	0.32	0.15-0.38	0.26-0.29	0.21-0.38	0.15-0.38	0.37	0.38
Valine	0.21-0.53	0.39-0.52	0.53	0.49-0.85	0.38-0.45	0.30-0.61	0.21-0.85	0.77	0.63
Leucine	0.79-1.54	0.98-1.38	1.29	1.04-2.41	1.05-1.14	0.84-1.84	0.79-2.41	1.45	1.54
Arginine	0.29-0.60	0.36-0.51	0.52	0.22-0.64	0.42-0.43	0.27-0.57	0.22-0.64	0.54	0.62
Phenylalanine	0.29-0.58	0.39-0.54	0.52	0.37-0.58	0.43-0.44	0.32-0.64	0.29-0.64	0.62	0.62
Glycine	0.26-0.47	0.32-0.41	0.43	0.49	0.38	0.29-0.45	0.26-0.49	0.53	0.51
Non-essential amino acids									
Alanine	0.65-1.00	0.59-0.79	0.79	0.88-0.95		0.56-1.04	0.56-1.04	1.23	0.94
Aspartic acid	0.59-0.73	0.52-0.71	0.73	0.67-0.72		0.48-0.85	0.48-0.85	1.01	0.87
Glutamic acid	1.25-1.98	1.46-2.01	1.97	1.99-2.15		1.26-2.58	1.25-2.58	2.65	2.35
Proline	0.67-1.04	0.71-0.99	0.92	1.06-1.36		0.63-1.16	0.63-1.36	1.21	1.09
Serine	0.42-0.56	0.35-0.49	0.50	0.57-0.61	0.42	0.37-0.91	0.35-0.91	0.64	0.60
Tyrosine	0.29-0.47	0.22-0.34	0.43	0.22-0.79	0.28-0.34	0.12-0.48	0.12-0.79	0.51	0.51

Notes: 1. Values calculated from a given percentage of total amino acids (10.1% total protein). 2. Values calculated from a given percentage of total amino acids (8.74% total protein). 3. Possibly including GMO-varieties. 4. Values calculated from given values on total weight basis. 5. Values calculated from given values on total weight basis, using average moisture content of 12.50%. 6. Values taken from NRC (1996, 1998, 2001). Values from NRC (1994 and 1998) calculated from given values on total weight basis, using reported moisture content of 12.00% and 11.00%, respectively. Values from NRC (2001) were calculated from reported percentage of crude protein, using given crude protein content of 9.4% on dry basis. 7. Commercial range on non-GMO controls, compiled from data from AgrEvo (1995, 1998), Dow AgriSciences LLC (2000), Monsanto (1995, 1997, 2000) and Pioneer Hybrid International (1998). 8. Values for air-popped kernels.

Sources: Watson (1982); White and Pollack (1995); USDA (2001); Souci et al. (2000); NRC (1994, 1998, 2001).

Table 5.7. **Fatty acid composition of maize kernels in percentage of kernel dry weight**

Reference	Field maize					Sweet maize	Popcorn maize
	USDA[1,2]	Souci et al.[1,3]	NRC[1,4]	Comm. range[5]	Range	USDA[1,2]	USDA[1,2,6]
16:0 Palmitic	0.63	0.29-0.79	0.70	0.30-0.37	0.29-0.79	0.71	0.52
18:0 Stearic	0.084	0.04-0.17	0.11	0.05-0.08	0.04-0.17	0.046	0.073
18:1 incl. Oleic	1.39	1.26	1.31	0.70-1.03	0.70-1.39	1.44	1.15
18.2 incl. Linoleic	2.34	0.67-2.81	2.04	1.80-2.21	0.67-2.81	2.25	1.92
18.3 incl. Linolenic	0.073	0.03-0.08	0.10	0.03-0.04	0.03-0.10	0.067	0.063

Notes: 1. Possibly including GMO-varieties. 2. Values calculated from given values on total weight basis. 3. Values calculated from given values on total weight basis, using average moisture content of 12.50%. 4. Values taken from NRC (1994) are calculated from given values on total weight basis, using average moisture content of 11.00%. 5. Commercial range compiled from data from Aventis crop Science (1999) and Monsanto (1996b, 2000). 6. Values for air-popped kernels.

Sources: USDA (2001); Souci et al. (2000); NRC (1994).

Oil

Oil is produced from the field maize germ by wet milling. Maize oil in the germ consists mostly of triglycerides (TG) (75-92%). Crude maize oil contains 95.6% TG and 1.7% free fatty acids (FFA). Refined oil contains 98.8% TG and 0.03% FFA (oleic acid) (Anderson and Watson, 1982). The fatty acids, linoleic acid, oleic acid and palmitic acid form the major part of the TG (Watson, 1987). In Table 5.8, fatty acids consistently present at levels below 1% are not included. Maize oil is used in salad- and cooking oil, mayonnaise and margarine, baking and frying fat and in sauces and soups. In the production process for refined maize oil, protein is reportedly reduced to amounts below 100 micrograms per ml (SCF, 1999), or to amounts below the level of detection (Federal Register, 2000; EPA, 2001).

Table 5.8. **Fatty acid composition of refined maize oil in percentage of total fatty acids**

	USDA[1,2]	Codex Alimentarius[1]	Anderson and Watson	Orthoefers and Sinram
16:0 Palmitic	11.4	8.6-16.5	11.5	11.0 ± 0.5
18:0 Stearic	1.9	0-3.3	2.0	1.8 ± 0.3
18:1 incl. Oleic	25.3	20.0-42.2	24.1	25.3 ± 0.6
18.2 incl. Linoleic	60.7	34.0-65.5	61.9	60.1 ± 1.0
18.3 incl. Linolenic	0.73	0-2.0	0.7	1.1 ± 0.3

Notes: 1. Possibly including GMO-varieties. 2. Values calculated from a given percentage of oil.

Sources: USDA (2001); Codex Alimentarius (1999); Anderson and Watson (1982); Orthoefer and Sinram (1987).

Grits, meal, flour, bran

Grits, meals and flours are products of the dry-milling process of field maize with de-germination. Bran is a by-product of this process. Grits are used to make cereals and snacks and also to produce alcoholic beverages; meal is used for bread and muffins, flour for pancakes and snacks. Bran is used as a dietary source of fibre. Typical composition (percentage as-is basis) of dry milled corn products is 7-8% protein; less than 1% fat, ash or fibre; and 77-79% starch (88-90%, dry basis) (Alexander, 1987).

Table 5.9. **Proximate analysis of grits, flour and meal in percentage of dry weight**

Reference	Grits			Flour				Meal	
	Alexander	Anderson and Watson	USDA[1]	Alexander	Anderson and Watson	USDA[1]	Souci et al.[1]	Anderson and Watson	USDA[1]
Moisture[2]	11.5	12	10	13	12	9.81	12.00	12	11.59
Protein	8.47	9.9	9.78	6.0	8.9	6.13	7.65-11.38	9.0	9.59
Carbohydrates	90.2	88.8	88.4	90.7	87.3	90.7		89.1	87.9
Fat	0.79	0.91	1.33	2.3	3.0	1.52	1.77-4.43	1.36	1.87
Crude fibre	0.23	0.45		0.57	0.80				
Total dietary fibre[3]			1.78			2.08	0.7	0.68	8.4
Ash	0.34	0.45	0.44	0.46	0.91	0.50	1.30-1.36	0.57	0.68

Notes: 1. Possibly including GMO-varieties. 2. Values are a percentage of total weight (all other values are a percentage of d.w., calculated from a given percentage of total weight). 3. Measured according to the AOAC method.

Sources: Alexander (1987); Anderson and Watson (1982); USDA (2001); Souci et al. (2000).

Table 5.10. **Levels of minerals and vitamins in grits, flour and meal**

Reference		Grits	Flour		Meal
		USDA	USDA	Souci et al.	USDA
Sodium (Na)	mg/100 g	1.1	1.1	0.80	3.4
Potassium (K)	mg/100 g	152	99	136	183
Calcium (Ca)	mg/100 g	2.2	2.2	11-30	5.7
Phosphorus (P)	mg/100 g	81	66		95
Magnesium (Mg)	mg/100 g	30	20	53	45
Iron (Fe)	mg/100 g	1.1	1.0	2.7	1.2
Copper (Cu)	mg/100 g	0.083	0.16		0.088
Selenium (Se)	mg/100 g	0.19	0.088		0.088
Zinc (Zn)	mg/100 g	0.46	0.41		0.81
Vitamin A	mg/kg RE[1]	4.9	0.55	0.57	4.6
Vitamin B1 (thiamin)	mg/kg	1.44	0.81	4.3-5.6	1.58
Vitamin B2 (riboflavin)	mg/kg	0.44	0.64	1.3-1.9	0.57
Vitamin B6 (pyridoxine)	mg/kg	1.63	1.06	0.68	2.91
Vitamin C (ascorbic acid)	mg/kg	0	0	0	0
Vitamin E	mg/kg ATE[1]	2.9	3.6		3.7
Folate, total	mg/kg	0.56	5.3	0.11 mg/kg folic acid	5.4
Niacin (nicotinic acid)	mg/kg	13.3	29.1	19-23.5 mg/kg nicotinamide	11.3

Notes: All values are expressed on dry weight basis, calculated from given values per 100 g of total weight; data possibly include GMO-varieties. 1. RE: retinol equivalents; ATE: alpha tocopherol equivalents.

Sources: USDA (2001); Souci et al. (2000).

Starch

Starch is derived from field maize by the wet-milling process. About 60% of the starch is converted (by acid or enzyme hydrolysis) to sweeteners (syrups) and ethanol. The remaining 40% is used for foods and industrial uses. The lipids in starch are mainly free fatty acids (Anderson and Watson, 1982). Starch is used in a variety of products that include bakery products, baby foods, sauces, dressings and soups. Typically, maize starch contains residual protein at 0.4% (SCF, 1999) or 0.6% (Federal Register, 2000), whereas starch hydrolysates contain 100-200 ppm of protein (SCF, 1999).

Table 5.11. **Proximate analysis of corn starch in percentage of dry weight**

	Anderson and Watson	USDA[1]	Souci et al.[1]
Moisture (% of total weight)	11	8.32	11-12.6
Protein	0.39	0.28	0.30-0.78
Lipids	0.61	0.055	0-0.23
Carbohydrate	98.9	99.55	
Fibre	0.11[2]	0.98[3]	
Ash	0.11	0.098	0.07-0.34

Notes: 1. Possibly including GMO-varieties. 2. Measured as crude fibre. 3. Measured as total dietary fibre (AOAC method).

Sources: Anderson and Watson (1982); USDA (2001); Souci et al. (2000).

Feed

Gluten meal and gluten feed are by-products of the wet processing of maize. Hominy feed and distillers' grain with solubles are products of the maize dry-milling industry. Gluten meal is high in protein (65-69%) and carotenoids. Gluten feed is medium in protein (24-25%) and is higher in fibre. Hominy feed is lower in protein with about the same fibre content as gluten feed. Distillers' grain with solubules is a medium protein (29%) higher fibre product.

All maize products are relatively low in the amino acid lysine and in calcium. When maize and maize products are used in diets containing soybean meal, the amino acid composition of the feed meets nutritional requirements of most domestic animals.

Table 5.12. **Proximate of common maize animal feed products**

Parameter		Gluten meal	Gluten feed	Hominy feed	Distillers grain w/solubules	Maize silage	Maize grain[1]
Moisture	% of fw	86-90	90	90	90.2-93.0	62-78	7-23
Protein	% of dw	65.0-68.9	23.98-24.4	11.4-11.56	29.7-29.5	4.7-9.2	6-12.7
Neutral detergent fibre	% of dw	9.17-14.00	33.5-37.0	23.0-38.8	45.0	40-48.2	8.3-10.8
Acid detergent fibre	% of dw	5.00-5.11	11.89-12.1	6.2-9.0	17.53-19.7	25.6-34	3.0-4.3
Fat	% of dw	2.5-3.22	2.77-3.33	5.7-8.89	9.03-10.00	1.5-3.2	3.1-5.8
Ash	% of dw	1.9	6.8-6.9	2.2-2.7	5.2-7.7	2.9-5.7	1.1-3.9

Note: 1. Values taken from Table 5.2.

Sources: BNF (includes Monsanto (1995, 1996a, 1996b, 1997, 1999, 2000); Aventis Crop Science (1999); Dow Agrisciences LLC (2000)); NRC (1994, 1998, 2001); Ensminger et al. (1990).

Anti-nutrients and allergens in maize

Phytic acid

Phytic acid (myo-Inositol 1,2,3,4,5,6-hexakis [dihydrogen phosphate]) is present in maize and binds about 60-75% of the phosphorus in the form of phytate (NRC, 1998). Because of phytate binding, bioavailability of phosphorus in maize is less than 15% for non-ruminant animals.

Ruminants utilise considerably more phosphorus since the rumen microbes produce the enzyme phytase that breaks down phytate and releases phosphorus (Ensminger et al., 1990). It is becoming common for feed formulators to add phytase to swine and poultry diets to improve the utilisation of phosphorus. Phytic acid levels in maize grain vary from 0.45% to 1.0% of dry matter (Monsanto, 1995; Watson, 1982).

Table 5.13. **Levels of minerals, amino acids and fatty acids in common maize animal feed products**

% of dry weight

Parameter	Gluten meal	Gluten feed	Hominy feed	Distillers grain w/solubules	Maize silage	Maize grain[1]
Calcium	0.06-0.08	0.04-0.27	0.05-0.06	0.22-0.32	0.15-0.31	0.003-0.15
Phosphorus	0.49-0.56	0.55-1.00	0.48-0.57	0.83-1.40	0.20- 0.27	0.23-0.75
Argenine	2.02-2.14	0.91-1.16	0.52-0.62	1.21-1.22	0.17-0.34	0.22-0.64
Histidine	1.33-1.42	0.68-0.79	0.22-0.31	0.74-0.74	0.16-0.17	0.26-0.37
Isoleucine	2.67-2.76	0.74-0.98	0.40-0.44	1.10-1.11	0.29-0.34	0.22-0.71
Leucine	10.9-11.3	2.10-2.44	0.93-1.09	2.76 -2.85	0.75-0.76	0.79-2.41
Lysine	1.10-1.14	0.65-0.71	0.42-0.44	0.67-0.67	0.22-0.33	0.05-0.55
Methionine	1.54-1.66	0.38-0.50	0.14-0.20	0.54-0.54	0.135-0.15	0.10-046
Phenylalanine				1.44-1.45	0.34-0.40	0.29-0.64
Threonine	2.20-2.31	0.82-0.99	0.44-0.44	1.01-1.02	0.28-0.37	0.27-0.58
Tryptophan	0.34-0.40	0.08-0.13	0.11-0.12	0.26-0.27	0.04- 0.09	0.04-0.13
Valine	3.02-3.10	1.06-1.16	0.54-0.58	1.40 -1.40	0.39-0.47	0.48-0.59
Cysteine	1.21-1.22	0.51-0.57		0.55-0.56	0.118-0.12	0.08-0.32
Glycine						0.26-0.49
Palmitic 16:0						0.29-0.79
Stearic 18:0	0.07					0.04-0.17
Oleic 18:1	0.68					0.70-1.39
Linoleic 18:2	1.29-1.30					0.67-2.81
Linolenic 18:3						0.03-0.10

Note: 1. Values for calcium and phosphorus taken from Table 5.4; those for amino acids from Table 5.6; and those for fatty acids from Table 5.7.

Sources: Monsanto (1995, 1996); NRC (1994, 1998, 2001); Ensminger et al. (1990).

Dimboa

2,4-Dihydroxy-7-methoxy-2H-1,4-benzoxazin-3(4H)-one (DIMBOA) belongs to a group of metabolites, hydroxamic acids and benzoxazinoids, commonly found in cereal plants. The glycoside of DIMBOA, DIMBOA-glc, is the most prominent of these compounds in green aerial- and root- tissues of maize during initial plant development (Cambier et al., 2000).

Levels of DIMBOA and related compounds in green- and root- tissues of maize seedlings vary by orders of magnitude (approximately 0-1 000 ppm fresh weight) among maize varieties (Xie et al., 1992). High levels are associated with elevated resistance of conventional maize varieties against insects, such as European Corn Borer (Sicker et al., 2000). In addition, these levels change in the course of green tissue development, reaching a maximum within several days after germination and then declining to a fraction within weeks (Cambier et al., 2000).

DIMBOA-glc is enzymatically deglycosylated in injured plant tissues to DIMBOA, which is toxic to insects. The mechanism of DIMBOA's toxicity to insects has not been elucidated yet. In addition, data on the possible toxic and physiological effects of DIMBOA and related compounds on humans and domestic animals are scarce. One report, for example, describes the *in vitro* mutagenicity of DIMBOA in the Ames test (Hashimoto et al., 1979). In addition, a number of reports document hormonal effects of MBOA, a metabolite of DIMBOA, in wild rodents (Korn, 1988). Data on hormonal effects of MBOA in domestic animals are, however, fragmentary.

Analysis of DIMBOA in maize silage is not recommended in Table 5.15 because of the high variability of its levels among maize varieties and the fragmentary knowledge on its toxicology.

Raffinose

Raffinose is a non-digestible oligosaccharide (NDO), i.e. it cannot be broken down by enzymes in the gastro-intestinal tract. Raffinose is considered an anti-nutrient due to gas production and resulting flatulence caused by its consumption (Maynard et al., 1979). A daily dose of 15 g NDO is considered to be safe (Voragen, 1998). Raffinose is not a toxicant but may cause discomfort. It can be removed from food and feed by soaking, cooking, enzyme or solvent treatment and by irradiation.

Percentages of raffinose in field maize are 0.21-0.31%, and in sweet maize 0.1%. (Naczk et al., 1997; Aung et al., 1993; NOTIS Plus, 1999).

Other anti-nutrients

Maize contains low levels of trypsin and chymotrypsin inhibitors, neither of which is considered nutritionally significant (White and Pollak, 1995).

Identification of allergens

Maize is not a common allergenic food, although in some case studies, allergic reactions were reported (Hefle, 1996). These reported allergic effects for maize include skin, gastrointestinal and respiratory complaints.[1]

Secondary plant metabolites

Secondary plant metabolites are neither nutrients nor anti-nutrients. They are important though for compositional analysis and the comparative approach (OECD, 1997). As part of the comparative approach, selected secondary plant metabolites, for which characteristic levels in the species are known, are analysed as further indicators of the absence of unintended effects of the genetic modification on the metabolism. Characteristic plant metabolites in maize are furfural and phenolic acids (ferulic acid and p-coumaric acid). The biological function is not always known, but furfural might play a role in toxicity and the phenolic acids might influence digestion, while other data suggest beneficial effects.

Furfural

Furfural is a heterocyclic aldehyde. It occurs in several vegetables, fruits and cereals. It is used as a pesticide, but also in foodstuff as flavouring. Furfural is generally recognised as safe (GRAS) by FEMA under conditions of intended use as a flavour ingredient, i.e. at levels 100 times lower than the occurrence of furfural as a natural ingredient in traditional foods. Field maize contains <0.01 ppm (mg/kg) furfural (Adams et al., 1997).

The acute toxicity of furfural is moderate, with $LD_{50\ (oral)}$ 50-149 mg/kg bw (rats), 250-500 mg/kg bw (mice) and 650-950 mg/kg bw (dogs) (Adams et al., 1997). In acute and sub-chronic studies in rodents, effects were seen mainly in the liver. Evidence of genotoxicity and carcinogenic activity after oral administration is limited. Furfural is considered an oral genotoxic carcinogen of low potency. An increase of the furfural level in food stuff should be avoided (Feron et al., 1991). Furfural can partly be removed from products by heating.

Ferulic acid and p-coumaric acid

The phenolic acids, ferulic acid and p-coumaric acid are structural and functional components of plant cells (Kroon and Williamson, 1999). Their function is, amongst others, to act as a natural pesticide against insects and fungi. Ferulic acid and p-coumaric acid are found in vegetables, fruit and cereals. They are also used as flavouring in foods, as supplements and in traditional Chinese herbal medicine.

Daily intake of phenolic acids by humans is estimated to be 0.2-5.2 mg/day (Clifford, 1999; Radtke et al., 1998). There are indications that phenolic acids may play a role in the beneficial health effects of vegetables and fruits. The anti-oxidative action of phenolic acids might be involved in prevention of chronic diseases. Ferulic acid and p-coumaric acid are weak anti-oxidants. *In vitro* tests are equivocal as to whether ferulic acid enhances or inhibits the effects of mutagenic substances (Sasaki et al., 1989; Stich, 1992).

Reported concentrations of ferulic acid in field maize kernels are 0.02-0.03% (NOTIS Plus, 1999), 0.02-0.1% (Classen et al., 1990; Rosazza, 1995) or 0.3% (Dowd and Vega, 1996). Concentrations of p-coumaric acid in field maize kernels are reported to be 0.003-0.03% (Classen et al., 1990; NOTIS Plus, 1999).

Food use

Identification of key maize products consumed by humans

In the EU, 2.9 million tonnes of field maize is consumed as food along with 21 million tonnes as feed (Eurostatistics, 1994). Field maize products (starch, oil, grits, meal and flour) are used in many foods. Starch is mostly fermented to sweeteners (syrups) and ethanol. It is also used for foods, such as bakery products, baby foods, sauces, dressings and soups. Maize oil is used in salad and cooking oil, mayonnaise, margarine, baking and frying fat, and in sauces and soups. Grits are used to make cereals and snacks and also to produce alcoholic beverages. Meal is used for bread and muffins and flour is used for pancakes and snacks. Bran is used as a dietary source of fibre. Field maize is also used as a raw material for the production of paper, fuel, glue, textiles, pharmaceuticals and soap.

During 1995, consumption of sweet maize (mostly the whole kernel is consumed as vegetable) amounted to 76 000 tonnes frozen, 298 000 tonnes canned and 45 000 tonnes fresh in the EU (AGPM, 1996).

Popcorn maize kernels are used (in dried form) as popcorn and as a basis for confections (Juggenheimer, 1976).

Identification of key products and suggested analysis for new varieties

Since all maize-derived food products are produced from kernels, analysis of the composition of kernels is the most appropriate test for food use. If only agronomical traits are influenced by the genetic modification, derived products need not be analysed separately. In other cases, the additional analysis of derived products can be useful, depending on the nature and purpose of the modification (e.g. deliberately changing the oil composition). This can apply to the following products: maize oil, starch, grits, meal and flour. The parameters to be analysed were discussed in detail above.

Table 5.14. **Suggested nutritional and compositional parameters to be analysed in maize matrices for human food use**

Parameter	Oil	Starch	Grits/meal/flour	Kernels (field maize, sweet maize, popcorn)
Proximate analysis[1]		X	X	X
Minerals				X
Vitamins				X
Amino acids			(X)	X
Fatty acids	X		X	X
Phytic acid				X
Raffinose				X
Furfural				X
Ferulic acid				X
p-coumaric acid				X

Note: 1. Proximate includes protein, fat, total dietary fibre, ash and carbohydrates.

Feed use

Identification of key maize products consumed by animals

Maize is the preferred feedstuff in livestock production either as a processed whole grain, as a by-product of the milling industry or as a whole plant silage (Newcomb, 1995). The preference results from its high nutrient value and relative low cost. Yellow dent maize and flint dent maize are the primary types that are fed, though other types of maize such as white, waxy or popcorn, may be fed under certain economically feasible circumstances. The maize kernel contains the most energy of all the grains used for livestock feed, but also has the lowest crude protein content (9-11%) (Ensminger et al., 1990). However, since maize grain is usually included in a high percentage in animal diets, a substantial amount of protein-containing essential amino acids is provided by corn. The corn milling industry, as previously mentioned, produces several animal feed products, such as gluten feed, gluten meal, distillers grains, distillers solubles, germ meal and hominy, that are economically attainable in specific areas. The products of major significance are maize gluten feed and maize gluten meal. Most corn gluten feed is fed to ruminants, but some is fed to swine. The major use of gluten meal is in poultry diets because the gluten contains carotenoid pigments that express themselves in skin and eggs of poultry.

Identification of key products and suggested analysis for new varieties

Maize grain is fed to animals as a source of energy from carbohydrates and oils and provides a source of essential and non-essential amino acids. From the oil, essential fatty acids are also provided. The kernel (grain) is generally fed at moisture levels of 10-15%, which is considered safe for storage. Corn grain is sometimes fed to cattle and swine at moisture levels up to 30-35% where the maize has either been ensiled or treated with an organic acid. The kernel contains about 83% carbohydrate that is in the form of starch, pentosans, dextrins, sugars, cellulose and hemicellulose. Starch makes up the biggest part of the carbohydrate fraction and provides most of the energy. The fibre portion includes the cellulose and hemicellulose portions that are generally unavailable to non-ruminants. Maize grain is rich in linoleic acid, one of the essential fatty acids needed by swine and poultry. Maize also has a favourable content of essential amino acids, with the exception of lysine and tryptophan which are the most limiting amino acids in corn, particularly for swine. Maize provides an important source of methionine which is the most limiting amino acid in poultry. In cattle and sheep, where microbial protein from the rumen is considered the primary protein source for the animal, there is increased interest in proteins that escape rumen fermentation, particularly in high-producing dairy cattle. Thus, nutritionists are taking a closer look at the potential for cattle to also have certain limiting essential amino acids. Methionine and lysine have been found to be the two most limiting amino acids for lactating dairy cattle fed corn-based diets (NRC, 2001). The ten traditional essential amino acids are arginine, histidine, isoleucine, leucine, lysine, methionine, phenylalanine, threonine, tryptophan and valine. Glycine is considered essential for poultry. Cystine, tyrosine and serine are also important amino acids as they can partially substitute for methionine, phenylalanine and glysine, respectively. Proline has also been shown to be an important amino acid for young chicks (NRC, 1994). Calcium and phosphorus are important minerals in animal nutrition. Maize grain is extremely low in calcium, and thus not a big contributor to the calcium in animal diets. Maize, on the other hand, is a fair source of phosphorus, yet a substantial amount of the phosphorus is bound in the form of phytic acid – a form of phosphorus that is of little value to nonruminant animals such as swine and poultry (Ensminger et al., 1990). However, many producers are now adding the enzyme phytase to the diet to release some of the bound phosphorus from the phytic acid. Other minerals such as selenium are also important, but the amount in plants has been shown to reflect the amount of the mineral in the soil. Nutritionists incorporate supplemental sources of calcium, phosphorus, sodium chloride, magnesium, iron, zinc, copper, manganese, iodine and selenium as needed to balance diets. Maize grain is a source of vitamins A, E, thiamin, riboflavin, pantothenic acid and pyridoxine. While niacin occurs in relatively high concentration, it is in the form of niocytin that is biologically unavailable. Again, nutritionists supplement swine diets with vitamins A, D, E, K, B12, riboflavin, niacin and pantothenic acid (NRC, 1998); and ruminant diets with vitamins A, D, E and K.

Maize silage is a very important feed ingredient for feedlot cattle and dairy cattle. In the United States, approximately 10% of the maize crop is harvested as silage. It is regarded highly as a palatable energy source (Newcomb, 1995). The whole corn plant contains about 1.5 times the nutrients of the grain, and the ensiling process preserves more than 90% of the nutrients (Ensminger et al., 1990). In that silage is fed to lactating dairy cows, nutritionists are becoming more interested in the amino acid content of silage, particularly for high-producing animals. Concerning minerals and vitamins in silage, a similar situation exists as described for maize grain; although silage contains more calcium, levels are not enough to meet an animal's needs and should be supplemented.

As previously mentioned, most corn gluten feed is fed to ruminants, but some is fed to swine. The major use of gluten meal is in poultry rations because the gluten contains high amounts of protein and carotenoid pigments that express themselves in the skin and eggs of poultry. Cattle nutritionists are including corn gluten meal and dried corn distillers in diets because they are thought to contain by-pass rumen protein (Newcomb, 1995). Thus, the amino acid composition of these maize products has become important in addition to total protein content.

Proximate analyses are commonly conducted on animal feedstuffs, including the amounts of nitrogen, ether extract, ash and crude fibre. Carbohydrates are measured as starch or nitrogen-free extract. Nitrogen-free extract, which includes starch, sugars, some cellulose, hemicellulose and lignin, is calculated by subtracting the total of the determinates from 100. Crude protein is calculated by multiplying the nitrogen content by 6.25, a conversion factor based on the average amount of nitrogen in protein. Fat is considered to be acid ether extractable material (Ensminger et al., 1990). In the case of ruminants and swine, the traditional analysis for crude fibre is considered obsolete and has been replaced by analyses for acid detergent fibre and neutral detergent fibre. For amino acids, the ten essential amino acids plus glysine, cystine, tyrosine, serine and proline are the key nutrients. Linoleic is the fatty acid of key importance for the kernel, while the fatty acid spectrum is more important for the oil.

In considering the anti-nutrients and natural toxins in maize, only phytic acid is significant to the animal feed. With the use of the enzyme phytase, it is possible to break down part of the phytic acid and release bound phosphorus and calcium. Hence, the phytic acid content of the grain is beneficial to know.

When one considers the remainder of the maize products that might be fed to animals, their nutrient content would not be expected to change if the whole maize plant and the maize kernel are not changed. Hence, only the whole plant (silage) and the kernel are suggested to be analysed (Table 5.15).

Table 5.15. **Suggested nutritional and compositional parameters to be analysed in maize matrices for animal feed**

Parameter	**Kernel**	**Silage**
Proximate	X	X
Amino acids	X	
Fatty acids	X	
Calcium	X	X
Phosphorus	X	X
Phytic acid	X	

Notes

1. Using sera from 22 maize allergic patients, Pastorello et al. (2000) identified two proteins as the major food allergens in maize, i.e. a 9-kd lipid transfer protein (LTP) and a 16-kd trypsin inhibitor. The 9-kd LPT represents a significant fraction of the amount of soluble protein in maize and has a high physicochemical stability, thus possessing important general characteristics of food allergens. In another report, zeins, water-insoluble proteins from maize, were implicated as causative agents of allergic responses to a hypoallergenic, cow's milk-based infant formula containing maize starch (Frisner et al., 2000). The clinical relevance of these findings is, however, uncertain.

References

Adams, T.B. et al. (1997), "The FEMA GRAS assessment of furfural used as a flavour ingredient", *Food and Chemical Toxicology*, Vol. 35, No. 8, August, pp. 739-51.

AGPM-Congres 95/96- Section Mais Doux.

AgrEvo USA (1998), "Safety and nutritional aspects of Bt Cry9C insect resistant, glufosinate tolerant corn transformation event CBH-351", U.S. FDA/CFSAN BNF 41.

Alexander, R.J. (1987), "Corn dry milling: Processes, products and applications", in: *Corn: Chemistry and Technology*, pp. 351-376.

Anderson, R.A. and S.A. Watson (1982), "The corn milling industry", in: Wolffe, I.A. (ed.), *CRC Handbook of Processing and Utilisation in Agriculture*, Vol. II, Part I: Plant Products, pp. 31-78, CRC Press, Inc, Florida.

Aung, L.H. et al. (1993), "Effects of imbibition and modified atmospheres on the soluble sugar content of supersweet sweet corn embryo and endosperm", *Journal of Horticultural Sciences*, Vol. 68, No. 1, pp. 37-43.

Aventis Crop Science (1999), "Corn with transformation event MS6", U.S. FDA/CFSAN BNF 66.

Cambier, V. et al. (2000), "Variation of DIMBOA and related compounds content in relation to the age and plant organ in maize", *Phytochemistry*, Vol. 53, No. 2, pp. 223-229, January.

Classen, D. et al. (1990), "Correlation of phenolic acid content of maize to resistance to Sitophilus zeamais, the maize weevil, in CIMMYT's collections", *Journal of Chemical Ecology*, Vol. 16, No. 2, pp. 301-15.

Clifford, M.N. (1999), "Chlorogenic acids and other cinnamates: Nature, occurrence and dietary burden", *Journal of the Science of Food and Agriculture*, Vol. 79, No. 3, pp. 362-72.

Codex Alimentarius (1999), "Codex standard for named vegetable oils", CODEX-STAN 210-1999.

Corn Refiners Association (2001), website www.corn.org, Washington, DC at *Source:* U.S. Department of Agriculture, Foreign Agricultural Service. Based on local marketing years in thousands of metric tons. Updated May 17, 2001

Dow AgroSciences LLC (2000), "Safety, compositional, and nutritional aspects of B.t. Cry 1F insect resistant, glufosinate tolerant maize", U.S. FDA/CFSAN BNF 73.

Dowd, P.F. and F.E. Vega (1996), "Enzymatic oxidation products of allelochemicals as a basis for resistance against insects: Effects on the corn leafhopper Dalbulus maidis", *Natural Toxins*, Vol. 4, No. 2, pp. 85-91.

Eckhoff, S.R. and M.R. Paulsen (1996), "Cereal chemistry", Chapter 3 in: *Cereal Grain Quality*, Chapman and Hull Publishers.

Ensminger, M.E. et al. (1990), *Feeds and Nutrition*, 2nd edition, The Ensminger Publishing Co., Clovis, California.

EPA (2001), *White paper on the possible presence of Cry9C protein in processed human foods made from food fractions produced through the wet milling of corn*, United States Environmental Protection Agency, Washington, DC, available at: www.epa.gov/scipoly/sap/meetings/2001/july/wetmilling.pdf.

Eurostatistics (1994), "Crop production", Quarterly statistics, No. 3.

FAO (1996), *Report of a Joint FAO/WHO Expert Consultation on Biotechnology and Food Safety*, Food and Agriculture Organization, Rome, Italy, 20 September- 4 October 1996 http://www.fao.org/ag/agn/food/pdf/biotechnology.pdf (accessed 11 February 2015)

FAO (1992), *Maize in Human Nutrition*, FAO Food and Nutrition Series, No. 25. Food and Agriculture Organization, Rome.

Federal Register (2000), "Assessment of scientific information concerning StarLink corn Cry9C Bt corn plant-pesticide", *Notice*, Vol. 65, No. 211.

Feron, V.J. et al. (1991), "Aldehydes: Occurrence, carcinogenic potential, mechanism of action and risk assessment", *Mutation Research*, Vol. 259, No. 3-4, March-April, pp. 363-85.

Frisner, H. et al. (2000), "Identification of immunogenic maize proteins in a casein hydrolysate formula", *Pediatric Allergy and Immunology*, Vol. 11, No. 2, pp. 106-10.

Hashimoto, Y. et al. (1979), "Mutagenicities of 4-hydroxy-1,4-benzoxazinones naturally occurring in maize plants and of related compounds", *Mutation Research*, Vol. 66, pp. 191-194, http://dx.doi.org/10.1016/0165-1218(79)90066-1.

Hefle, S.L. et al. (1996), "Allergenic foods", *Critical Review in Food Science and Nutrition*, Vol. 36S, pp. 69-90.

Jugenheimer, R.W. (1976), *Corn Improvement, Seed Production, and Uses*, John Wiley & Sons, Inc, New York.

Korn, H. (1988), "A feeding experiment with 6-methoxybenzoxazolinone and a wild population of the deer mouse (Peromyscus maniculatus)", *Canadian Journal of Zoology*, Vol. 67, No. 9, pp. 2 220-2 224, http://dx.doi.org/10.1139/z89-313.

Kroon, P.A. and G. Williamson (1999), "Hydroxycinnamates in plants and food: Current and future perspectives", *Journal of the Science of Food and Agriculture*, Vol. 79, Issue 3, pp. 355-361.

Maynard, L.A. et al. (1979), *Animal Nutrition*, 7th edition, Mcgraw-Hill Book Co., New York.

Monsanto (2000), "Safety and nutritional assessment of corn rootworm protected corn event MON 863", U.S. FDA/CFSAN BNF 75.

Monsanto (1999), "Information to support the human food/animal feed safety of roundup read corn line NK603", U.S. FDA/CFSAN BNF 71.

Monsanto (1997), "Information to support the human food/animal feed safety of roundup read corn line GA21", U.S. FDA/CFSAN BNF 51.

Monsanto (1996a), "Information to support the human food/animal feed safety of roundup read corn lines MON 809 and MON 810", U.S. FDA/CFSAN BNF 34.

Monsanto (1996b), "Information to support the human food/animal feed safety of roundup read corn lines MON 802, MON 805, MON 830, MON 831, and MON 832", U.S. FDA/CFSAN BNF 35.

Monsanto (1995), "Safety, compositional and nutritional aspects of insect-protected corn line MON 801: Conclusion based on studies and information evaluated according to FDA's policy on foods from new plant varieties", U.S. FDA/CFSAN BNF 18.

Naczk, M. et al. (1997), "a-Galactosides of sucrose in foods: Composition, flatulence-causing effects and removal", in: *Anti-Nutrients and Phytochemicals in Food*, American Chemical Society.

NCGA (1999), *The World of Corn*, National Corn Growers Association, St. Louis, Missouri.

NEVO (2001), Dutch food composition table 2001, Nederlands Voedingsstoffenbestand.

Newcomb, M.D. (1995), "Corn and animal nutrition in the United States", BNF, United States Food and Drug Administration.

NOTIS Plus database analysis (1999).

NRC (2001), *Nutrient Requirements of Dairy Cattle*, Seventh Revised Edition, National Research Council, National Academies Press, Washington, DC, available at: http://www.nap.edu/openbook.php?record_id=9825.

NRC (2000), *Nutrient Requirements of Beef Cattle*, Seventh Revised Edition: Update 2000, National Research Council, National Academies Press, Washington, DC, available at: www.nap.edu/catalog/9791/nutrient-requirements-of-beef-cattle-seventh-revised-edition-update-2000.

NRC (1998), *Nutrient Requirements of Swine*, Tenth Revised Edition, National Research Council, National Academies Press, Washington, DC, available at: www.nap.edu/catalog/6016/nutrient-requirements-of-swine-10th-revised-edition.

NRC (1996), *Nutrient Requirements of Beef Cattle*, 7th edition, National Research Council, National Academies Press, Washington, DC.

NRC (1994), *Nutrient Requirements of Poultry*, Ninth Revised Edition, National Research Council, National Academies Press, Washington, DC, available at: http://books.nap.edu/catalog/2114.html.

OECD (1997), "Report of the OECD workshop on toxicological and nutritional testing of novel foods", SG/ICGB(1998)1, OECD, Paris, final version available at: www.oecd.org/officialdocuments/publicdisplaydocumentpdf/?doclanguage=en&cote=sg/icgb(98)1/final.

Orthoefer, F.T. and R.D. Sinram (1987), "Corn oil: Composition, processing and utilization", in: *Corn Chemistry and Technology*, pp. 535-551.

Panel on the Definition of Dietary Fibre and the Standing Committee on the Scientific Evaluation of Dietary Reference Intakes, Food and Nutrition Board (2001), *Dietary Reference Intakes: Proposed Definition of Dietary Fibre*, National Academies Press, Washington, DC, available at: www.nap.edu/catalog/10161.html.

Pastorello, E.A. et al. (2000), "The maize major allergen, which is responsible for food-induced allergic reactions, is a lipid transfer protein", *Journal of Allergy and Clinical Immunology*, Vol. 106, No. 4, pp. 744-51.

Pingali, P.L. (ed.) (2001), *CIMMYT 1999-2000 World Maize Facts and Trends. Meeting World Maize Needs: Technological Opportunities and Priorities for the Public Sector*, CIMMYT, Mexico.

Pioneer Hybrid International, Inc. (1998), "Safety, compositional, and nutritional aspects of male sterile corn", U.S. FDA/CFSAN BNF 36.

Radtke, J. et al. (1998), "Phenolic intake of adults in a Bavarian subgroup of the national food consumption survey", *Zeitschrift fur Ernahrungswissenschaft*, Vol. 37, No. 2, pp. 190-7.

Rosazza, J.P.N. et al. (1995), "Review: Biocatalytic transformations of ferulic acid: An abundant aromatic natural product", *Journal of Industrial Microbiology*, Vol. 15, pp. 457-71, Society for Industrial Microbiology.

Sasaki, Y.F. et al. (1989), "Modifying effects of components of plant essence on the induction of sister-chromatid exchanges in cultured Chinese hamster ovary cells", *Mutation Research*, Vol. 226, No. 2, June, pp. 103-10.

SCF (1999), Opinion concerning the scientific basis for determining whether food products, derived from genetically modified soya and from genetically modified maize, could be included in a list of food products which do not require labelling because they do not contain (detectable) traces of DNA or protein, SCF/CS/LABEL/13 Rev. 4, E.U. SCF.

Sicker, D. et al. (2000), "Role of natural benzoxazinones in the survival strategy of plants", *International Review of Cytology*, Vol. 198, pp. 319-346, http://dx.doi.org/10.1016/S0074-7696(00)98008-2.

Souci, S.W. et al. (2000), *Food Composition and Nutrition Tables*, 6th ed., Medpharm Scientific Publishers – CRC Press, Stuttgart, Germany, online database at: www.sfk-online.net.

Stich, H.F. (1992), "Teas and tea components as inhibitors of carcinogen formation in model systems and man", *Preventive Medecine*, Vol. 21, No. 3, May, pp. 377-84.

UPOV (1994), "Guidelines for the conduct of tests for distinctness, uniformity and stability", International Union for the Protection of New Varieties of Plants, Geneva.

USDA Agricultural Research Service (2001), *USDA National Nutrient Database for Standard Reference*, Release 14, Nutrient Data Laboratory, available at: http://ndb.nal.usda.gov.

Voragen, A.G.J. (1998), "Technological aspects of functional food-related carbohydrates", *Trends in Food Science and Technology*, Vol. 9, Issue 8-9, August, pp. 328-35, http://dx.doi.org/10.1016/S0924-2244(98)00059-4.

Watson, S.A. (1987), "Structure and composition", in: Watson, S.A. and R.E. Ramstad (eds.), *Corn Chemistry and Technology*, pp. 53-80, American Society of Cereal Chemists Inc., St. Paul, Minessota.

Watson, S.A. (1982), "Corn, amazing maize, general properties", in: Wolff, I.A. (ed.), *CRC Handbook of Processing and Utilisation in Agriculture*, Vol II, Part I: Plant Products, pp. 3-29, CRC Press, Inc, Florida.

White, P.J. and L.M. Pollak (1995), "Corn as a food source in the United States: Part II: Processes, products, composition and nutritive values", *Cereal Foods World*, Vol. 40, No. 10, pp. 756-762.

Xie, Y. et al. (1992), "Variation to hydroxamic acid content in maize roots in relation to geographic origin of maize germ plasm and resistance to Western corn rootworm (Coleoptera: Chrysomelidae)", *Journal of Economic Entomology*, Vol. 85, No. 6, December, pp. 2 478-2 485.

Chapter 6

Wheat (*Triticum aestivum*)

*This chapter, prepared by the OECD Task Force for the Safety of Novel Foods and Feeds with Australia as the lead country, deals with the composition of bread wheat (*Triticum aestivum*). It contains elements that can be used in a comparative approach as part of a safety assessment of foods and feeds derived from new varieties. Background is given on wheat production, classification, uses and processing, followed by quality criteria and elements for comparative analyses. Nutrients in wheat and its products, anti-nutrients, allergens and other compounds are then detailed. The final sections suggest the key products and constituents for analysis of new varieties for food use and for feed use.*

Background

Production of wheat[1]

Wheat is grown as a commercial crop in over 120 countries worldwide (FAO, 2002), which makes it the most widely grown crop in the world today. Table 6.1 shows world production and export figures for wheat. The People's Republic of China, the European Union (EU), India and the United States are the major wheat producers, accounting for nearly 60% of the total world production. The major wheat exporters, accounting for 86% of the total exports, are the United States, Canada, Australia, the EU and Argentina.

Table 6.1. **Production and export of wheat, 2001-02**

Country/region	Production (million tonnes)		Exports (million tonnes)	
	2000	2001 estimate	2000/01 estimate	2001/02 forecast
Argentina	16.0	15.5	11.0	11.0
Australia	23.8	23.3	16.5	18.0
Canada	26.8	21.3	16.8	16.0
China (Peopple's Republic of)	99.6	94.2	0.4	0.3
European Union	105.2	92.0	14.5	11.0
India	75.6	68.5	2.3	2.5
Kazakhstan	9.1	13.5	3.7	4.2
Pakistan	21.1	19.0	0.3	1.0
Russian Federation	34.4	46.9	0.7	2.5
Turkey	18.0	16.0	1.6	0.4
Ukraine	10.2	21.3	0.1	4.5
United States	60.8	53.3	27.9	27.5
World total	598.3	591.1	100.4	106.0

Source: FAO (2002).

Classification of wheat

The commercially relevant crops of wheat are limited to four species of the genus *Triticum*. These are: *T. monococcum*, *T. turgidum*, *T. timopheevi* and *T. aestivum*. Of these, *T. aestivum* and *T. turgidum* are the most widely grown. *T. aestivum* includes the common bread wheats and *T. turgidum* includes the durum wheats. This chapter only considers those constituents relevant to the common bread wheats.

Extensive cultivation, breeding and selection have resulted in many thousands of commercial varieties of bread wheats. This had led to yield improvement and to the development of wheats with the required milling and flour-processing qualities. For commercial purposes, the common wheats are classified into broad classes that are used as a basis of world trade. The major factors used to distinguish wheats are hardness or softness of the kernel, winter or spring growing habit, red or white seed coat, and protein content (Orth and Shellenberger, 1988).

Uses of wheat

Of the wheat that is produced in the world, about 74% is destined for human food use, 16% for animal feed use, 5.5% for seed, with the remaining 4.5% for use in industrial applications (International Grain Council, 1996). As these figures show, the vast majority of wheat is used for human food, although it is also popular as an animal feed, particularly in years where there is a grain surplus and its price becomes competitive with other feed grains. The wheat plant is also popular as an animal feed where it is used for forage as well as hay and silage production.

Processing of wheat

The processing of wheat can be divided into two categories: *i)* dry milling; and *ii)* wet milling with aqueous solvents. The two processes produce distinctly different products. In addition to milling, wheat is also used in fermentation processes to produce industrial alcohol as well as beverages such as beer.

Dry milling

The major products resulting from the dry milling of wheat are flour, bran and wheat germ. These products result from separating the grain (kernel) into its three distinct parts: the mealy or starchy endosperm (composed of the endosperm but lacking the aleurone layer), which is subsequently processed into fine particles (flour); the bran (composed of the pericarp, the seed coat and the aleurone layer); and the germ (composed of the embryonic axis and the scutellum). Wheat germ is further processed into wheat germ meal and oil. Wheat grains from current commercial varieties typically comprise about 2-3% germ, 13-17% bran and 80-85% mealy endosperm, on a dry matter basis (Belderok, 2000).

Of the three major products resulting from the processing of wheat, flour is by far the most valued and versatile. Flour is used to produce a wide range of products, including pan bread, flat bread, noodles/pasta, cakes, pastries and biscuits.

Figure 6.1 shows the typical steps employed in the dry milling of wheat. The objective during milling is to separate the bran and germ of the wheat kernel from the endosperm. Initial steps in the milling involve the cleaning and sifting of the grain. This is achieved using a separator, which is a series of screens that remove stones, sticks and other foreign material. From the separator, the wheat passes through an aspirator where jets of air remove many of the light impurities, such as dust and chaff. The next step involves a disc separator – the surface of the discs are indented so that wheat is caught but larger and smaller particles, such as foreign grains, are rejected. After passing through the disc separator, the wheat is scoured to remove the beard from the individual grains and passes through a magnetic separator to the washer-stoner (a machine that uses high-speed rollers and water to remove stones). Prior to grinding, the cleaned wheat is tempered by adding water. This hydrates the outer bran coat so that it becomes more elastic and will not splinter during grinding and contaminate the flour, and also mellows the endosperm so that it breaks easily off the bran during grinding, resulting in less power required to reduce large, pure particles of flour. The tempered wheat passes into an Entoleter machine, which breaks and removes unsound wheat. The sound wheat then passes through a series of grinders, sifters and purifiers to separate the various parts of the grain. This process is repeated over and over again until the maximum amount of flour is separated.

None of the kernel fractions coming out of the mill are entirely pure, and each will contain some part of the other fractions. The level of purity of each product at the end of the process is one of the measures of milling efficiency. Generally, modern day millers remove about 80% of the wheat kernel for wheat flour, with the other 20% of the millstream going into the production of animal feeds as well as dietary fibre ingredients for human foods (Orth and Shellenberger, 1988).

The remaining 20% of the millstream is known collectively as "millfeed". Several different names, such as middlings, shorts and red dog, have been assigned to various combinations of these millstreams but each name refers to a rather poorly defined material whose composition can be varied by changing roller settings, purification conditions and the way in which the millstreams are combined (Matz, 1991). In some cases, all the non-flour streams are combined to yield a single product called "millrun". Wheat middlings consists mostly of the layers of the wheat kernel just inside the bran, bran particles, some flour and some non-wheat material. About 45% of the millfeed is middlings, the exact amount depending on the efficiency of the process, the type of wheat and the miller's decision as to how the millstreams are to be combined. Wheat shorts typically contain more flour than do middlings and red dog contains more flour than the other millfeeds.

Figure 6.1. **Dry milling of wheat**

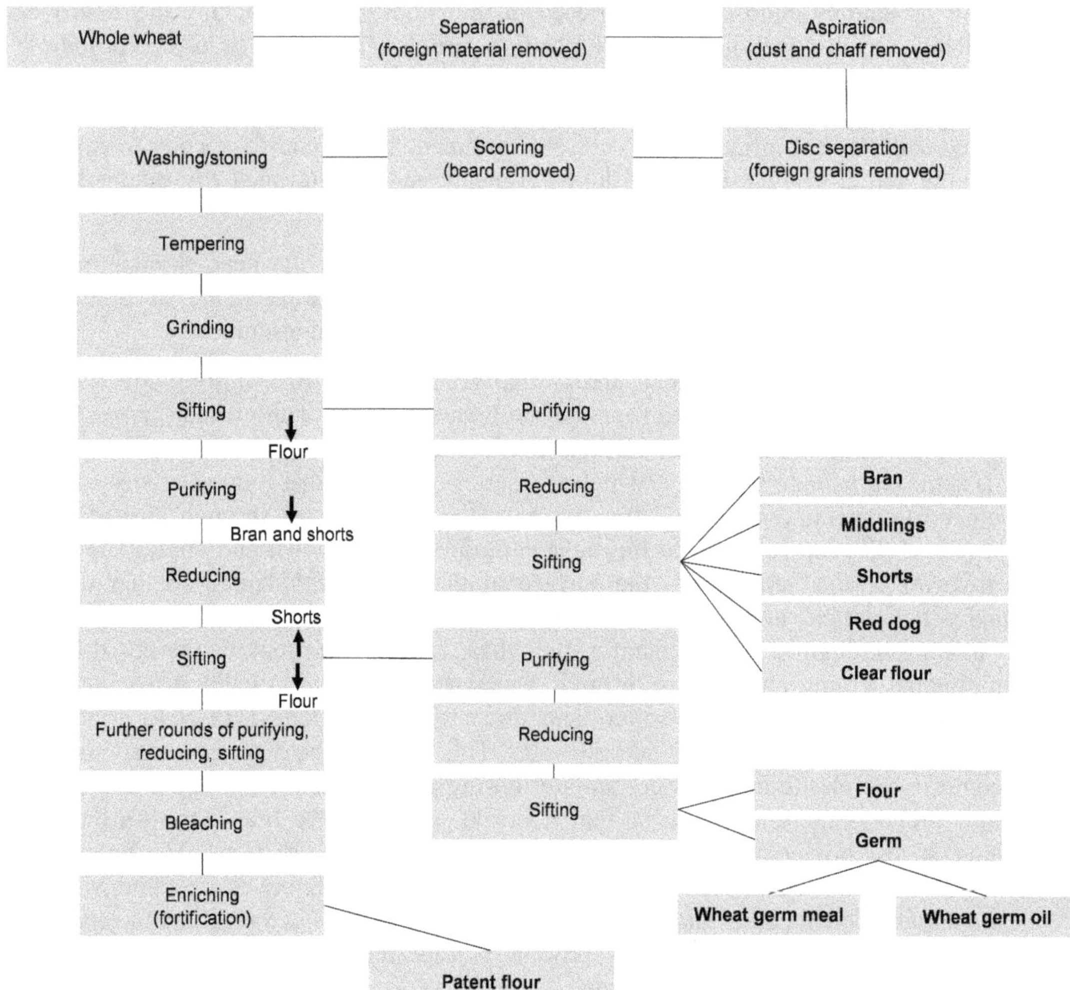

Source: Adapted from Matz (1991).

Wet milling

Wet milling is the process used to separate starch from gluten, both of which are contained within the wheat endosperm. Wheat starch has physiochemical properties similar to cornstarch, although viscosity and gel strength are usually lower. When used in baking, it can replace portions of wheat flour, increase cake volumes and tenderness, and reduce fat absorption in doughnuts, as well as having applications in the confection and canning industries (Becker and Hanners, 1991). Gluten is primarily used as an additive to improve the quality of flour for bread making (Matz, 1991). Gluten is also used in hamburger buns and hot dog rolls to give the desired low density and texture. Small amounts of gluten are also processed into liquid dietary supplements and flavouring hydrolysates.

There are a number of different wet-milling techniques that are used, with both the whole kernel and flour-milling fractions able to be used as the starting material (Rao, 1979). Whole wheat as the starting material has several advantages over flour-milling fractions, including: it is a readily available raw material that is not tied to the supply of dry-milled flour products; less starch damage can be expected because of the absence of the high shear effects of dry milling therefore yielding higher quantities of prime starch; and increased vitality in the gluten because the entire endosperm protein is recovered (i.e. including the gluten protein that would normally be "lost" to the high-quality patent flour) (Fellers, 1973, Rao, 1979).

The most well-known method for separating starch from gluten, using flour as the starting product, is the Martin Process (Fellers, 1973). It involves the continuous mixing of the flour with about 40-60% of its weight of water. The resulting dough is then washed in a continuous kneader with additional water, which washes away the starch, leaving a coherent mass of gluten. The starch milk is collected and passed through a centrifugal extractor fitted with special slotted screens to remove fibre and pentosan material. The starch milk is then concentrated, refined, washed and dried.

When whole wheat is used, the wheat is first tempered with water, and then flaked before further water is added to obtain hydrated dough. Separation of the dough is accomplished by washing with water under high pressure. The starch, bran, germ and other non-gluten components are removed, leaving the hydrated elastic gluten. Vital gluten obtained after purification and drying contains about 75-80% protein.

Fermentation processes

Although barley and corn grains are the most common substrates used for brewing and the production of industrial alcohol, respectively, a small amount of wheat is also used for these purposes (Wu, 1989).

Wheat malt has only had limited use in the brewing industry, primarily because of its higher price but also because of the brewer's traditional preference for barley malt (Matz, 1991). Despite this, a significant amount of wheat beers are brewed in Europe, with smaller amounts also being brewed in North America and Australia. Brewers' grains, which are typically a blend of the spent grain and hops, are one of the major by-products of the brewing process and are a popular feedstuff for cattle. Brewers' grains are supplied either wet or dry and are an excellent source of high-quality bypass protein and digestible fibre.

Fermentation of cereal grains to make ethanol for industrial uses results in a protein-rich material (stillage) after the ethanol is distilled. The fermentation process predominantly consumes the starch in cereal grains, and the other nutrients, particularly protein, are concentrated (Wu, 1989). The optimum recovery and use of stillage is important for the commercial success of the fermentation process. Stillage is usually centrifuged to yield a solid fraction (distillers' grains) and a soluble fraction (stillage solubles). Manufacturers typically dry the distillers' grains to yield dehydrated distillers' grains and sell them as an ingredient for animal feed. The soluble fraction can be concentrated, blended with dehydrated distillers' grains and co-dried yielding distillers' dried grains with solubles (DDGS). Wheat, however, is generally only an economical choice as a fermentation substrate if the dehydrated distillers' grains or DDGS can be sold as a human food ingredient rather than as a component of animal feed. Distillers' grains, including those from wheat, have been used as ingredients in baked goods and other foods to enhance protein and dietary fibre content.

Typical criteria used to determine wheat quality

Grain hardness

Bread wheat varieties can vary greatly in grain hardness and are usually classified into one of two categories – hard wheat or soft wheat. The terms "hard wheat" and "soft wheat", as used in this chapter, do not have the same meaning as described in EU regulations. In the EU, the term "hard wheat" is used exclusively for durum wheats, and the term "soft wheat" for *aestivum*, or bread wheats, irrespective of the hardness of their grains. In this chapter, the terms "hard" and "soft" refer to the hardness of the grain.

The hardness of the grain is of particular relevance to the milling and baking industries. Hard grains exhibit more resistance to grinding than soft wheat grains and thus produce more damaged starch than soft wheat in the milling process (Belderok, 2000). A certain degree of starch damage is desirable in bread baking as it contributes to the soft texture and pleasant mouthfeel of the crumb and also has a retarding effect on the bread going stale. Soft wheat flours have less damaged starch and are more suited to the production of biscuits, cakes, crackers, wafers, etc.

Moisture content

Moisture content is thought to be one of the most important considerations in judging the quality of wheat because it is fundamental to the keeping and milling quality of the grain (Rasper and Walker, 2000). The moisture content of commercial lots of wheat may vary between 8.0% and 18%, depending on the weather during harvest (Belderok, 2000). Prior to milling, the moisture content of the grain is optimised by either the addition or removal of moisture. This ensures maximum milling efficiency and optimum performance in the final product (Bass, 1988).

Protein content

The protein content of wheat typically ranges between 10.0% and 16.0% (air dried matter), depending on the variety and the environmental conditions during the growth of the crop (see Tables 6.3 and 6.6). Hard wheats typically have higher protein contents than soft wheats. Abundant rainfall during kernel development usually results in low protein content. Available soil nitrogen also has considerable influence on protein content. It is the protein content of kernel that generally dictates the end use of the flour produced.

Protein quality

Varieties of wheat having the same total protein content can produce flours that behave quite differently in baking operations. In many instances these differences are attributed to qualitative differences in the gluten proteins. Gluten quality is largely a varietal characteristic, although high temperatures and low relative humidity during the period when the wheat is maturing in the field have a marked deleterious effect on the quality of the gluten.

Alpha-amylase activity

Wet weather after wheat has matured in the field but before harvest may cause some of the kernels to sprout. These kernels are very high in α-amylase activity. Even if visible sprouting does not occur, the α-amylase level may be considerably elevated as a result of a wet harvest season. Although some α-amylase activity is optimal to sustain the production of sugars required for proper fermentation and subsequent gas production for bread making, excessively high levels of α-amylase can impair the quality of both the dough and the final baked product because of the rapid degradation of the starch molecules and subsequent reduction in viscosity of the dough (Rasper and Walker, 2000).

Fat acidity

Fat acidity refers to the breakdown of fats by lipases and the release of free fatty acids in the grain. Under most practical storage conditions, fat acidity in wheat increases after several years to levels considerably above those associated with freshly harvested sound wheat, even though the wheat shows no appreciable physical evidence of deterioration. Such wheat may still be useful for milling purposes but the keeping and baking quality of the flour may be adversely affected. Low temperature and moisture content during storage markedly reduce the rate of increase in fat acidity.

Crude fibre and ash

Both crude fibre and the ash content in wheat are related to the amount of bran in the wheat and hence have a rough inverse relationship to flour yield. Small or shrivelled kernels have more bran on a percentage basis and therefore more crude fibre and ash than large, plump kernels and consequently yield less flour. Wheat usually contains 2.0-2.7% crude fibre and 1.4-2.0% ash, both calculated to a 14% moisture basis.

Comparative analyses

This chapter suggests parameters that wheat developers should measure when undertaking comparative analyses of new varieties of wheat. Data from the new variety should be compared to those obtained from the conventional counterpart and may also be compared to the literature values presented in this chapter. Wheat composition is known to vary quite markedly from one area to another as well as from year to year within any given area (Matz, 1991) therefore for effective comparison it is important that the new variety and its comparator (i.e. the control) are grown at the same site (preferably in adjacent plots) and at the same time. Also, given the variation that can occur in some constituents between different classes of wheat (e.g. in protein content between hard and soft wheats), when comparing results to the literature values for a particular constituent it is important that the comparison is made to data derived from the same class of wheat.

Nutrients in wheat and wheat products

Whole grain wheat is a major source of nutrients for humans as well as for livestock. Although often seen mainly as a source of highly digestible carbohydrate, whole wheat is also recognised as a significant source of protein, B vitamins, as well as a number of minerals, particularly iron, phosphorus, zinc, potassium and magnesium (Orth and Shellenberger, 1988). Overall, wheat contributes slightly less than 20% of the world's total energy and protein (Betschart, 1988), making it a significant staple food for the world population.

Table 6.2. **Key wheat nutrients and their location in the kernel**

Fraction	% kernel (by weight)	Key nutrients
Bran	8	– Dietary fibre, protein, potassium, phosphorus, magnesium, iron and zinc
Aleurone layer	7	– Protein, niacin, thiamine, folate, minerals – especially phosphorus (mainly as phytate), potassium, magnesium, iron and zinc
Endosperm	82	– Starch, protein, minerals
Germ		
Embryo	1	– Fats and lipids, protein and sugars
Scutellum	2	– B vitamins (especially thiamine), phosphorus

Source: Orth and Shellenberger (1988).

Whole kernel and fractions

Typical values for proximate composition of wheat are presented in Table 6.3 and the relative distribution of the major components in the various kernel fractions are presented in Table 6.4.

Most of the mealy endosperm, used to derive the flour, consists of food reserves in the form of carbohydrate (mainly starch), whereas the bran contains high levels of fibre and comparatively more minerals and fat than the endosperm. The germ also contains comparatively high levels of fat and minerals as well as significant amounts of fibre and carbohydrate and is also very rich in protein.

Table 6.3. **Typical values for the proximate composition of whole wheat**

Constituent	% air dried matter
Moisture	8.0-18.0
Protein	10.0-16.0
Ash	1.2-3.0
Carbohydrate	65.4-78.0
Fat	1.5-2.0
Energy	1 377-1 431 kJ/100 g
Crude fibre	2.0-2.7
Acid detergent fibre	3.6-4.0
Neutral detergent fibre	12.0-13.5

Sources: Compiled from USDA (1999); Matz (1991); Belderok (2000); and Ensminger et al. (1990).

Table 6.4. **Chemical composition (% dry matter) of whole wheat and its various fractions**

	Kernel	Flour	Bran	Germ
Protein	16	13	16	22
Fat	2	1.5	5	7
Carbohydrate	68	82	16	40
Dietary fibre	11	1.5	53	25
Ash	1.8	0.5	7.2	4.5
Other	1.2	1.5	2.8	1.5
Total	100	100	100	100

Source: Belderok (2000), with kind permission from Springer Science+Business Media B.V.

Carbohydrates

Carbohydrates constitute the bulk of the total dry matter of the wheat kernel and are normally classified into three categories on the basis of their different monomeric and polymeric forms. The three categories are: sugars, composed of the monosaccharides (glucose, fructose, galactose) and disaccharides (sucrose, maltose); oligosaccharides (e.g. raffinose, stachyose); and polysaccharides, composed of starch (amylose, amylopectin) and the non-starch polysaccharides (cellulose, pentosans, β-glucans) (FAO, 1998).

The non-starch polysaccharides make up the bulk of what is termed dietary fibre, which also includes lignin (a non-carbohydrate component) plus resistant oligosaccharides and resistant starch (FAO, 1998). The carbohydrate, and more particularly the dietary fibre component of whole wheat, confers significant health benefits to humans (Kritchevsky and Bonfield, 1995).

The majority of the carbohydrate in wheat, of which most is found in the endosperm, is composed almost entirely of starch and serves as the energy source for the germ upon germination. As the most abundant carbohydrate component of wheat and wheat flour, wheat starch is also an important macronutrient for humans (Shelton and Lee, 2000) as well as for other animals. Wheat starch is made up of two main fractions, amylose and amylopectin, which make up approximately 25% and 75% of the total starch mass, respectively. In addition to being an important energy source, starch also serves a number of important roles in bread making, such as providing a framework to which gluten can adhere, regulating the distribution of water in a loaf and filling up spaces that are created as the loaf changes shape during baking (Belderok, 2000).

Other carbohydrate components, such as the sugars (glucose, fructose, galactose, sucrose, maltose) and the non-starch polysaccharides (cellulose, pentosans, and β-glucans), are present in lesser amounts and are located primarily in the bran and germ fractions. The fibre and carbohydrate components of the germ arise from contamination by the bran during milling.

Table 6.5 gives typical ranges for the various carbohydrate components of wheat and its fractions. The carbohydrate content of wheat is subject to variation due to variety, environmental conditions, and also due to processing/milling conditions (Becker and Hanners, 1991).

Table 6.5. **Carbohydrate composition of wheat**

Dry matter basis, in %

Constitutent	Kernel	Flour	Bran	Germ
Total dietary fibre	11-14.6	2.3-5.6	43-53	13.2-15
– Pentosans	1.4-6.7	1.1-1.4	21-43	6.6
– Cellulose	2.0-2.7	0.3-0.6	7.2-8.0	2.7
Free sugars	2.1-2.6	1.2-2.1	7.6	16.0
Starch	59-72	65-74	14.1	28.7

Sources: Compiled from: FAO (1998); Matz (1991); Becker and Hanners (1991); USDA (1999); Shelton and Lee (2000); Belderok (2000).

Proteins

In addition to its high energy content, wheat is also a good source of protein and contains considerably more protein on average than other cereals. The proteins of wheat are complex. They can be divided into two broad categories based on their biological functions: the biologically active enzymes (albumins and globulins) and the biologically inactive storage proteins (gliadins and glutenins) (Lookhart and Bean, 2000). The gliadins and glutenins are referred to collectively as the gluten proteins, and are mainly located within the mealy endosperm of the grain, whereas the albumins and globulins are concentrated in the bran (the aleurone layer) and the germ. The gluten proteins play a key role in the formation of dough for bread making.

The protein content of wheat is affected by both the genetic makeup of the plant and by environmental conditions during growth of the plant, and development of the seed, therefore protein content can vary quite markedly, a typical range being 10.0-16.0%. It is possible that in the normal course of events many samples will be found that fall outside this range because of unusual weather patterns, heavy fertiliser applications, disease or characteristics of a particular variety. For this reason it is important that an appropriate comparator is used for the comparative analysis. The typical protein contents of wheat, by class, are listed in Table 6.6.

Table 6.6. **Protein content of wheat by class**

Wheat class[1]	% dry matter
Hard red spring	13.6-15.8
Hard red winter	12.6-14.1
Soft red winter	10.4-13.0
Soft white winter	10.0-12.4
Soft white spring	13.5-14.2
Hard white winter	11.5-12.1
Hard white spring	12.3-13.4

Note: 1. Using the US classification.

Sources: Davis et al. (1981); Ensminger et al. (1990); USDA (1999); NRC (1998).

Although wheat can be a significant source of protein, the nutritional quality of the protein for humans and other monogastric animals (for example, pigs and poultry) is limited by the low content of two essential amino acids, lysine and threonine

(Matz, 1991; Shewry et al., 1994). Of these, lysine is the more limiting. Because of this, it is necessary to mix wheat with more lysine-rich proteins to provide a balanced diet or to provide lysine as a supplement in animal feeds. The low levels of lysine in the kernel result from the low proportions of lysine in the gluten proteins, which are located mainly in the endosperm, whereas the lysine-rich albumins and globulins are located mainly in the bran and the germ (Lookhart and Bean, 2000). Wheat germ is the grain fraction that contains the most lysine (Matz, 1991). Table 6.7 lists the amino acid content of wheat and its fractions.

Vitamins

In the wheat kernel, the vitamin content can vary from one part of the grain to another (see Table 6.2) with vitamins being found in high concentrations in the germ and bran. The removal of these kernel structures during the milling process can result in significant loss of certain vitamins. The vitamin content of wheat is also known to be highly variable (Pomeranz, 1988). Significant differences in vitamin content may occur due to variety, crop year, crop site, fertilisation practices, soil type, wheat class and the analytical techniques used.

Table 6.7. **Typical amino acid composition of wheats (% total protein)**

Amino acid	Kernel	Flour	Bran	Germ
Tryptophan	1.0-2.1	0.7-1.0	1.6-1.8	1.0-1.3
Threonine	2.4-3.2	2.2-3.0	2.6-3.5	3.4-4.2
Isoleucine	3.0-4.3	3.4-4.1	3.1-3.8	3.5-3.9
Leucine	5.0-7.3	6.5-7.2	5.5-6.8	5.7-6.8
Lysine	2.2-3.0	1.8-2.4	3.5-4.5	5.3-6.3
Methionine	1.3-1.7	0.9-1.5	1.1-1.6	1.7-2.0
Cystine	1.7-2.7	1.6-2.6	1.5-2.4	1.0-2.0
Phenylalanine	3.5-5.4	4.5-4.9	3.2-4.0	3.4-4.0
Tyrosine	1.8-3.7	1.8-3.2	2.1-2.8	2.8-3.0
Valine	4.4-4.8	3.7-4.5	4.0-5.1	4.7-5.2
Arginine	4.0-5.7	3.1-3.8	5.5-7.0	6.9-8.1
Histidine	2.0-2.8	1.9-2.6	2.1-2.8	2.3-2.8
Alanine	3.4-3.7	2.8-3.0	4.6-4.9	5.2-6.4
Aspartic acid	4.8-5.6	3.7-4.2	6.6-7.3	7.5-8.9
Glutamic acid	29.9-34.8	34.5-36.9	16.2-20.8	14.0-17.3
Glycine	3.8-6.1	3.2-3.5	5.0-7.1	5.2-6.2
Proline	9.8-11.6	11.4-11.7	5.7-6.9	5.0-5.3
Serine	4.3-5.7	3.7-4.8	4.4-4.6	4.5-4.8

Sources: Pomeranz (1988); Ensminger et al. (1990); USDA (1999); Lookhart and Bean (2000); Posner (2000).

Wheat has relatively low levels of lipids and hence tends to only contain low amounts of the fat-soluble vitamins – provitamin A, and vitamins D, E and K. The exception to this is the germ fraction, which contains relatively high levels of lipids and hence the tocopherols, which are responsible for the vitamin E activity of plant tissues, are most abundant in this fraction. Wheat germ oil is considered a particularly rich source of vitamin E. Wheat has only four major tocol derivatives, namely α-tocopherol (α-T), α-tocotrienol (α-T-3), β-tocopherol (β-T) and β-tocotrienol (β-T-3), with α-T being the major form in wheat (Morrison, 1981). The γ- and δ-tocopherols and tocotrienols either are absent or are only present in trace amounts. Among cereal grains, the wheat tocopherols are considered to have particularly good vitamin E activity and antioxidant properties (Morrison, 1981).

Table 6.8. **Tocol derivative content (mg/100 g) of whole wheat and its fractions**

Grain fraction	α-T	α-T-3	β-T	β-T-3	Total
Whole grain	0.9-1.8	0.2-0.7	0.4-0.9	1.9-3.6	4.9-5.8
Germ	22.1-25.6	<0.2-0.3	8.6-11.4	<0.2-1.0	..
Bran	1.6-3.3	1.1-1.5	0.8-1.3	2.9-5.6	..
Endosperm	0.007	0.045	0.01	1.4	1.4

Note: .. not available.

Source: Adapted from Chung and Ohm (2000).

Compared to maize, the carotenoids are considered to be very minor constituents of wheat. Their very low levels mean that wheat is not a significant source of vitamin A precursors and thus wheat carotenoids are not considered to have any nutritional importance (Bock, 2000). The colour due to carotenoids, however, is an important factor in the use of cereal grains in food production, particularly in durum wheat used to make pasta. The major carotenoids of bread wheat are carotene, xanthophyll and xanthophyll ester. The carotenoids are not homogenously distributed in the wheat kernel. Bran contains 0.9-0.95 mg/kg; germ contains 7.2-11.0 mg/kg; and endosperm contains 1.6-2.2 mg/kg (Chung and Ohm, 2000). Carotenoid composition also differs among wheat classes and wheat fractions within a given class of wheat.

Whole wheat is considered to be a particularly good source of the B vitamins, especially thiamine, riboflavin, niacin and pyridoxine (vitamin B6), and is also a moderate source of folic acid. These vitamins are concentrated in the bran (aleurone layer) and germ. Typical ranges in vitamin content are shown in Tables 6.9 and 6.10.

Table 6.9. **Vitamin content (mg/100 g, dry weight basis) of whole wheat**

Vitamin	Range
Thiamine	0.13-0.99
Riboflavin	0.06-0.31
Niacin	2.20-11.10
Pyridoxine	0.09-0.79
Folic acid	0.02-0.09

Source: Adapted from Davis et al. (1984a).

Table 6.10. **Vitamin content (mg/100 g, dry weight basis) of whole wheat by class[1]**

Vitamin	Hard red winter	Hard red spring	Soft red winter	Hard white winter	Soft white spring	Soft white winter
Thiamine	0.334-0.57	0.416-0.50	0.411-0.51	..	0.46-0.50	0.411-0.46
Riboflavin	0.11-0.14			..	0.10-0.15	
Niacin	4.95-7.4	4.97-6.25	4.84-6.70	4.33-4.90	4.68-6.00	5.19-5.59
Pyridoxine	0.092-0.53	0.202-0.53	0.169-0.38	..	..	..

Notes: .. not available. 1. According to the US classification.

Source: Davis et al. (1984a).

Minerals

The average mineral content of a given wheat grain varies significantly from one part of the world to another. This appears to be a function of a number of factors, including the wheat variety, the growing and soil conditions, and fertiliser application (Davis et al., 1984b; Bock, 2000). The mineral composition of wheat has more to do with environmental conditions than varietal characteristics. Major constituents of the mineral fraction of wheat are magnesium, phosphorus and potassium. There are also significant amounts of copper, iron, manganese and zinc present (Davis et al., 1984b). As with the vitamins, minerals are especially concentrated in the bran (aleurone layer), therefore the milling process can also result in significant losses of minerals, especially copper, iron, manganese and zinc.

In wheat, phosphorus is mostly present in the form of calcium, potassium or magnesium phytate (Hazell, 1985), primarily in the bran fraction. Whole wheat, wheat germ and wheat bran are classified as high sources of phosphorus (Bock, 2000), although most of this is biologically unavailable to monogastric animals, including humans. Whole cereal grains and particularly wheat are the main sources of magnesium for humans. Magnesium is located primarily in the bran fraction and also binds with phytic acid. Whole wheat is considered to be a moderately good source of magnesium, with wheat bran and wheat germ considered to be high sources (Bock, 2000). The iron in wheat is located in the outer endosperm and bran. Both wheat germ and wheat bran are considered to be good sources of dietary iron (Bock, 2000). Wheat germ and bran are also excellent sources of dietary zinc and are also the only cereal products that serve as good sources of copper (Bock, 2000).

Lipids

Lipids are relatively minor constituents of the kernel; however, they are important nutritionally as well as for grain storage and processing. Measured lipid content and composition depend largely on extraction and purification procedures and to a lesser extent on the samples, therefore care should be taken when comparing lipid content or composition data reported in the literature. The lipid content of wheat typically ranges from 1.5% to 2.0% but is not dispersed evenly throughout the grain, with between 34% and 42% of the lipid being in the germ fraction (Zeringue and Feuge, 1980).

The majority of the lipids in wheat are acyl lipids containing the fatty acids commonly found in higher plants, that is, palmitic (16:0), stearic (18:0), oleic (18:1, *n-9*), linoleic (18:2, *n-6*) and linolenic (18:3, *n-3*) acids. The typical fatty acid composition of whole wheat is presented in Table 6.11.

Table 6.11. **Typical fatty acid composition (% total fatty acids) of wheat**

Fatty acid	Kernel	Germ
Palmitic acid	11-32	18-19
Stearic acid	0-4.6	..
Oleic acid	11-29	8-17
Linoleic acid	44-74	57-62
Linolenic acid	0.7-4.4	7-11

Note: .. not available.

Sources: Davis et al. (1980); Barnes (1982), cited in Pomeranz (1988).

The main nonsaponifiable components of lipids in wheat are the tocol derivatives and carotenoids (discussed above), plus the sterols. Of the sterols, β-sitosterol is the primary sterol in all cereal grains and comprises 41-53% of the total sterols found in wheat. Campesterol is the next most abundant sterol found in wheat. The total sterol content has been estimated as 0.5% of the germ (Pomeranz, 1988).

Other components

Wheat also contains a number of other constituents, some of which, at higher intakes, are suggested to be implicated in protection against disease (Thompson, 1994; Slavin et al., 1997). These include phenolic acids, lignans and the flavonoids.

The most abundant phenolic acid is ferulic acid, followed by vanillic, *p*-coumaric, protocatechuic, syringic, *p*-hydroxybenzoic, caffeic and genitisic acids. Ferulic acid is ester-linked to specific polysaccharides (the arabinoxylans), which form 65% of the aleurone cell walls. Bacterial enzymes in the human colon slowly and partially degrade the aleurone cell walls. This degradation results in the release of feruloylated oligosaccharides, which can then be further degraded to release ferulic acid. Ferulic acid is a good antioxidant (Rice-Evans et al., 1997).

The flavonoids are a large group of phenolic compounds that occur widely in plants. Many of them have good antioxidant properties (Ferguson and Harris, 1999). The highest concentration of flavonoids in whole wheat is in the germ, followed by the bran. Flavonoids have structures based on a C15 nucleus and are usually grouped into classes (e.g. flavones). The flavone tricin, as well as two glycosides of the flavone apigenin, have been identified from wheat bran. Wheat bran also contains small amounts of the flavonol catechin and proanthocyanidin (also known as condensed tannins), which are oligomers or polymers based on flavonol units.

The lignans are phenolic dimers, and are predominantly present in the bran (Nilsson et al., 1997). They are converted by fermentation in the large intestine to mammalian lignans.

Whole plant

In addition to the production of grains, which contain large quantities of carbohydrates, the entire wheat plant can be used for forage – pasture, hay or silage – for grazing animals. In addition, by-products of the harvested grains, such as chaff, stover and straw, can be used as low-quality forages for ruminant animals. The typical constituents measured in wheat forage are shown in Table 6.12.

Table 6.12. **Typical constituents measured in wheat forage**

Analysis	Forage type			
	Hay, sun-cured	Chaff	Straw	Immature, fresh
Proximate analysis				
Dry matter	89	93	90	22
Ash	7.0	15.6	6.9	3.0
Neutral detergent fibre	60.5	..	70.3	10.2
Acid detergent fibre	36.5	..	47.7	6.3
Crude fat	2.0	1.9	1.8	1.0
N-free extract	46.4	39.5	40.4	8.3
Crude protein	7.7	5.4	3.2	6.1
Minerals				
Calcium	0.13	0.19	0.16	0.09
Phosphorus	0.18	0.08	0.05	0.09

Note: .. not available.

Source: Adapted from Ensminger et al. (1990).

Other wheat products

Dehydrated distillers grains

Wheat DDGS and products prepared by fractionation of DDGS have a very high protein content (29-59%) as well as dietary fibre content (40-55%). Wheat DDGS also provide higher levels of calcium, iron and zinc than whole wheat. The level of thiamine, however, is significantly lower in DDGS (0.09-0.19 mg/100 g dry weight) than in wheat flour (0.162-0.168 mg/100 g dry weight) or whole grain (see Tables 6.9 and 6.10) but the level of riboflavin is comparable (0.17-0.50 mg/100 g dry weight).

Anti-nutrients, allergens and other compounds in wheat and wheat products

Compared to the legumes, the content of common anti-nutrients in cereals, including wheat, is considered to be quite low (Klopfenstein, 2000).

Anti-nutrients

Protease and amylase inhibitors

Protease inhibitors, especially trypsin inhibitors, may decrease the digestibility and biological value of ingested protein and retard growth when sufficient amounts are present in the diet and amylase inhibitors may affect the digestibility of starch. Both protease and amylase inhibitors have been identified in wheat; however, they do not appear to be responsible for any serious anti-nutritional activity in humans (Klopfenstein, 2000), probably because both inhibitor types tend to be heat labile.

The type of amylase inhibitor found in wheat is fairly common and although found to be present at quite high concentrations is relatively heat labile (Wiseman et al., 1998). The inhibitor is mainly associated with the starch granules in the endosperm with very little present within the bran portion. Wheat amylase inhibitor is found to be active against chicken pancreatic amylases, although in practice it is unlikely to have much nutritional significance in chickens since it is largely inactivated by pepsin in the gizzard (Macri et al., 1977).

Lectins

Lectins, sometimes called phytohemagglutinins, are glycoproteins that bind to certain carbohydrate groups on cell surfaces, such as intestinal epithelial cells, where they cause lesions and severe disruption and abnormal development of the microvilli (Liener, 1989). One of the major consequences of the lectin damage to the intestinal mucosa appears to be serious impairment in the absorption of nutrients across the intestinal wall.

Although more commonly associated with legumes, cereal grains, including wheat, are also known to contain lectins, although their possible physiological significance is unknown because of the absence of suitable studies (Liener, 1989). As lectins are usually inactivated by heat treatment, they are really only of interest when raw or inadequately cooked food or feed is consumed. Therefore, in the case of wheat they are more likely to be an animal feed concern. Although lectins have been detected in wheat, as well as in wheat germ, virtually no evidence exists of any significant anti-nutritional effect of these lectins (Klopfenstein, 2000).

Phytic acid

Phytic acid (myo-inositol hexaphosphate) chelates minerals such as iron, zinc, phosphate, calcium, potassium and magnesium. The bioavailability of trace elements such as zinc and iron can thus be reduced by the presence of phytic acid in monogastric animals, although in humans phytic acid does not seem to have a major affect on potassium, phosphorus or magnesium assimilation. Ruminants, on the other hand, are more readily able to utilise phytate-complexed minerals such as phosphorus because they have abundant amounts of microbial phytase, which degrades phytate, in the rumen. The typical levels of phytate in various wheat fractions are given in Table 6.13.

Table 6.13. **Phytate content of wheat**

Food	mg phytate/100 g edible portion
Wheat flour, all purpose	282
Wheat flour, whole wheat	845
Wheat bran, crude	3 011
Wheat germ	4 071

Source: Adapted from Harland (1993).

Wheat allergens

Wheat is one of the most common allergenic foods associated with IgE-mediated reactions in the world (FAO, 1995) but has only rarely been reported to cause anaphylaxis (Bousquet et al., 1998; Takizawa et al., 2001). Wheat is most commonly associated with the (IgE-mediated) conditions known as baker's asthma, resulting from the inhalation of wheat flour, and atopic dermatitis. The ingestion of wheat flour has also produced anaphylaxis in rare instances in children.

A diversity of allergens appears to be implicated in each of these conditions, with some in common between conditions, although thus far the specific allergens involved have not been identified. In baker's asthma, components of both the water/salt-soluble fraction (globulins and albumin) and the water/salt-insoluble fraction (gliadins and glutenins) have been reported to be allergens (Sutton et al., 1984; Franken et al., 1994; Sandiford et al., 1997). In cases of food-dependent, exercise-induced anaphylaxis, gliadin has been reported to be an allergen prominently involved (Palosuo et al., 1999). In cases of non-exercise-induced anaphylaxis in young children resulting from the ingestion of wheat flour, two or more protein components of the wheat proteins appear to be implicated and some of the proteins characterised are in common with those implicated in cases of atopic dermatitis (Takizawa et al., 2001).

Wheat, along with other gluten-containing cereals such as rye and barley, is also associated with a condition known as gluten-sensitive enteropathy (also called coeliac disease), which affects genetically predisposed individuals (FAO, 2001). The response is triggered by gliadin (Howdle et al., 1984).

Other compounds

DIBOA and DIMBOA

Hydroxamic acids and benzoxazinoids belong to a group of metabolites commonly found in the roots and leaves, but not in the seeds, of cereal plants. In young plantlets (seedlings) of wheat, DIBOA (2,4-dihydroxy-2*H*-1,4-benzoxazin-3(4*H*)-one) and DIMBOA (2,4-dihydroxy-7-methoxy-2*H*-1,4-benzoxazin-3(4*H*)-one) are prominent representatives of these metabolites, both in glycosylated and non glycosylated form (Nagakawa et al., 1995). High tissue levels of DIBOA and DIMBOA are associated with insect resistance, for example aphid resistance, of wheat obtained through conventional breeding.

After reaching their maximum during the first days of plant development, levels of DIBOA and DIMBOA decrease in the time thereafter, and are relatively low in flag leaves and ears at late growth stages (Copaja et al., 1999; Nicol and Wratten, 1997). In addition, levels vary between varieties. For example, young plantlets of screened wheat varieties contained 0-1.1 mmol (kg fw)$^{-1}$ DIBOA and 1.4-10.9 mmol (kg fw)$^{-1}$ DIMBOA (Copaja et al., 1991).

The mechanism of DIBOA's and DIMBOA's toxicity to insects has not been elucidated yet. In addition, data on the possible toxic and physiological effects of DIBOA, DIMBOA and related compounds on humans and domestic animals are scarce. One report, for example, describes the *in vitro* mutagenicity of DIBOA and DIMBOA in the Ames test (Hashimoto et al., 1979). In addition, a number of reports document hormonal effects of MBOA, a metabolite of DIMBOA, in wild rodents (Korn, 1988). Data on hormonal effects of MBOA in domestic animals are, however, fragmentary.

Food use

Identification of the key wheat products consumed by humans

Nearly 600 million tonnes of wheat are produced annually worldwide (FAO, 2002), with the majority of this destined for human consumption. Wheat and wheat foods are a major source of nutrients for people in many regions of the world. Overall, cereal grains contribute 50% and 45% of the world's dietary calories and protein, respectively, with wheat providing slightly less than 20% of total calories and protein (Betschart, 1988). The importance of wheat as a human food is primarily due to the fact that the wheat kernel can be ground into flour, which forms the basic ingredient of bread and other products, such as breakfast cereals, biscuits, cakes, pastry and noodles. Other products of the dry-milling process – the bran and the germ – are also highly valued as food: both are good sources of vitamins and minerals and the bran is also a good source of dietary fibre, and the germ is rich in vitamin E (measured as α-tocopherol).

Identification of key constituents and suggested analysis for food use

The suggested key constituents to be analysed for human food use are shown in Table 6.14. As all the food products are derived from the whole grain it is considered sufficient, in most circumstances, to analyse key constituents for the kernel only and it will not be necessary to separately analyse key constituents for the derived fractions, that is, the flour, bran or germ. Depending on the nature and purpose of the specific modification, however, additional analyses of the various derived fractions may also be useful.

Table 6.14. **Suggested constituents to be analysed in wheat for human food use**

Constituents	Kernel	Flour	Bran	Germ
Proximate	√	√	√	√
Amino acids	√			
Fatty acids				√
α-Tocopherol	√			√
B vitamins	√	√	√	
Phytate	√		√	

Feed use

Traditionally, wheat is considered a good animal feed, with approximately 16% of the world wheat production going into animal feeds. Wheat is said to compare favourably with corn in feed value and is regarded as superior to barley (Matz, 1991). Although corn has the higher energy value, wheat has the highest amount of crude protein (Ensminger et al., 1990).

Identification of the key wheat products consumed by animals

The key wheat products in animal feeding can be divided into three categories: *i)* whole and minimally processed grain; *ii)* processing by-products; and *iii)* forages derived from the whole plant.

Whole and minimally processed grain

Whole and minimally processed grain is fed to animals primarily for its high energy content and also because it is a valuable source of protein, vitamins and minerals. The grain also contains relatively low levels of fibre and thus is a highly digestible feed for both non-ruminants and ruminants. The grain, however, is not used very extensively in animal rations because of its high cost. In years in which harvests are adversely affected by rain and significant quantities of the grain are made unsuitable for milling because of sprouting, significant quantities of grain may be used for feed. The use of wheat for animal feeding varies from country to country. For example, most US wheat is used for human food whereas in Europe, where there is often a surplus of wheat, large quantities are fed to livestock (Ensminger et al., 1990). Most of the European feed wheat is thus of the hard or medium-hard type, whereas in the United States, most wheat for animal feed is of the soft type, which is not as valuable for milling (Matz, 1991).

Due to its low crude fibre content and high digestibility, wheat is a valuable starch source for various animals. Limitations on the amount of grain that can be included in the diet are generally not necessary, although digestive upsets in some young animals are known. For this reason, it is usually recommended that the proportion of wheat grain in the concentrate mix not exceed 50% in diets for piglets and 20% for chickens and broilers. The wheat grain is usually cracked, crushed or coarsely ground to improve palatability and digestibility. Heat processing (hot extrusion, steam flaking and popping) does not appear to improve the palatability or performance of wheat used as feed over minimally processed wheat (Matz, 1991).

Processing by-products

The amount of milling by-products used in animal feeds is almost entirely a function of the demand for flour (Matz, 1991). In the United States, wheat is cultivated primarily as a food grain for human consumption. As a result, most of the wheat fed to livestock is in the form of mill by-products.

By-products of the dry milling of wheat have long been employed as ingredients of animal feeds. Generally, millers remove about 80% of the kernel for wheat flour and the other 20% goes into the production of livestock feeds, generally described as wheat middlings, wheat bran, wheat shorts, wheat red dog, wheat screenings, wheat germ meal and wheat germ oil. With the exception of wheat germ meal and oil, such individual by-products have largely lost their identity in the milling industry. Many flourmills combine all by-product streams with the screenings, merchandising a single product (generally termed "millfeed") to the feed industry, with individual by-products generally not being available (Dale, 1996).

Wheat millfeeds are suitable for many types of livestock rations. Millfeeds are usually combined with other cereal grains and various supplements when fed to cattle and are often used in swine feeds but are rarely fed to sheep, although wheat bran is a favoured supplement for use in gestating sheep rations. Most horse rations contain an average of about 10% wheat bran (Matz, 1991).

By-products from fermentation and brewing processes, such as DDGS and brewers grains, are also important constituents of animal feeds. They serve as cheap energy sources and are particularly valuable for their protein content, as well as their vitamins and mineral content.

Whole plant

In addition to the production of grains, the entire wheat plant can be used for forage – pasture, hay or silage – which are all important categories in animal feeding (Matz, 1991). Winter wheat is an excellent source of pasture in the autumn and early spring, particularly for cattle. The by-products of the harvested grain, such as chaff, stover and straw, can also be used as low-quality forages for ruminant animals (Ensminger et al., 1990). They are generally high in fibre and low in protein and can be used as filler and also to provide some nutrients for cattle.

Identification of key constituents and suggested analysis for feed use

The suggested nutritional and compositional parameters to be analysed for animal feed use are shown in Table 6.15. Analysis of processing by-products should not be necessary as all the animal feed products are derived from either the whole grain or the whole plant.

Table 6.15. **Suggested constituents to be analysed in wheat for feed use**

Constituents	Kernel	Whole plant
Proximate	√	√
Amino acids	√	
Fatty acids	√	
Phytate	√	

The key analysis for animal feeds is the proximate analysis. Feeds are typically evaluated in terms of six components – moisture (dry matter), ash (mineral matter), crude protein (N X 6.25), ether extract (fat, organic acids, pigments, alcohols and fat soluble vitamins), crude fibre (cellulose, hemicellulose and lignin) and carbohydrate (starch, sugars, some cellulose, hemicellulose and lignin). For proximate analysis of animal feeds, acid-detergent fibre (ADF) and neutral-detergent fibre (NDF) should be substituted for crude fibre analysis. These give an indication of the digestibility of the feed and are particularly important for forage analysis.

Notes

1. For information on the environmental considerations for the safety assessment of wheat, see OECD (1999).

References

Barnes, P.J. (1982), "Lipid composition of wheat germ and wheat germ oil", *Fette. Seifen. Anstrichm*, Vol. 84, pp. 256-269.

Bass, E.J. (1988), "Wheat flour milling", in: Pomeranz, Y. (ed), *Wheat Chemistry and Technology*, Third Edition, Vol. II, American Association of Cereal Chemists, Inc., Minnesota, pp. 1-68.

Becker, R. and G.D. Hanners (1991), "Carbohydrate composition of cereal grains", in: Lorenz, K.J. and K. Kulp (eds.), *Handbook of Cereal Science and Technology*, Marcel Dekker, Inc., New York, pp. 469-496.

Belderok, B. (2000), "Developments in bread-making processes", *Plant Foods for Human Nutrition*, Vol. 55, pp. 1-86, Kluwer Academic Publishers, Netherlands.

Betschart, A.A. (1988), "Nutritional quality of wheat and wheat foods" in: Pomeranz, Y. (ed.), *Wheat Chemistry and Technology, Third Edition*, Vol. II, American Association of Cereal Chemists, Inc., Minnesota, pp. 91-130.

Bock, M.A. (2000), "Minor constituents of cereals", in: Kulp, K. and J.G. Ponte, Jr. (eds.), *Handbook of Cereal Science and Technology*, Second Edition, Marcel Dekker, Inc., New York, pp. 479-504.

Bousquet, J. et al. (1998), "Scientific criteria and the selection of allergenic foods for product labelling", *Allergy*, Vol. 53, Supplement s47, November, pp. 3-21, http://dx.doi.org/10.1111/j.1398-9995.1998.tb04987.x.

Chung, O.K. and J.-B. Ohm (2000), "Cereal lipids", in: Kulp, K. and J.G. Ponte, Jr. (eds.), *Handbook of Cereal Science and Technology*, Second Edition, Marcel Dekker, Inc., New York, pp. 417-477.

Copaja, S.V. et al. (1999), "Accumulation of hydroxamic acids during wheat germination", *Phytochemistry*, Vol. 50, Issue 1, January, pp. 17-24, http://dx.doi.org/10.1016/S0031-9422(98)00479-8.

Copaja, S.V. et al. (1991), "Hydroxamic acid levels in Chilean and British wheat seedlings", *Annals of Applied Biology*, Vol. 118, Issue 1, February, pp. 223-227, http://dx.doi.org/10.1111/j.1744-7348.1991.tb06100.x.

Dale, N. (1996), "The metabolizable energy of wheat by-products", *Journal of Applied Poultry Research*, Vol. 5, pp. 105-108, available at: http://japr.oxfordjournals.org/content/5/2/105.full.pdf .

Davis, K.R. et al. (1984a), "Variability of the vitamin content in wheat", *Cereal Foods World*, Vol. 29, No. 6, pp. 364-370.

Davis, K.R. et al. (1984b), "Evaluation of the nutrient composition of wheat. III. Minerals", *Cereal Foods World*, Vol. 29, pp. 246-248.

Davis, K.R. et al. (1981), "Evaluation of the nutrient composition of wheat. II. Proximate analysis, thiamine, riboflavin, niacin and pyridoxine", *Cereal Chemistry*, Vol. 58, pp. 116-120.

Davis, K.R. et al. (1980), "Evaluation of the nutrient composition of wheat. I. Lipid constituents", *Cereal Chemistry*, Vol. 57, pp. 178-184.

Ensminger, M.E. et al. (1990), *Feeds & Nutrition*, 2nd edition, Ensminger Publishing Company.

FAO (2002), *Food Outlook*, No. 1, February, Food and Agriculture Organization, Rome.

FAO (2001), "Evaluation of allergenicity of genetically modified foods", Report of a Joint FAO/WHO Expert Consultation on Allergenicity of Foods Derived from Biotechnology, 22-25 January 2001, Food and Agriculture Organization, Rome, available at: www.fao.org/3/a-y0820e.pdf.

FAO (2000), *Report of a Joint FAO/WHO Expert Consultation on Foods Derived from Biotechnology*, World Health Organization, Geneva, 29 May-2 June 2000, http://www.fao.org/fileadmin/templates/agns/pdf/topics/ec_june2000_en.pdf (accessed 11 February 2015).

FAO (1998), *Carbohydrates in Human Nutrition*, Report of the Joint FAO/WHO Expert Consultation 14-18 April 1997, Food and Nutrition Paper 66, Food and Agriculture Organization, Rome.

FAO (1996), *Report of a Joint FAO/WHO Expert Consultation on Biotechnology and Food Safety*, Food and Agriculture Organization, Rome, Italy, 20 September- 4 October 1996 http://www.fao.org/ag/agn/food/pdf/biotechnology.pdf (accessed 11 February 2015)

FAO (1995), "Report of the FAO Technical Consultation on Food Allergies", Rome, 13-14 November 1995, Food and Agriculture Organization, Rome.

Fellers, D.A. (1973), "Fractionation of wheat into major components", in: *Industrial Uses of Cereal Grains*, Symposium Proceedings of the 58th Annual Meeting of the American Association of Cereal Chemists, 4-8 November 1973, Minnesota.

Ferguson, L.R. and P.J. Harris (1999), "Protection against cancer by wheat bran: Role of dietary fibre and phytochemicals", *European Journal of Cancer Prevention*, Vol. 8, No. 1, February, pp. 17-25.

Franken, J. et al. (1994), "Identification of alpha-amylase inhibitor as a major allergen of wheat flour", *International Archives of Allergy and Applied Immunology*, Vol. 104, No. 2, pp. 171-174.

Harland, B.L. (1993), "Phytate contents of foods", in: Spiller, G.A. (ed.), *CRC Handbook of Dietary Fibre in Human Nutrition*, CRC Press, Boca Raton, Florida, pp. 617-623.

Hashimoto, Y. et al. (1979), "Mutagenicities of 4-hydroxy-1,4-benzoxazinones naturally occurring in maize plants and of related compounds", *Mutation Research*, Vol. 66, Issue 2, February, pp. 191-194, http://dx.doi.org/10.1016/0165-1218(79)90066-1.

Hazell, T. (1985), "Minerals in foods: Dietary sources, chemical forms, interactions, bioavailability", *World Review of Nutrition and Dietetics*, Vol. 46, pp. 1-123.

Howdle, P.D. et al. (1984), "Are all gliadins toxic in coeliac disease? An *in vitro* study of alpha, beta, gamma, and w gliadins", *Scandinavian Journal of Gastroenterology*, Vol. 19, No. 1, January, pp. 41-47.

International Grain Council (1996), *World Grain Statistics 1995/1996*, London.

Klopfenstein, C.F. (2000), "Nutritional quality of cereal-based foods", in: Kulp, K. and J.G. Ponte, Jr. (eds.), *Handbook of Cereal Science and Technology*, Second Edition, Marcel Dekker, Inc., New York, pp. 705-723.

Korn, H.A. (1988), "Feeding experiment with 6-methoxybenzoxazolinone and a wild population of the deer mouse (*Peromyscus maniculatus*)", *Canadian Journal of Zoology*, Vol. 67, No. 9, pp. 2 220-2 224, http://dx.doi.org/10.1139/z89-313.

Kritchevsky, D. and C. Bonfield (1995), *Dietary Fiber in Health and Disease*, Eagan Press, St. Paul.

Liener, I.E. (1989), "The nutritional significance of lectins", in: Kinsella, J.E. and W.G. Soucie (eds.), *Food Proteins*, American Oil Chemists' Society, Champaigne, Illinois, pp. 329-353.

Lookhart, G. and S. Bean (2000), "Cereal proteins: Composition of their major fractions and methods for identification", in: Kulp, K. and J.G. Ponte, Jr. (eds.), *Handbook of Cereal Science and Technology*, Second Edition, Marcel Dekker, Inc., New York, pp. 363-383.

Macri, A. et al. (1977), "Adaptation of the domestic chicken, *Gallus domesticus*, to continuous feeding of albumin amylase inhibitors from wheat flour as gastro-resistanct microgranules", *Poultry Science*, Vol. 56, No. 2, March, pp. 434-441.

Matz, S.A. (1991), *The Chemistry and Technology of Cereals as Food and Feed*, Second Edition, Van Nostrand Reinhold, New York.

Morrison, W.R. (1981), "Lipids", in: Pomeranz, Y. (ed.), *Wheat Chemistry and Technology*, Third Edition, American Association of Cereal Chemists, Inc., Minnesota, pp. 373-439.

Nagakawa, E. et al. (1995), "Non-induced cyclic hydroxamic acids in wheat during juvenile stage of growth", *Phytochemistry*, Vol. 38, Issue 6, April, pp. 1 349-1 354, http://dx.doi.org/10.1016/0031-9422(94)00831-D.

Nicol, D. and S.D. Wratten (1997), "The effect of hydroxamic acid concentration at late growth stages of wheat on the performance of the aphid *Sitobion avenae*", *Annals of Applied Biology*, Vol. 130, No. 3, pp. 387-396.

Nilsson, M. et al. (1997), "Content of nutrients and lignans in roller milled fractions of rye", *Journal of the Science of Food and Agriculture*, Vol. 73, Issue 2, February, pp. 143-148, http://dx.doi.org/10.1002/(SICI)1097-0010(199702)73:2<143::AID-JSFA698>3.0.CO;2-H.

NRC (1998), *Nutrient Requirements of Swine*, Tenth Revised Edition, National Research Council, National Academies Press, Washington, DC, available at: www.nap.edu/catalog/6016/nutrient-requirements-of-swine-10th-revised-edition.

OECD (2000), "Report of the Task Force for the Safety of Novel Foods and Feeds", prepared for the G8 Summit held in Okinawa, Japan on 21-23 July 2000, C(2000)86/ADD1, OECD, Paris, available at: www.oecd.org/chemicalsafety/biotrack/REPORT-OF-THE-TASK-FORCE-FOR-THE-SAFETY-OF-NOVEL.pdf (accessed 28 Jan. 2015).

OECD (1999), "Consensus document on the biology of *Triticum aestivum* (bread wheat)", ENV/JM/MONO(99)8, OECD, Paris, available at: www.oecd.org/env/ehs/biotrack/46815608.pdf.

OECD (1997), "Report of the OECD workshop on toxicological and nutritional testing of novel foods", SG/ICGB(1998)1, OECD, Paris, final version available at: www.oecd.org/officialdocuments/publicdisplaydocumentpdf/?doclanguage=en&cote=sg/icgb(98)1/final.

OECD (1993), *Safety Evaluation of Foods Derived by Modern Biotechnology, Concepts and Principles*, OECD, Paris, available at: www.oecd.org/science/biotrack/41036698.pdf.

Orth, R.A. and J.A. Shellenberger (1988), "Origin, production, and utilization of wheat", in: Pomeranz, Y. (ed.), *Wheat Chemistry and Technology*, Third Edition, American Association of Cereal Chemists, Inc., Minnesota, pp. 1-14.

Palosuo, K. et al. (1999), "A novel wheat gliadin as a cause of exercise-induced anaphylaxis", *Journal of Allergy and Clinical Immunology*, Vol. 103, Issue 5, May, pp. 912-917, http://dx.doi.org/10.1016/S0091-6749(99)70438-0.

Pomeranz, Y. (1988), "Chemical composition of kernel structures", in: Pomeranz, Y. (ed.), *Wheat Chemistry and Technology*, Third Edition, American Association of Cereal Chemists, Inc., Minnesota, pp. 97-158.

Posner, E.S. (2000), "Wheat", in: Kulp, K. and J.G. Ponte, Jr. (eds.), *Handbook of Cereal Science and Technology*, Second Edition, Marcel Dekker, Inc., New York, pp. 1-30.

Rao, G.V. (1979), "Wet wheat milling", *Cereal Foods World*, Vol. 24, pp. 334-335.

Rasper, V.F. and C.E. Walker (2000), "Quality evaluation of cereals and cereal products", in: Kulp, K. and J.G. Ponte, Jr. (eds.), *Handbook of Cereal Science and Technology*, Second Edition, Marcel Dekker, Inc., New York, pp. 505-537.

Rice-Evans, C.A. et al. (1997), "Antioxidant properties of phenolic compounds", *Trends in Plant Science*, Vol. 2, Issue 4, April, pp. 152-159, http://dx.doi.org/10.1016/S1360-1385(97)01018-2.

Sandiford, C.P. et al. (1997), "Identification of the major water/salt insoluble wheat proteins involved in cereal hypersensitivity", *Clinical & Experimental Allergy*, Vol. 27, No. 10, October, pp. 1 120-1 129.

Shelton, D.R. and W.J. Lee (2000), "Cereal carbohydrates", in: Kulp, K. and J.G. Ponte, Jr. (eds.), *Handbook of Cereal Science and Technology*, Second Edition, Marcel Dekker, Inc., New York, pp. 385-416.

Shewry, P.R. et al. (1994), "Opportunities for manipulating the seed protein composition of wheat and barley in order to improve quality", *Transgenic Research*, Vol. 3, No. 1, January, pp. 3-12.

Slavin, J. et al. (1997), "Whole-grain consumption and chronic disease: Protective mechanisms", *Nutrition and Cancer*, Vol. 27, No. 1, pp. 14-21.

Sutton, R. et al. (1984), "The diversity of allergens involved in baker's asthma", *Clinical Allergy*, Vol. 14, pp. 93-107.

Takizawa, T. et al. (2001), "Identification of allergen fractions of wheat flour responsible for anaphylactic reactions to wheat products in infants and young children", *International Archives of Allergy and Immunology*, Vol. 125, No. 1, May, pp. 51-56.

Thompson, L.U. (1994), "Antioxidants and hormone-mediated health benefits of whole grains", *Critical Reviews in Food Science and Nutrition*, Vol. 34, No. 5-6, pp. 473-497.

USDA (1999), "Food group 20, cereal grains and pasta", *USDA National Nutrient Database for Standard Reference*, Release 13, available at: http://ndb.nal.usda.gov.

Wiseman, J. et al. (1998), "Variability in the chemical composition of wheat and its utilisation by young poultry", *Home-Grown Cereals Authority (HGCA) Project Report*, No. 177, HGCA, London.

WHO (1991), *Strategies for Assessing the Safety of Foods Produced by Biotechnology*, Report of a Joint FAO/WHO Consultation, World Health Organization of the United Nations, Geneva, out of print.

Wu, Y.V. (1989), "Utilization of by-products of wheat-based alcohol fermentation", in: Pomeranz, Y. (ed.), *Wheat is Unique: Structure, Composition, Processing, End-Use Properties, and Products*, American Association of Cereal Chemists, Inc., Minnesota, pp. 657-674.

Zeringue, H.J., Jr. and R.O. Feuge (1980), "A comparison of the lipids of triticale, wheat and rye grown under similar ecological conditions", *Journal of the American Oil Chemists' Society*, Vol. 57, Issue 11, pp. 373-376.

Chapter 7

Rice (*Oryza sativa*)

*This chapter, prepared by the OECD Task Force for the Safety of Novel Foods and Feeds with Japan as the lead country, deals with the composition of rice (*Oryza sativa*). It contains elements that can be used in a comparative approach as part of a safety assessment of foods and feeds derived from new varieties. After some background on rice production and processing, nutrients and anti-nutrients are detailed. The final sections suggest the key products and parameters for analysis of new varieties for food use and for feed use.*

Note: This chapter is currently being revised by the Task Force for the Safety of Novel Foods and Feeds; an updated document on rice composition is expected to be available in the course of 2015. The updated version will be made available on the BioTrack Website at www.oecd.org/biotrack

Background

Production of rice

Rice is cultivated in more than 100 countries around the world and is a staple food for about a half of the world's population. The worldwide production area for rice is 150 million hectares (ha), and the annual production of rice (paddy rice) is about 590 million tonnes (FAO, 2004a). Asia is the main producer of rice with 92% of total world production. The country with the highest production is the People's Republic of China, with 186 million tonnes, or 31% of the total production. India is the second with 131 million tonnes, or 22%. India has the largest production area with about 43 million hectares (Table 7.1). Yield (tonnes/ha) has rapidly increased since the second half of the 1960s as the semi-short (short-stem) and high-yield varieties became widespread. Rice is mostly consumed in each producing country. The trade amount of rice is approximately 25 million tonnes (Table 7.2), which is less than 5% of the world production.

Table 7.1. **Word rice production, average production per year, 1999-2002**

Rank	Country	Production area ('000 ha)	Yield (tonne/ha)	Production ('000 tonnes)
1	China (People's Republic of)	29 815	6.25	186 519
2	India	42 724	3.06	130 606
3	Indonesia	11 724	4.37	51 207
4	Bangladesh	10 809	3.37	36 658
5	Viet Nam	7 572	4.29	32 489
6	Thailand	9 928	2.59	25 670
7	Myanmar	6 228	3.42	21 312
8	Philippines	4 037	3.12	12 600
9	Japan	1 738	6.58	11 441
10	Brazil	3 446	3.16	10 871
11	United States	1 323	7.06	9 334
12	Korea	1 069	6.61	7 065
	World	**151 385**	**3.94**	**596 989**

Note: The yield and production values are expressed as paddy rice. The countries are listed in order of production quantity.

Source: FAO (2004a).

Most of the rice varieties grown in the world belong to the species *Oryza sativa*, which has its origin in Asia. Another species grown in western Africa, *Oryza glaberrima*, is considered to have been domesticated in the Niger River delta. Varieties of the species *Oryza glaberrima* are cultivated in limited regions and detailed production data are scarcely available. For these reasons, this chapter deals only with *Oryza sativa* which occupies the great majority of the production and consumption in the world.

Rice is consumed in the world, mostly in Asia, as shown in Table 7.3 (FAO, 2001). Rice accounts for over 20% of global caloric intake (FAO, 2001).

Terminology

In this chapter, a number of technical and scientific terms that are specific to the rice industry are used. In order to facilitate common understanding, these rice-specific terms and their definitions are listed in Table 7.4.

Table 7.2. **World rice exports**

Thousands of tonnes

Country/year	1999	2000	2001	2002
Argentina	659	467	366	231
Australia	669	622	615	331
China (People's Republic of)	2 819	3 071	2 011	2 068
Egypt	307	393	656	464
India	1 895	1 533	2 194	5 053
Italy	667	666	563	593
Myanmar	54	251	939	730
Pakistan	1 791	2 016	2 424	1 684
Thailand	6 839	6 141	7 685	7 338
United States	2 668	2 736	2 622	3 267
Uruguay	699	741	811	652
Viet Nam	4 508	3 477	3 721	3 241
World	25 250	23 560	26 827	27 372

Note: Rice export quantities are calculated on the basis of the following multiplication factors: paddy rice, 0.65; husked rice, 1.00; milled/husked rice, 1.00; milled paddy rice, 1.00; and broken rice, 1.00.

Source: FAO (2004b).

Table 7.3. **Production and consumption of milled rice**[1]

Region	Production ('000 tonne)	Consumption (kg/caput/year)
Asia	363 255	83.8
North and Central America	8 061	11.2
South America	13 225	29.6
Africa	11 070	19.1
Europe	2 109	4.4
Oceania	1 187	15.8
World	398 907	56.5

Note: A milled rice equivalent.

Source: FAO (2001).

Cooking of rice

Rice is eaten as brown rice or milled rice after being cooked in grain form (for the processing of rice into brown or polished rice, see the following paragraph.). There are many recipes for cooked rice in which rice is boiled, steamed, boiled into porridge or mixed with other grain flours. Boiled or steamed rice can be further baked or fried.

Processing of rice

Paddy rice is processed as shown in Figure 7.1. Parboiled rice is obtained by boiling paddy rice as it is. Brown rice is produced by hulling, namely removing the hulls, from paddy rice. Milled rice is derived from brown rice by milling to remove all or most of the bran which primarily consists of seed coat, aleurone layer and germ. Germ seed is separated through bolting of the by-products of milling. Milled rice is processed by polishing to remove residual bran on the surface to give a smoother finish; and may further be polished to obtain the inner part of rice grain containing less protein for further processing. Most of the rice used for food is polished rice. Rice flour, which is partly used for rice wine (Sake) fermentation, is a pulverized product of the outer part or the whole part of milled rice. Rice bran oil is made from rice bran, which is used as cooking oil. Defatted bran (cake of rice bran) can be further utilised for feed and fertilizer.

Table 7.4. **Definitions used in this chapter**

Term	Definition in this document	Synonyms
Bran	Germ and several histologically identifiable soft layers (pericarp, seedcoat, nucellus and aleurone layer).	
Broken rice	Milled broken rice grains, subdivided into second heads (one-half to three-quarters), screenings (one-quarter to one-half) and brewer's rice (less than one-quarter) by the grain length, compared with that of the whole rice.	
Brown rice	Paddy rice from which the hull only has been removed; the process of hulling and handling may result in some loss of bran.	Caryopsis, cargo rice, hulled rice, husked rice, dehusked rice
Endosperm	Starchy tissue covered by the aleurone layer; divided into two regions: the subaleurone layer and the central core region containing mainly starch.	
Germ	The part consisting of scutellum, plumule, radicle and epiblast.	Embryo
Glutinous rice	Rice of which amylose content is less than 5%.	Waxy rice
Head rice	Milled whole rice kernels, exclusive of broken rice that is smaller than three-quarters of the grain length of the whole rice.	Head yield
Hull	Outermost layer of paddy rice.	Husk, shell, chaff
Hulling	Removal of the hull from paddy rice (note: sometimes referred to the removal of both hulls and bran).	Dehulling, husking, dehusking, shelling
Milled rice	Rice grain with removed germ and outer layer such as pericarp , seed coat and a part of aleulone layer by milling.	
Milling	Removal of all or most of the bran to produce the milled rice that is white.	Scouring, whitening
Paddy rice	Rice grain after threshing and winnowing and retains its hull.	Rice grain, rough rice
Parboiled rice	Hulled or milled rice processed from paddy or hulled rice which has been soaked in water and subjected to a heat treatment so that the starch is fully gelatinized, followed by a drying process.	
Polished rice	Rice grain with removed outer layer by polishing.	
Polishing	Abrasive removal of traces of bran on the surface of milled rice to give a smoother finish.	
Polishings	The by-product from polishing rice, consisting of the inner bran layers of the kernel with part of the germ and a small portion of the starchy interior.	

Figure 7.1. **Rice processing and the resulting products**

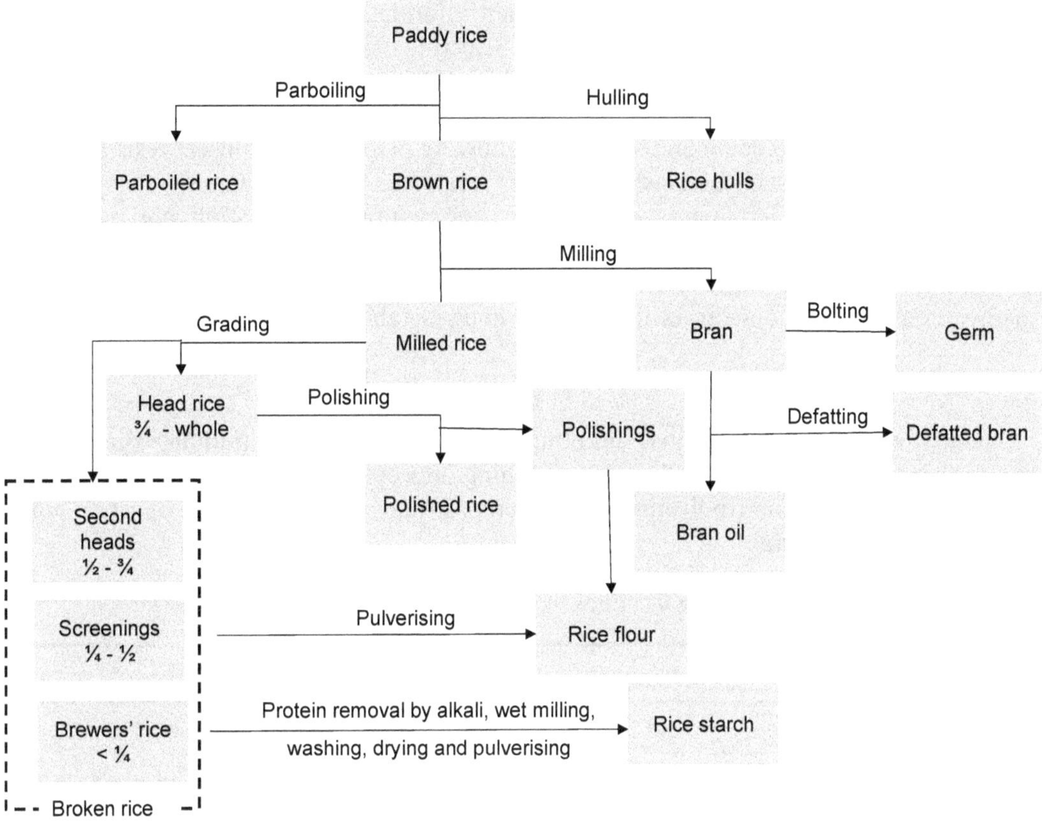

Only a relatively small amount of rice is consumed as prepared rice products worldwide. However, prepared rice products are widely found and consumed in Asia as rice noodle, rice cake, rice cracker, rice sweets and alcoholic beverages. For example, rice noodles are found in different shapes, flat or tubular, thick or thin, and given local names in Asian countries including China and Thailand. Rice sweets and cakes are also common in Asia. As for alcoholic beverages, there are rice wines and distilled rice wines in China, Japan and Korea. Those products that are undesirable for human consumption, such as poor grade paddy rice, broken rice, hulls, bran, rice flour and hulls/polishings of parboiled rice, are used in animal feed.

Appropriate comparators for testing new varieties

This chapter suggests parameters that rice developers should measure. Measurement data from the new variety should ideally be compared to those obtained from the near isogenic non-modified line grown under identical conditions. A developer can also compare values obtained from new varieties with the literature values of conventional counterparts presented in this chapter. Critical components include key nutrients and key toxicants for the food source in question. Key nutrients are those components in a particular product which may have a substantial impact in the overall diet. These may be major constituents (carbohydrates, proteins and lipids) or minor ones (minerals and vitamins). Key toxicants are those toxicologically significant compounds known to be inherently present in the species, that is, compounds whose toxic potency and level may impact on human and animal health. Similarly, the levels of known anti-nutrients and allergens should be considered.

Traditional characteristics screened by rice developers

Phenotype characteristics provide important information related to the suitability of new varieties for commercial distribution. Selecting new varieties is based on data from parental lines. Plant breeders developing new varieties of rice evaluate many parameters at different stages in the developmental process. In the early stages of growth, breeders evaluate stand count and seedling vigour. As plants mature, insect resistance and resistance to disease such as blast disease are evaluated. At near maturity or maturity, heading, maturation, lodging, shedding and pre-matured germination are evaluated. The matured plant is measured for plant height, ear height, number of ears and yield. The harvested grain is measured for yield, moisture, test weight, shape, size, visual quality, component's contents, milling quality and palatability.

Nutrients in rice

Paddy rice can be separated into hull and brown rice by hulling. Brown rice can further be separated by milling and polishing into polished rice and another fraction that consists of bran and polishings. The composition of each fraction of rice ranges widely as shown in Table 7.5.

Table 7.5. **Rice fractions by hulling, milling and polishing**

Fraction	Ratio (on a weight basis)
Hull	16-28% (average 20%) of paddy rice
Brown rice	72-84% (average 80%) of paddy rice
Polished rice	90% of brown rice
Bran + polishings	10% of brown rice

Source: Adapted from Juliano and Bechtel (1985).

Key nutrients in grain

Table 7.6 lists the key nutrients in rice products for food use.

Carbohydrates

Most of the available carbohydrates such as starch are found in the endosperm of rice grain. Milled rice mainly consists of starch with a few other carbohydrates, including free sugars and non-starch polysaccharides. The hull is comprised of a small amount of starch and mostly non-starch polysaccharides, such as cellulose and hemicellulose. The bran and germ are comprised mainly of non-starch polysaccharides, such as cellulose and hemicellulose and partly of free sugars as well as a small amount of starch.

Starch

Starch, the principal component of rice, consists of amylose and amylopectin. Starch in non-glutinous rice is composed of 15-30% amylose and 70-85% amylopectin. Starch in glutinous rice contains less than 5% of amylose and consists mostly of amylopectin (Juliano and Villareal, 1993).

Amylose content in rice grown in Asia ranges widely from 0% to 32% (Nakagahra et al., 1986).

Recently in Japan, a low-amylose variety of rice was developed, whose amylose content in starch is between those of non-glutinous rice and glutinous rice (Okuno et al., 1983). Many types of rice exist across the range from non-glutinous to glutinous varieties.

Table 7.6. **Proximate content (% of dry matter) of rice products used as food**[1]

Nutrient	Paddy	Brown	Milled	Bran	Germ	Polishings
Protein (N x 5.95)	5.8-7.7	7.1-8.3	6.3-7.1	11.3-14.9	14.1-20.6	11.2-12.4
Crude fat	1.5-2.3	1.6-2.8	0.3-0.5	15.0-19.7	16.6-20.5	10.1-12.4
Crude ash	2.9-5.2	1.0-1.5	0.3-0.8	6.6-9.9	4.8-8.7	5.2-7.3
Carbohydrates						
Available carbohydrates	63.6-73.2	72.9-75.9	76.7-78.4	34.1-52.3	34.2-41.4	51.1-55.0
Starch	53.4	66.4	77.6	13.8	2.1	41.5-47.6
Free sugars	0.5-1.2	0.7-1.3	0.22-0.45	5.5-6.9	8.0-12.0	
Neutral detergent fibre	16.4	3.9	0.7-2.3	23.7-28.6	13.1	
Crude fibre	7.2-10.4	0.6-1.0	0.2-0.5	7.0-11.4	2.4-3.5	2.3-3.2
Cellulose				5.9-9.0	2.7	
Hemicelluloses			0.1	9.5-16.9	9.7	...
Pentosans	3.7-5.3	1.2-2.1	0.5-1.4	7.0; 8.3	4.9; 6.4	3.6-4.7
Lignin	3.4		0.1	2.8-3.9	0.7-4.0	2.8
Energy (kJ/g)	15.8	15.2-16.1	14.6-15.6	16.7-19.9		17.9

Note: 1. Sample contains 14% of moisture.

Sources: Juliano and Bechtel (1985); Nyman et al. (1984); Dikeman et al. (1981); Kennedy et al. (1974); Houston (1972).

Dietary fibre

Although dietary fibre is an important nutrient, it is low in cooked rice such as cooked polished rice and polished rice porridge. It is lost by milling and polishing as can be seen in Table 7.7.

Table 7.7. **Dietary fibre in rice products**

Weight percentage

Product	Total dietary fibre	Soluble fibre	Insoluble fibre
Brown rice (15.5)[1]	3.0	0.7	2.3
Milled rice with the germ (15.5)[1]	1.3	0.3	1.0
Milled rice (15.5)[1]	0.5	Trace	0.5
Cooked milled rice (60.0)[1]	0.3	0.0	0.3
Milled rice porridge (83.0)[1]	0.1	0.0	0.1

Note: 1. Water content of the product.

Source: Resources Council, Science and Technology Agency of Japan (2000).

Protein

Total protein content in rice is calculated by multiplying total nitrogen content by the rice-specific Kjeldahl conversion factor of 5.95, which is based on the nitrogen content of glutelin, the major protein in rice. The protein content in brown rice based on the analysis of about 8 000 samples ranged from 5% to 17% (on a dry matter basis) (Juliano, 1968). The protein content of rice fluctuates according to the variety grown and can also be affected by growing conditions such as early or late maturing, soil fertility and water stress.

Rice proteins are classified by their solubility in albumin (soluble in pure water), globulin (soluble in salt water), prolamin (soluble in alcohol) and glutelin (soluble in aqueous alkaline solution) (Hoseney, 1986). The proportion of each protein type compared with the total protein is shown in Table 7.8. Albumin and globulin have a balanced composition of amino acids. They are found mostly in the outer layer of brown rice, and less in the inner layer of milled rice. Prolamin and glutelin are considered to be the storage proteins of rice, and exist in the outer layer and the inside of milled rice. Thus, the protein composition of bran and germ differs greatly from that of milled rice.

Table 7.8. **Typical proportions of osborne protein fractions in total rice protein**

Protein fraction	% of total protein
Albumin	2-5
Globulin	2-10
Prolamin	1-5
Glutelin	75-90

Note: Proteins were fractionated by the method of Osborne (Hoseney, 1986).

Source: Adapted from Simmonds (1978).

Amino acid composition

Protein content and amino acid composition varies in different fractions of rice kernel (Table 7.9). The key protein in rice is glutelin (oryzenin), and the most limiting amino acid is lysine. However, compared to other cereal grains, rice has nutritionally a more complete balance of amino acids.

To evaluate the nutritional value of each protein, amino acid score is calculated as follows: 100 x (mg of essential amino acid in the protein) / (mg of the essential amino acid in the reference protein ideal for human) (FAO/WHO, 1973; WHO, 1985). Rice (amino acid score of 61) has more balanced amino acid composition than those of other major cereals such as wheat (medium flour: amino acid score of 39) and corn (corn grits: amino acid score of 31) due to its higher contents of lysine and sulphur-containing amino acids.

Table 7.9. **Mean amino acid composition (% protein) of rice products used as food**

Amino acid	Paddy rice	Brown	Milled	Bran	Germ	Polishings
Alanine	4.6-6.7	5.8	5.6-5.8	6.2-6.7	6.6-7.2	6.2; 6.3
Arginine	4.2-10.0	8.5-10.5	8.6-8.7	8.2-8.7	9.7-10.4	8.5; 8.6
Aspartic	7.2-11.0	9.0; 9.5	9.1-9.6	9.5-10.5	9.1-10.6	9.2; 10.2
Cysteine	1.2-3.0	2.2-2.4	1.8-2.6	2.4-2.7	2.6-2.8	2.6; 2.7
Glutamic	15.4-20.5	16.9; 17.6	18.3-18.5	13.9-14.3	15.1-17.3	15.3; 16.8
Glycine	4.1-5.7	4.7; 4.8	4.5-4.8	5.5-5.9	6.0-6.6	5.3; 5.4
Histidine	1.6-2.9	2.4; 2.6	2.3-2.7	2.8-3.5	3.4-3.8	2.7; 2.8
Isoleucine	3.2-5.0	3.6-4.6	3.7-4.8	2.8-4.3	3.2-3.8	2.8; 4.0
Leucine	7.2-9.2	8.3-8.9	8.4-8.6	7.2-8.0	6.9-7.0	6.9; 8.0
Lysine	3.4-4.9	3.9; 4.3	3.4-4.2	5.0-5.7	6.2-7.4	4.4; 4.9
Methionine	1.6-3.6	2.3; 2.5	2.3-3.0	1.8-2.4	1.4-1.9	2.3; 2.9
Phenylalanine	3.3-6.1	5.0; 5.3	5.3-5.5	4.7-5.0	4.0-4.5	4.4; 4.8
Proline	3.9-6.3	4.8; 5.1	4.6-5.1	4.4-5.8	4.3-5.0	4.0; 5.4
Serine	4.2-6.0	4.8-5.8	5.3-5.9	4.9-5.7	4.8-5.4	4.7; 5.6
Threonine	3.2-4.7	3.9-4.0	3.7-3.9	4.0-4.4	4.2-4.5	3.7; 4.2
Tryptophan	1.3-2.1	1.3-1.5	1.3-1.8	0.6; 1.3	1.0-1.4	1.3
Tyrosine	4.0-5.7	3.8-4.6	4.4-5.5	3.3-3.6	3.3-3.7	3.6; 4.1
Valine	4.8-7.4	5.0-6.6	4.9-6.8	5.1-6.3	5.1-6.3	4.6; 5.9
Ammonia	1.4-6.8	2.8; 6.8	3.0-7.0	1.8-7.2	1.8-9.7	2.1; 6.2
alb/glo/pro/glu ratio[1]		6:10:3:81	5:9:3:83	37:36:5:22	24:14:8:54	30:14:5:51

Note: 1. alb/glo/pro/glu: albumin/globulin/prolamin/glutelin.

Sources: Juliano and Bechtel (1985); Kennedy and Schelstraete (1974); Houston et al. (1969).

Lipids

Rice lipid is contained mainly in the germ, aleurone layer and sub-aleurone layer. Within a cell, lipids exist in the form of a lipid globule with a diameter of 0.7-3 μm, which is called a spherosome. Some exist as starch-lipid complexes.

Most of the rice lipids are neutral. They are triglycerides in which glycerol is esterified with three fatty acids, primarily oleic, linoleic and palmitic. Besides triglycerides, free fatty acids, sterol and diglycerides are also found in rice. Rice also contains lipid-conjugates like acylsterolglycoside and sterolglycoside, glycolipids, such as cerebroside, and phospholipids, such as phosphatidylcholine and phosphatidylethanolamine (Table 7.10).

Lipids in a starch-lipid complex are not extracted by such organic solvent as ether, but by water-saturated butanol and others. The percentage of these lipids contained in non-glutinous brown rice is 0.5-0.7% and approximately 0.2% in glutinous brown rice. The major components are phospholipids followed by neutral lipids and glycolipids. Among fatty acids, palmitic and linoleic acids make up a large proportion, and oleic acid makes up a lesser amount (Choudhury and Juliano, 1980a, 1980b).

Fatty acid composition is dependent on the growing season and the ecogeographical varieties. Cultivated rice is classified into four varieties: Indian, Chinese, Japanese and Javanese. The amount of palmitic acid found in this order: Indian > Chinese > Japanese > Javanese (Taira et al., 1988). In terms of the fatty acid content, there is a strong negative correlation between oleic acid and linoleic acid, both of which are the key fatty acids of rice. In early crops, in which the ripening temperature is high, oleic acid content is high, while in late crops, linoleic acid content is high.

Table 7.10. **Mean composition of lipids in hull, brown rice and its fractions**[1]

Property	Free lipids[2]						Complexed lipids in non-glutinous starch	
	Hull	Brown	Milled	Bran	Germ	Polishings	Brown	Milled
Lipid content (weight %)	0.4	2.7	0.8	18.3	30.2	10.8	0.6	0.5
Fatty acid composition[3] (weight % of total)								
Palmitic (16:0)	18	23	33	23	24	23	46	45
Oleic (18:1)	42	35	21	37	36	35	12	11
Linoleic (18:2)	29	38	40	36	37	38	38	40
Others[4]	12	4	6	4	3	4	4	4
Neutral lipids[3] (% of total lipids)	64	86	82	89	91	87	28	26
Triglycerides		71	58	76	79	72	4	2
Free fatty acids		7	15	4	4	5	20	21
Glycolipids (% of total lipids)	25	5	8	4	2	5	19	16
Phospholipids (% of total lipids)	11	9	10	7	7	8	53	58
Phosphatidylcholine		4	9	3	3	3	4	4
Phosphatidylethanolamine		4	4	3	3	3	5	5
Lysophosphatidylcholine		<1	2	<1	<1	<1	21	23
Lysophosphatidylethanolamine			1			<1	22	25

Notes: 1. Based on 6% bran germ, 4% polishings and 90% milled rice from brown rice. 2. Free lipids stand for the lipids which are not involved in starch-lipid complexes. 3. Mean of two non-glutinous and one glutinous rices for free lipids; mean of the two non-glutinous rices only for complexed lipids in non-glutinous starch (Choudhury and Juliano, 1980b); and values of IR42 only for the hull (Choudhury and Juliano, 1980a). 4. Trace to 3% myristic acid, 2-4% stearic acid and 1-2% linolenic acid. Source: Adapted from Juliano and Bechtel (1985).

Minerals

Mineral content is greatly influenced by cultivation conditions – including fertilization – and soil conditions.

Among the inorganic elements contained in rice, silicon is dominant in paddy rice. In brown and milled rice, phosphorus is the principal mineral but comparable amounts of potassium, magnesium and silicon are also found (Table 7.11).

Table 7.11. **Range of mean content of elements in paddy rice and milling fractions**

Element	Paddy	Brown	Milled	Hull	Bran	Germ	Polishings
			Macroelements (mg/g dry matter)				
Calcium	0.1-0.9	0.1-0.6	0.1-0.3	0.7-0.5	0.4-1.4	0.2-1.2	0.6-0.8
Magnesium	0.7-1.7	0.2-1.7	0.2-0.6	0.4	5.8-15.1	5-15	7-8
Phosphorus	2.0-4.5	2.0-5.0	0.9-1.7	0.4-0.8	13-29	11-24	12-26
Phytin phosphorus	2.1-2.4	1.5-3.1	0.4-0.8	0	11-26	8-19	14-20
Potassium	1.7-4.3	0.7-3.2	0.8-1.5	1.7-8.7	12-23	13-17	8;13
Silicon	12.6	0.7-1.6	0.1-0.5	74-110	3-6	0.5-1.1	1.3;1.9
Sulfur	0.5-0.7	0.3-2.2	0.9	0.5	2.0	.	1.9
			Microelements (µg/g dry matter)				
Copper	2-13	1-7	2-3	35-45	11-40	11-40	6-30
Iron	16-70	2-60	2-33	45-110	100-500	70-210	50-180
Manganese	20-110	2-42	7-20	116-337	110-270	106-140	
Sodium	62-940	20-400	6-100	78-960	83-390	162-740	Trace - 160
Zinc	2.0-36	7-33	7-27	11-47	50-300	66-300	20;70

Sources: Juliano and Bechtel (1985); Dikeman et al. (1981); Kennedy and Schelstraete (1975).

Minerals are unevenly distributed in a brown rice grain. By milling stepwise from the outer layer of a brown rice with an abrasive rice mill, mineral contents in each layer fraction can be measured.

Mineral contents in a brown rice grain tend to decrease toward the endosperm. Endosperm contains much less minerals than germ and the outer bran layer fractions (Table 7.12).

Table 7.12. **Distribution of minerals[1] in brown rice grain**

Fractions	Phosphorus	Potassium	Magnesium	Calcium	Manganese	Iron	Silica
Bran layer fractions							
100-98.5[2]	100	100	100	100	100	100	100
98.5-97.0[2]	109	108	111	98	90	100	66
97.0-95.5[2]	117	108	112	90	81	79	49
95.5-94.0[2]	108	95	100	76	58	76	34
94.0-92.5[2]	100	81	83	61	40	54	24
92.5-91.0[2]	82	61	65	41	29	46	17
91.0-88.0[2]	42	39	40	35	18	29	13
88.0-85.0[2]	20	19	19	23	11	23	10
85.0-82.0[2]	12	10	10	14	7	16	
Endosperm							
82.0-0[2]	2.2	1.9	0.8	6.6	2.9	2.0	0.6
Germ	100	102	67	78	91	56	41

Notes: 1. Mineral contents for each layer fractions and products are expressed in weight ratio in comparison with those of the most exterior layer of the seed coat as 100. 2. Each value shows the weight ratio (%) of the milled rice to the whole grain, and each layer fraction was collected between two weight ratios indicated.Sources: Kubo (1960); Ohtsubo and Ishitani (1995).

Vitamins

Rice contains water-soluble vitamins including thiamine (B1), riboflavin (B2), pyridoxine (B6), nicotinic acid, inositol and cyanocobalamin (B12), and alpha-tocopherol (E). It does not contain significant amounts of hydrophobic vitamins A and D. Vitamins mainly exist in the endosperm and bran layer, thus milled rice contains less vitamins compared with brown rice (Table 7.13).

Table 7.13. **Vitamin content (µg/g dry matter) in paddy rice and milling fractions**

Vitamin	Paddy	Brown	Milled	Hull	Bran	Germ	Polishings
Retinol (A)	0-0.08	0-0.11	0-trace	0	0-3.6	0-1.0	0-0.9
Thiamine (B1)	2.6-3.3	2.9-6.1	0.2-1.1	0.9-2.1	12-24	17-59	3-19
Riboflavin (B2)	0.6-1.1	0.4-1.4	0.2-0.6	0.5-0.7	1.8-4.3	1.7-4.3	1.7-2.4
Niacin (nicotinic acid)	29-56	35-53	13-24	16-42	267-499	28-83	224-389
Pyridoxine (B6)	4-7	5-9	0.4-1.2		9-28	13-15	9-27
Pantothenic acid	7-12	9-15	3-7		20-61	11-28	26-56
Biotin	0.04-0.08	0.04-0.10	0.01-0.06		0.2-0.5	0.3-0.5	0.1-0.6
Inositol, total	800	1 000	90-110		4 000; 8 000	3 200; 5 500	3 700; 3 900
Choline, total	760-980	950	390-880		920-1 460	1 700; 2 600	860-1 250
p-Aminobenzoic acid	0.3	0.3	0.12-0.14		0.65	0.9	0.6
Folic acid	0.2-0.4	0.1-0.5	0.03-0.04		0.4-1.4	0.8-4.1	0.9-0.8
Cyanocobalamin (B12)	0-0.003	0-0.004	0-0.0014		0-0.004	0-0.01	0-0.003
alpha-Tocopherol (E)	9-20	9-25	Trace-3		26-130	76	54-86

Sources: Juliano and Bechtel (1985); Kennedy et al. (1975).

Key nutrients in animal feeds

The whole rice plant is sometimes used for animal feed. Table 7.14 provides proximate and major mineral content of the whole rice plant at different growth stages.

Nutritional composition of whole rice plant is dependent on its growth stage. Starch content increases as the rice kernel ripens. However, the nutritional value may decrease, as the rice kernel that is rich in nutrients could be lost if the harvest is delayed until its mature stage. Therefore, rice is generally harvested at its yellow ripe stage. Crude protein content of whole rice plant at that stage is low (about 7%).

The mineral content of rice plant is high; however, the contents of calcium and phosphorus are low as is the case with rice straw, because the silica content is more than the half of the mineral content.

Table 7.14. **Proximate and major mineral content (% of dry matter) of whole rice plant**

Ripening stage		Protein	Neutral detergent fibre	Acid detergent fibre	Nitrogen free extract	Ash	Calcium	Phosphorus
Whole rice plant	Early bloom	6.5-8.8	60.0-60.1	37.0-40.4	40.1-43.2	14.7-14.9	0.12	0.16
	Milk stage	5.6-8.5	52.5	33.1	45.6-49.9	12.0-13.6		
	Dough stage	5.3-9.6	49.3	29.9-31.6	49.7-61.4	9.7-15.6		
	Yellow ripe	4.9-7.2	43.4-56.8	26.1-35.0	50.9	12.6-12.9		
	Mature	4.0-7.6	38.9-48.3	22.9-33.7	52.9-60.6	9.1-15.5	0.17-0.19	0.40-0.67

Sources: National Agricultural Research Organization, Japan (2001); Enishi et al. (1995); Enishi and Shijimaya (1998); Horiguchi et al. (1992); Itoh et al. (1975); Nakui et al. (1988); Quinitio et al. (1990); Rahal et al. (1997); Taji et al. (1991); Taji and Quinitio (1992).

As most of the valuable nutrients are transferred from the leaves and stems and are stored in the ripening seeds, the straw which consists of the mature stems and leaves contains relatively little protein, starch and fat. Rice straw is low in calcium, phosphorus and most vitamins, but high in manganese. The high content of fibre, lignin and silica are responsible for the low digestibility. By adding 1-3% ammonia on a dry matter basis, its crude protein content is increased by 2-3 times. The dry matter digestibility and preservability are also improved.

Table 7.15 provides the range of nutrient content of the major rice feed ingredients. Proximate and major minerals are provided for rough (paddy) and broken rice, hulls, bran, polishings, straw and ammoniated straw. Animal nutritionists prefer that fibre be measured as neutral detergent fibre (NDF) and acid detergent fibre (ADF). Both of these measures are used to calculate feed energy values. Crude fibre values are included because of existing databases, but are not encouraged as a comparative method for feed.

Only the major minerals are important since the mineral content of plants is highly influenced by the level of minerals in the soil, and animal diets are fortified with the important minerals. Amino acid composition is provided for rough and broken rice, and bran. The amino acids included are those that are essential to be added to the diet and those that can contribute to the conversion to essential amino acids. Fatty acids levels are provided in Table 7.10, but only linoleic acid is important in animal nutrition.

Table 7.15. **Proximate, major mineral and amino acid content (% of dry matter) of rice products used as feed**

Nutrient	Paddy[1]	Broken[2]	Hull[3]	Bran[4]	Polishings[5]	Straw[6]	Ammonia[6] treated straw
Dry matter	81-90	87.0-89.0	87.0-92.5	89-94	90	90.9	
Protein (N x 6.25)[7]	7.5-9.7	6.7-9.8	2.1-4.3	10.6-16.9	11.2-13.4	1.2-7.5	8.2-16.0
Crude fat	1.5-2.3	0.5-1.9	0.30-0.93	5.1-19.7	10.1-13.9	0.8-2.1	
Neutral detergent fibre	3.9	13.7-16.0		26.1-33.0		67.9-78.6	60.3-63.9
Acid detergent fibre		3.5		13.1-15.4		38.3-56.7	41.7-46.8
Crude fibre	7.2-20.2	0.6	30.0-53.4	7.0-18.9	2.3-3.6	33.5-68.9	
Ash	2.9-6.5	5.0	13.2-24.4	8.8-28.8	5.2-8.3	12.2-21.4	14.2-14.8
Carbohydrates	63.6-84.4		22.4-35.3	90	51.1-55.0	39.1-47.3	
Starch	53.4		1.5		41.5-47.6		
Calcium	0.01-0.11	0.09-0.19	0.04-0.21	0.08-1.4	0.05	0.30-0.71	
Phosphorus	0.22-0.32	0.03-0.04	0.07-0.08	1.3-2.9	1.48	0.06-0.16	
Arginine	0.50-0.64	0.56-0.83		0.72-1.59			
Glycine	0.27-0.37	0.38-0.56		0.63-0.81			
Histidine	0.15-0.25	0.18-0.29		0.23-0.47			
Isoleucine	0.25-0.34	0.34-0.41		0.40-0.66			
Leucine	0.51-0.63	0.65-0.76		0.70-1.17			
Lysine	0.25-0.30	0.30-0.36		0.49-0.91			
Methionine	0.10-0.20	0.21-0.36		0.23-0.43			
Cystine	0.10-0.17	0.11-0.24		0.10-0.33			
Phenylalanine	0.32-0.38	0.43-0.54		0.44-0.76			
Threonine	0.25-0.30	0.27-0.40		0.41-0.64			
Tryptophan	0.10-0.12	0.11		0.10-0.19			
Tyrosine	0.10-0.60	0.29-0.70		0.32-0.48			
Valine	0.36-0.50	0.46-0.85		0.64-1.14			

Notes: 1. AgrEvo (1999); Farrell and Hutton (1990); Ffoulkes (1998); FAO (2003); Herd (2003); Juliano and Bechtel (1985); Miller et al. (1991); NGFA (2003); NRC (1982). 2. Farrell and Hutton (1990); NGFA (2003); NRC (1982, 1994, 1998). 3. AgrEvo (1999); Farrell and Hutton (1990); Ffoulkes (1998); FAO (2003); Herd (2003); Juliano and Bechtel (1985); Miller et al. (1991); NGFA (2003). 4. AgrEvo (1999); Farrell and Hutton (1990); Ffoulkes (1998); FAO (2003); Herd (2003); Juliano and Bechtel (1985); Miller et al. (1991); NGFA (2003); NRC (1982, 1994, 1998, 2000, 2001). 5. Miller et al. (1991); NRC (1994, 1998). 6. Drake et al. (2002), Fadel and MacKill (2002); FAO (2003); Ffoulkes (1998); Wanapat et al. (1996); Nour (2003). 7. Animal scientists commonly use a conversion factor of N x 6.25 for crude protein (AOAC, 2002).Anti-nutrients in rice

Phytin

Phytin is an organic phosphorous compound contained primarily in the bran layer, and it exists as a mixture of calcium-magnesium salts of phytic acid. Free phytic acid (myo-inositol 1,2,3,4,5,6-hexakis dihydrogen phosphate) chelates nutritional metal ions such as calcium and iron ions, which reduces the absorbability of these ions into the body (Thompson and Weber, 1981).

It has recently been reported that phytic acid reduced platelet aggregation and had an inhibitory effect against blood clot formation which may cause thrombosis and atherosclerosis (Vucenik et al., 1999). Phytic acid is considered to be an anti-carcinogen influencing signal transduction pathways, cell cycle regulatory genes, differentiation genes or suppressor genes (Shamsuddin, 1999).

Allergens

While rice is not considered to be a common cause of food allergic reactions, allergic reactions have been documented, and certain proteins in rice have been identified as rice allergens. The first reported allergens in rice were 14-16kDa proteins which were detected using sera from patients allergic to rice (Matsuda et al., 1991). A 16 kDA protein was later recognised as a major rice allergen. This protein has significant amino acid homology to barley trypsin inhibitor and wheat alpha amylase inhibitor (Izumi et al., 1992). Subsequently, rice seed proteins with molecular masses of 26, 33 and 56 kDA have been recognised as being allergenic. The 33 kDA protein has been characterised and identified as the enzyme glyoxalase I (Usui et al., 2001).

There have been several attempts to produce hypoallergenic rice. Rice products of reduced allergenicity have been developed by specifically hydrolyzing or reducing allergenic proteins using protease, alkali and ultra-high pressure treatment (Yamazaki and Sasagawa, 1997). Some rice products of reduced allergenicity were proven to be effective for individuals hypersensitive to rice and with atopic dermatitis (Watanabe et al., 1990). Furthermore, transgenic rice lines with reduced expression levels of the 14-16 kDa allergens are under development.

Trypsin inhibitor

A trypsin inhibitor has been isolated from rice bran and characterised (Tashiro and Maki, 1979). These investigators reported a specific activity of 0.0227 units per mg protein in defatted rice bran. There seems to be no standard way of reporting the quantity of the inhibitor, and it does appear to be heat labile. AgrEvo (1999) detected no trypsin inhibitor in the grain or polished rice, but did detect it in the bran.

Oryzacystatin

Oryzacystatin has been isolated from rice bran (Abe et al., 1987) and is considered a cysteinyl proteinase inhibitor (cystatin). It is inactivated by heat above 120°C (Juliano, 1993).

Alpha-amylase subtilisin inhibitor

The amino acid sequence of the bifunctional α-amylase subtilisin inhibitor from rice has been published by Ohtsubo and Richardson (1992). Bifunctional inhibitors have been proposed to be associated with defence of the seed against insect pests and pathogenic microorganisms (Ryan, 1990).

Lectins

Lectins are carbohydrate-binding proteins which agglutinate cells and precipitate glycoconjugates or polysaccharides (Goldstein et al., 1980). The toxicity of lectins is due to their ability to bind to specific carbohydrate receptor sites on the intestinal mucosal cells and interference with the absorption of nutrients across the intestinal wall (Liener, 1986). Rice bran lectin, haemagglutinin, has been found to be associated with agglutination of human A, B and O group receptors with specific binding to 2-acetamido-2-deoxy-D-glucose (Poola, 1989). Rice bran lectin is heat labile at temperatures above 80°C (Ory et al., 1981; Poola, 1989). Mannnose-binding rice lectin is distributed in all parts of the rice plant, and it has a potential ability to agglutinate bacterial cells of *Xanthomonas campestris* pv. *oryzae*, the pathogen causing bacterial leaf blight in rice, and also spores and protoplasts of *Magnaporthe grisea*, the rice blast fungus (Hirano et al., 2000).

Food use

Brown, milled and polished rice are the major rice products consumed by humans in the form of grain after being cooked. Rice is also consumed as food ingredients which are part of food products. For example, rice flour is used in cereals, baby food and snacks. The primary nutrients provided by rice are carbohydrates and proteins. Rice bran also provides some vitamins, fat and fibre. Rice oil extracted from bran is valued as a high-quality cooking oil.

Although relatively little rice is consumed as prepared products, a variety of such products is available in the market, in particular in Asia. Examples of prepared rice products include: parboiled rice, rice bread, rice noodle, mixed crop flour, ready-to-eat cooked rice, cooked rice for medical use, infant formulae, rice products specifically designed for aged people, rice bran, rice bran oil, rice germ, rice pudding, rice sweets and crackers, rice paper, swollen rice, sticky rice cake, fermented soybean paste (made from rice *koji*), rice vinegar and rice wine. Table 7.16 shows suggested nutritional and compositional parameters to be analysed in rice matrices for food use.

Table 7.16. **Suggested nutritional and compositional parameters to be analysed in rice matrices for food use**

Parameter	Bran oil	Rice flour	Paddy rice
Proximate analysis[1]		X	X
Minerals			X
Vitamins			X
Amino acids		X	X
Fatty acids	X	X	X
Phytic phosphorus			X
Amylose content		X	X

Note: 1. Proximate includes protein, fat, total dietary fibre, ash and carbohydrates.

Feed use

Identification of key rice products consumed by animals

Animals are fed paddy rice and its by-products such as rice straw, rice hull and rice bran. Whole rice plants can be fed as whole crop silage.

Paddy rice

The use of paddy rice and brown rice is limited as animal feeds because of the cost. Paddy rice is mostly consumed by humans and fed to animals only when the quality is poor or off-grade. Because of the hull, paddy rice is higher in crude fibre content and lower in calorific content than brown rice.

Paddy rice can replace other grains in animal feeding. For dairy and beef cattle diets, paddy rice can replace maize at the maximum rates of 40% (hereafter, in weight percentage figures) and 65%, respectively (JSFA, 1979a, 1979b). For poultry and swine, paddy rice can replace maize up to 60-65% (JSFA, 1979a). As rice endosperm is hard and enclosed in hard rice hull, paddy rice should be ground for efficient feed use.

Brown rice is an excellent animal feed, but is usually too expensive. For swine and poultry feeds, brown rice can replace maize at a rate of 40% (JSFA, 1970). Brown rice should be ground before used as animal feed, except in the case of poultry. It is also an excellent poultry feed because of its high-energy and low-fibre content. As paddy rice is lacking in carotene, the colour of egg yolks will become paler as rice content of poultry feed increases (JSFA, 1970). Broken rice is commonly used particularly in pet foods in the United States. It is valued for its lack of significant allergens.

Rice provides a number of other by-products that are valuable feed stuffs through harvest and processing: rice straw, rice hull and rice bran.

Rice straw

As rice straw is high in fibre it can be fed to ruminants as roughage. In the tropical zone of monsoon Asia, rice straw is used as roughage especially in the dry season.

Ruminants cannot subsist only on rice straw because of the low protein content (Table 7.15). Thus, an adequate protein balance should be achieved by supplementing the straw.

Rice straw can only partly replace forage because of the low protein content and low digestibility. The straw contains oxalates that chelate calcium and decrease its absorption. Rice is coated with prickly hairs to which cattle need some time to adapt. Rice straw containing less than 50% acid detergent fibre (ADF) could be good forage. Rice straw treated with ammonia or urea improves crude protein content, digestibility and preservability (Itoh et al., 1975: Rahal et al., 1997).

Hull

The hull is not a very good feed, as it is very low in protein and high in fibre. The sharp edges of the hull that may irritate the digestive tract of cattle should be broken by sufficiently grinding the hull. Digestibility can be improved by specific processes which remove silica. Monocalcium phosphate is added to the hull, and the mixture is ammoniated under heat and pressure to make an acceptable sheep feed. The hull is commonly used as a carrier for mineral and animal drug premixes.

Bran

Rice bran is a good source of protein, thiamine (vitamin B1) and niacin. The quality of feed is dependent on the amount of the hull content. Fresh bran is fairly palatable. However, it often turns rancid during storage unless treated with heat, because of the high oil content and the release of enzymes during processing. Heating and drying at milling can improve the storage life (Morimoto et al., 1985).

Rice bran is a good feed for dairy cows unless the bran amount exceeds 20% of the concentrate mixture. In Japan, rice bran has been used as one of the most important feed ingredients for *WAGYU* (Japanese Black). Rice bran can be blended up to 20% of swine feed. When too much rice bran is fed to juvenile pigs, it may lead to serious scouring. Due to the fatty acid composition in bran, swine and dairy cattle fed with bran in excess may lead to both body fat and butter fat to an undesirable soft nature (Morimoto et al., 1985).

Rice bran can replace wheat bran or wheat middlings in poultry feed. The bran contains a high amount of phytate (3-5%), which reduces the availability of minerals, and particularly phosphorus (NRC, 1998). Compared with rice bran, defatted rice bran has a long storage life and a high content in crude protein, crude fibre and ash.

Rice polishings also find their way into animal diets. Rice polishings easily become rancid during storage, as in the case of the bran. Therefore, the polishings, an excellent source of thiamine and niacin, should be fed as fresh as possible. The polishings can be used as a part of the concentrate mixture for dairy and beef cattle, and are good feed for swine.

Rice screenings, a mixture of small and broken rice seeds, can be used for feed. However, the nutrient content of screenings is highly variable.

Whole rice plant

Whole rice plants can be fed to dairy and beef cattle as whole crop silage. Its nutritional value is almost equivalent to that of barley whole crop silages (Horiguchi et al., 1992). Rice whole crop silage is low in crude protein and calcium, which should be supplemented (Table 7.14). Rice whole crop silage is palatable for cows (Goto et al., 1991), and dry matter intake by dairy cows ranges 6.3-9.5 kg per day (Ishida et al., 2000). There is only limited compositional information on the whole rice plant.

Identification of key products and suggested analysis for new varieties

In addition to proximate analysis, calcium and phosphorus need to be analysed in the forage which is fed to ruminants. Moreover, when using rice grain and its by-products as feed for swine or poultry, amino acids and phytic acid (as phytic phosphorus) should also be analysed. The suggested nutritional and compositional parameters to be analysed in rice matrices for animal feed use are shown in Table 7.17.

Table 7.17. **Suggested nutritional and compositional parameters to be analysed in rice matrices for feed**

Parameter	Paddy rice/bran	Straw	Whole plant
Proximate analysis[1]	X	X	X
Amino acids	X		
Calcium	X	X	X
Phosphorus	X	X	X
Phytic phosphorus	X		

Note: 1. NDF (neutral detergent fibre) and ADF (acid detergent fibre) should be substituted for crude fibre.

References

Abe, K. et al. (1987), "Molecular cloning of a cysteine proteinase inhibitor of rice (oryzacystatin)", *Journal of Biological Chemistry*, Vol. 262, No. 35, pp. 16 793-16 797, available at: www.jbc.org/content/262/35/16793.full.pdf.

AgrEvo (1999), "Safety, compositional and nutritional aspects of LibertyLink rice transformation events LLRICE06 and LLRICE62", U.S. FDA/CFSAN BNF 63.

AOAC (2002), *Official Methods of Analysis of AOAC International*, 17th edition, Association of Official Analytical Chemists, Gaithersburg, Maryland, Chapter 4, pp. 20-27.

Choudhury, N.H. and B.O. Juliano (1980a), "Lipids in developing and mature rice grain", *Phytochemistry*, Vol. 19, Issue 6, pp. 1 063-1 069, http://dx.doi.org/10.1016/0031-9422(80)83057-3.

Choudhury, N.H. and B.O. Juliano (1980b), "Effect of amylose content on the lipids of mature rice grain", *Phytochemistry*, Vol. 19, Issue 7, pp. 1 385-1 389, http://dx.doi.org/10.1016/0031-9422(80)80179-8.

Dikeman, E. et al. (1981), "Distribution of elements in the rice kernel determined by X-ray analysis and atomic absorption spectroscopy", *Cereal Chemistry*, Vol. 58, No. 2, pp. 148-152, available at: www.aaccnet.org/publications/cc/backissues/1981/Documents/chem58_148.pdf.

Drake, D.J. et al. (2002), *Feeding Rice Straw to Cattle*, Publication 8 079, University of California, Division of Agriculture and Natural Resources, available at: http://anrcatalog.ucdavis.edu/pdf/8079.pdf.

Enishi, O. and K. Shijimaya (1998), "The fermentative patterns of rice whole crop silage", *Japanese Journal of Grassland Science*, Vol. 44, pp. 179-181 (in Japanese).

Enishi, O. et al. (1995), "The effect of varieties and strains on chemical composition and *in vitro* dry matter digestibility of sodium hydroxide and ammonia treated rice (*Oryza sativa* L.) straw", *Japanese Journal of Grassland Science*, Vol. 41, pp. 160-163 (in Japanese with English summary).

Fadel, J.G. and D.J. MacKill (2002), "Characterization of rice straw – 94", University of California, Davis, California, www.syix.com/rrb/94rpt/RiceStraw.htm.

Farrell, D.J. and K. Hutton (1990), "Rice and rice milling by-products", Chapter 24, in: Thacker, P.A. and R.N. Kirkwood (eds.), *Nontraditional Feed Sources for Use in Swine Production*, Butterworths Publishers, Stoneham, Massachusetts.

FAO (2004a), "Agricultural production, crops primary", *FAOSTAT-Agriculture Data*, http://faostat.fao.org.

FAO (2004b), "Agriculture & food trade, crops & livestock primary & processed", *FAOSTAT-Agriculture Data*, http://faostat.fao.org.

FAO (2003), "*Oryza sativa*", *Animal Feed Resources Information System*, www.fao.org/ag//AGA/AGAP/FRG/AFRIS/default.htm.

FAO (2001), "Food balance sheets", *FAOSTAT-Agriculture Data*, http://faostat.fao.org.

FAO (2000), *Report of a Joint FAO/WHO Expert Consultation on Foods Derived from Biotechnology*, Geneva, Switzerland, 29 May-2 June 2000, http://www.fao.org/fileadmin/templates/agns/pdf/topics/ec_june2000_en.pdf (accessed 11 February 2015)

FAO (1996), *Report of a Joint FAO/WHO Expert Consultation on Biotechnology and Food Safety*, Food and Agriculture Organization, Rome, Italy, 20 September- 4 October 1996 http://www.fao.org/ag/agn/food/pdf/biotechnology.pdf (accessed 11 February 2015)

FAO/WHO (1973), *Energy and Protein Requirements*, Report of a Joint FAO/WHO Ad Hoc Expert Committee, World Health Organization Technical Report Series, No. 724; FAO Nutrition Meeting Report Series, No. 52, reprinted in 1987 and available at: www.fao.org/docrep/003/aa040e/aa040e00.htm.

Ffoulkes, D. (1998), "Rice as a livestock feed", Agnote, No. J22, April, Northern Territory of Australia, available at: https://transact.nt.gov.au/ebiz/dbird/TechPublications.nsf/A7BA6A055B2C671169256EFE004F65FC/$file/273.pdf.

Goldstein, I.J. et al. (1980), "What should be called a lectin?", *Nature*, Vol. 285, No. 66, http://dx.doi.org/10.1038/285066b0.

Goto, M. et al. (1991), "Feeding value of rice whole crop silage as compared to those of various summer forage crop silages", *Animal Science and Technology*, Vol. 62, No. 1, pp. 54-57.

Herd, D. (2003), "Composition of alternative feeds – Dry basis", Texas A&M U., College Station Texas, available at: http://animalscience.tamu.edu/wp-content/uploads/sites/14/2012/04/nutrition-composition-alternative-feeds.pdf.

Hirano, K. et al. (2000), "Novel mannose-binding rice lectin composed of some isolectins and its relation to a stress-inducible *salT* gene", *Plant Cell Physiology*, Vol. 41, No. 3, pp. 258-267.

Horiguchi, K. et al. (1992), "Comparisons of nutritive value among whole crop silage of rice plant with V-type leaves and other forages", *Japanese Society of Grassland Science*, Vol. 38, pp. 242-245 (in Japanese with English summary).

Hoseney, R.C. (1986), "Proteins", in: *Principles of Cereal Science and Technology*, American Association of Cereal Chemists, St. Paul, Minnesota, pp. 69-88.

Houston, D.F. (1972), "Rice bran and polish", in: *Rice: Chemistry and Technology*, 1st edition, American Association of Cereal Chemists, Inc., St. Paul, Minnesota, pp. 272-300.

Houston, D.F. et al. (1969), "Amino acid composition of rice and rice by-products", *Cereal Chemistry*, Vol. 46, pp. 527-537, available at: www.aaccnet.org/publications/cc/backissues/1969/Documents/chem46_527.pdf.

Ishida, M. et al. (2000), "Preliminary observation on milk yield and nutrients utilization by Holstein cows fed the round baled silage of the newly developed variety of whole crop rice, 'Kanto-shi-206'", *Kanto Journal of Animal Science*, Vol. 50, No. 1, pp. 14-21 (in Japanese with English summary).

Itoh, H. et al. (1975), "Improving the nutritive values of rice straw and rice hulls by ammonia treatment", *Japanese Journal of Zootech Science*, Vol. 46, pp. 87-93.

Izumi, H. et al. (1992), "Nucleotide sequence of a cDNA clone encoding a major allergenic protein in rice seeds. Homology of the deduced amino acid sequence with members of alpha-amylase/trypsin inhibitor family", *FEBS Letters*, Vol. 302, No. 3, pp. 213-216.

JSFA (1979a), "FY1978 report on feed processing test of rough rice: Feeding tests using broiler, layer and swine and lactation test using dairy cows" (in Japanese), Japan Scientific Feed Association.

JSFA (1979b), "FY1978 report on feed processing test of rough rice: Feeding test using beef cattle" (in Japanese), Japan Scientific Feed Association.

JSFA (1970), "Report on feed processing test of rice and related issues" (in Japanese), Japan Scientific Feed Association.

Juliano, B.O. (1993), *Rice in Human Nutrition*, Food and Agriculture Organization, Rome, available at: www.fao.org/docrep/T0567E/T0567E0g.htm#Antinutrition%20factors.

Juliano, B.O. (1968), "Screening for high protein rice varieties", *Cereal Science Today*, Vol. 13, pp. 299-301.

Juliano, BO. and D.B. Bechtel (1985), "The rice grain and its gross composition", in: Juliano, B.O. (ed.), *Rice: Chemistry and Technology*, 2nd edition, American Association of Cereal Chemists, St. Paul, Minnesota, pp. 17-57.

Juliano, B.O. and C.P. Villareal (1993), *Grain Quality Evaluation of World Rices*, International Rice Research Institute (IRRI).

Kennedy, B.M. and M. Schelstraete (1975), "Chemical, physical, and nutritional properties of high-protein flours and residual kernel from the overmilling of uncoated milled rice. III. Iron, calcium, magnesium, phosphorus, sodium, potassium, and phytic acid", *Cereal Chemistry*, Vol. 52, pp. 173-182, available at: www.aaccnet.org/publications/cc/backissues/1975/Documents/chem52_173.pdf.

Kennedy, B.M. and M. Schelstraete (1974), "Chemical, physical, and nutritional properties of high-protein flours and residual kernel from the overmilling of uncoated milled rice. II. Amino acid composition and biological evaluation of the protein", *Cereal Chemistry*, Vol. 51, pp. 448-457, available at: www.aaccnet.org/publications/cc/backissues/1974/Documents/chem51_448.pdf.

Kennedy, B.M. et al. (1975), "Chemical, physical, and nutritional properties of high-protein flours and residual kernel from the overmilling of uncoated milled rice. VI. Thiamine, riboflavin, niacin, and pyridoxine", *Cereal Chemistry*, Vol. 52, pp. 182-188, available at: www.aaccnet.org/publications/cc/backissues/1975/Documents/chem52_182.pdf .

Kennedy, B.M. et al. (1974), "Chemical, physical, and nutritional properties of high-protein flours and residual kernel from the overmilling of uncoated milled rice. I. Milling procedure and protein, fat, ash, amylose, and starch content", *Cereal Chemistry*, Vol. 51, pp. 435-448, available at: www.aaccnet.org/publications/cc/backissues/1974/Documents/chem51_435.pdf.

Kubo, S. (1960), "Transmigration of chlorine in a rice grain V. Transmigration of several elements other than chlorine", *Nippon Nogeikagaku Kaishi*, Vol. 34, pp. 689-694 (in Japanese).

Liener, I.E. (1986), "Nutritional significance of lectins in the diet", Chapter 10, in: Liener, I.E. et al. (eds.), *The Lectins: Properties, Functions and Applications in Biology and Medicine*, Academic Press, New York, pp. 527-552.

Matsuda, T. et al. (1991), "Immunochemical studies on rice allergenic proteins", *Agricultural and Biological Chemistry*, Vol. 55, Issue 2, pp. 509-513, http://dx.doi.org/10.1080/00021369.1991.10870603.

Miller, E.R. et al. (1991), *Swine Nutrition*, Butterworth – Heinemann, Boston and London.

Morimoto, H. et al. (1985), *Shiryoh-gaku (Feed Science)*, 2nd edition, Yoken-Do, Tokyo, pp. 152-156 (in Japanese).

Nakagahra, M. et al. (1986), "Spontaneous occurrence of low amylase genes and geographical distribution of amylase content in Asian rice", *Rice Genetics Newsletter*, Vol. 3, pp. 46-49, available at: http://archive.gramene.org/newsletters/rice_genetics/rgn3/v3I5.html.

Nakui, T. et al. (1988), "The making of rice whole crop silage and an evaluation of its value as forage for ruminants. Dietary fiber content and composition in six cereals at different extraction rates (wheat, rye, barley, sorghum, rice, and corn)", *Bulletin of Tohoku National Agricultural Experimental Station*, Vol. 78, pp. 161-174 (in Japanese with English summary).

National Agricultural Research Organization, Japan (2001), *Standard Tables of Feed Composition in Japan.*

NGFA (2003), "Rice fractions", National Grain & Feed Association, Grafton, Wisconsin, www.Ingredients101.com/ricefrac.htm.

Nour, A.M. (2003), "Rice straw and rice hulls in feeding ruminants in Egypt", available at: www.fao.org/wairdocs/ilri/x5494e/x5494e07.htm.

NRC (2001), *Nutrient Requirements of Dairy Cattle*, Seventh Revised Edition, National Academies Press, Washington, DC, available at: http://books.nap.edu/catalog/9825/nutrient-requirements-of-dairy-cattle-seventh-revised-edition-2001.

NRC (2000), *Nutrient Requirements of Beef Cattle*, Seventh Revised Edition: Update 2000, National Academies Press, Washington, DC, available at: www.nap.edu/catalog/9791/nutrient-requirements-of-beef-cattle-seventh-revised-edition-update-2000.

NRC (1998), *Nutrient Requirements of Swine*, 10th Revised Edition, National Academies Press, Washington, DC, available at: www.nap.edu/catalog/6016/nutrient-requirements-of-swine-10th-revised-edition.

NRC (1994), *Nutrient Requirements of Poultry*, Ninth Revised Edition, National Academies Press, Washington, DC, available at: www.nap.edu/catalog/2114/nutrient-requirements-of-poultry-ninth-revised-edition-1994.

NRC (1982), *United States – Canadian Tables of Feed Composition: Nutritional Data for United States and Canadian Feeds*, Third Revision, National Academies Press, Washington, DC, available at: www.nap.edu/catalog/1713/united-states-canadian-tables-of-feed-composition-nutritional-data-for.

Nyman, M. et al. (1984), "Dietary fiber content and composition in six cereals at different extraction rates (wheat, rye, barley, sorghum, rice, and corn)", *Cereal Chemistry*, Vol. 61, Issue 1, pp. 14-19, available at: www.aaccnet.org/publications/cc/backissues/1984/Documents/chem61_14.pdf.

OECD (2000), "Report of the Task Force for the Safety of Novel Foods and Feeds", prepared for the G8 Summit held in Okinawa, Japan on 21-23 July 2000, C(2000)86/ADD1, OECD, Paris, available at:

www.oecd.org/chemicalsafety/biotrack/REPORT-OF-THE-TASK-FORCE-FOR-THE-SAFETY-OF-NOVEL.pdf (accessed 28 Jan. 2015).

OECD (1997), "Report of the OECD workshop on toxicological and nutritional testing of novel foods", SG/ICGB(1998)1, OECD, Paris, final version available at: www.oecd.org/officialdocuments/publicdisplaydocumentpdf/?doclanguage=en&cote=sg/icgb(98)1/final.

OECD (1993), *Safety Evaluation of Foods Derived by Modern Biotechnology: Concepts and Principles*, OECD, Paris, available at: www.oecd.org/science/biotrack/41036698.pdf.

Ohtsubo, K. and T. Ishitani (1995), *Kome no Kagaku*, Asakurashoten, Tokyo, pp. 18-48 (in Japanese).

Ohtsubo, K. and M. Richardson (1992), "The amino acid sequence of a 20kDa bifunctional subtilisin/a-amylase inhibitor from bran of rice (Oryza sativa L.) seeds", *FEBS Letters*, Vol. 309, pp. 68-72.

Okuno, K. et al. (1983), "A new mutant gene lowering amylose content in endosperm starch of rice, *Oryza sativa* L.", *Japanese Journal of Breeding*, Vol. 33, pp. 387-394.

Ory, R.L. et al. (1981), "Properties of hemagglutinins in rice and other cereal grains", in Ory, R.L. (ed.), *Antinutrients and Natural Toxicants in Foods*, Food & Nutrition Press, Westport, Connecticut.

Quinitio, L.F. et al. (1990), "Feeding value and starch digestibility in paddy rice silage at the milk and the dough stages", *Japanese Journal of Zootech Science*, Vol. 61, pp. 663-665.

Poola, I. (1989), "Rice lectin: Physicochemical and carbohydrate-binding properties", *Carbohydrate Polymers*, Vol. 10, pp. 281-288.

Rahal, A. et al. (1997), "Effect of urea treatment and diet composition on, and prediction of nutritive value of rice straw of different cultivars", *Animal Feed Science and Technology*, Vol. 68, pp. 165-182, http://dx.doi.org/10.1016/S0377-8401(97)00045-X.

Resources Council, Science and Technology Agency of Japan (2000), *Standard Tables of Food Composition in Japan*, Fifth Revised Edition.

Ryan, C.A. (1990), "Protease inhibitors in plants", *Annual Review of Phytopathology*, Vol. 28, September, pp. 425-449, http://dx.doi.org/10.1146/annurev.py.28.090190.002233.

Shamsuddin, A.M. (1999), "Metabolism and cellular functions of IP6: A review", *Anticancer Research*, Vol. 19(5A), September-October, pp. 3 733-3 736.

Simmonds, D.H. (1978), "Structure, composition and biochemistry of cereal grains", in: Pomeranz, Y. (ed.), *Cereals '78: Better Nutrition for the World's Millions*, American Association of Cereal Chemists, St. Paul, Minnesota, pp. 105-137.

Taira, H. et al. (1988), "Fatty acid composition of indica, sinica, javanica, and japonica groups of nonglutinous brown rice", *Journal of Agricultural and Food Chemistry*, Vol. 36, pp. 45-47.

Taji, K. and L.F. Quinitio (1992), "Feeding value, digestibility and effect of combining hay cube in whole paddy rice silage and ammoniated whole crop paddy rice at the flowering stage", *Japanese Society of Grassland Science*, Vol. 38, Issue 1, pp. 16-25.

Taji, K. ct al. (1991), "Feeding value of paddy rice silage at the ripe stage in sheep and beef cattle", *Memoirs of the College of Agriculture, Ehime University*, Vol. 35, No. 2, pp. 171-177.

Tashiro, M. and Z. Maki (1979), "Purification and characterization of a trypsin inhibitor from rice bran", *Journal of Nutritional Science and Vitaminology*, Vol. 25, No. 3, pp. 255-264.

Thompson, S.A. and C.W. Weber (1981), "Effect of dietary fiber sources on tissue mineral levels in chicks", *Poutry Science*, Vol. 60, No. 4, April, pp. 840-845.

Usui, Y. et al. (2001), "A 33-kD allergen from rice (*Oryza sativa* L. *Japonica*). cDNA cloning, expression, and identification as a novel glyoxalase 1", *Journal of Biological Chemistry*, Vol. 276, No. 14, pp. 11 376-11 381.

Vucenik, I. et al. (1999), "Antiplatelet activity of inositol hexaphosphate (IP6)", *Anticancer Research*, Vol. 19(5A), pp. 3 689-3 693.

Wanapat, M. et al. (1996), "Cottonseed meal supplementation of dairy cattle fed rice straw", *Livestock Research for Rural Development*, Vol. 8, No. 3, available at: www.lrrd.org/lrrd8/3/metha83.htm.

Watanabe, M. et al. (1990), "Production of hypoallergenic rice by enzymatic decomposition of constituent proteins", *Journal of Food Science*, Vol. 55, No. 3, May, pp. 781-783, http://dx.doi.org/10.1111/j.1365-2621.1990.tb05229.x.

WHO (1991), *Strategies of Assessing the Safety of Foods Produced by Biotechnology*, Report of a Joint FAO/WHO Consultation, World Health Organization of the United Nations, Geneva, out of print.

WHO (1985), *Energy and Protein Requirements*, Report of a Joint FAO/WHO Ad Hoc Expert Committee, World Health Organization Technical Report Series, No. 724, World Health Organization, Geneva.

Yamazaki, A. and A. Sasagawa (1997), "Story of development of high-pressure processed foods: Part 3: A method of development of low allergenic cooked rice and bread", *Review of High Pressure Science and Technology*, Vol. 6, No. 4, pp. 236-241.

Chapter 8

Barley (*Hordeum vulgare*)

*This chapter, prepared by the OECD Task Force for the Safety of Novel Foods and Feeds with Finland, Germany and the United States as lead countries, deals with the composition of barley (*Hordeum vulgare*). It contains elements that can be used in a comparative approach as part of a safety assessment of foods and feeds derived from new varieties. Background is given on barley production, classification, uses and processing, followed by quality criteria, elements for comparative analyses and characteristics screened by breeders. Then nutrients in barley and its products, anti-nutrients and other compounds are detailed. The final sections suggest key products and constituents for analysis of new varieties for food use and for feed use.*

Background

Production of barley

Barley (*Hordeum vulgare* L.) is grown as a commercial crop in some 100 countries worldwide and is one of the most important cereal crops in the world. Barley assumes the fourth position in total cereal production in the world after wheat, rice and maize, each of which covers nearly 30% of the world's total cereal production (FAO, 2004). The Russian Federation, Canada, Germany, Ukraine and France are the major barley producers, accounting for nearly half of the total world production. Data on the total production and major producers in 2001 are shown in Table 8.1.

Table 8.1. **Barley production, 2001**

Country/region	Area harvested (million ha)	Grain yield (tonnes/ha)	Total production (million tonnes)
Australia	3.4	1.9	7.5
Canada	4.4	2.6	11.4
France	1.7	5.7	9.8
Germany	2.1	6.4	13.6
Russian Federation	7.7	2.5	19.5
Spain	3.0	2.1	6.2
Turkey	3.6	1.9	6.6
Ukraine	3.9	2.6	10.2
United Kingdom	1.2	5.4	6.7
United States	1.7	3.1	5.4
World total	54.3	2.6	141.2

Source: FAO (2004).

The major barley grain importers in 2000 were Saudi Arabia, the People's Republic of China, Japan and Belgium (Table 8.2).

Table 8.2. **Import of barley grain, 2000**

Region	Million tonnes	Share in total (%)
Algeria	0.6	2.6
Belgium	1.2	5.5
China (People's Republic of)	2.1	9.5
Germany	0.7	2.9
Iran	1.0	4.7
Italy	0.7	3.0
Japan	1.7	7.4
Morocco	0.9	3.9
Netherlands	0.6	2.5
Saudi Arabia	5.3	24.0
Tunisia	0.4	1.8
United States	0.6	2.6
Others	6.6	29.6
World total	22.4	100.0

Source: FAO (2004).

Classification of barley

Barley is one of the most ancient crops cultivated already some 10 000 years ago. Barley cultivated for food and feed belongs to the species *Hordeum vulgare* L. (Harlan, 1995). Although the barley crop is distributed throughout the world, its supposed progenitor, *Hordeum vulgare* L. ssp. *spontaneum* C. Koch, occurred in a more restricted area, namely the Middle East and adjacent regions of North Africa (Ellis, 2002). After domestication, unrecorded migration and trade would have rapidly distributed the barley crop outside the region of its origin. The result is the development of landraces adapted to northern and western European environments and later to north American, Australian and southern African environments (Ellis, 2002).

Barley is well adapted to a wide range of environments. It is grown in different latitudes from the equator up to the 65th latitude in the north and to the 50th latitude in the south, as well as from sea level up to mountain slopes. Consequently, the list of agronomic criteria used in breeding consists of at least increased and stable yield, early flowering and harvest, winter hardiness, resistance to extremes of temperature, edaphic factors and water stress, resistance to drought and soil acidity, salt tolerance, resistance to diseases and insect pests, and lodging. Quality criteria for breeding are determined according to the respective uses (processing characteristics and nutritional value) of barley.

Extensive cultivation, intensive breeding and selection have resulted in thousands of commercial varieties of barley. For commercial purposes, barley varieties are classified into broad classes that are used as a basis for world trade. The major factors used to distinguish barley varieties are feed or malting barley, winter or spring growth habit, six-, four- or two-row varieties, covered or naked/hulled barley, and starch amylose/amylopectin ratio.

Uses of barley

It is estimated that about 85% of the world's barley production is destined for feeding animals, while the rest is used for malt production, seed production and food consumption but also for production of starch either for food use or for the chemical industry (Fischbeck, 2002).

Barley grain-based feeds are used on pig and cattle farms. Barley is a valuable grain for finishing beef cattle in the United States and is also used in swine diets particularly in geographic regions where maize cannot be economically produced. In these climates, it competes with wheat as a feed, though it is considered to have a poorer nutritive value because of its higher fibre content.

Although malt from barley can be used for a number of purposes, the brewing industry utilises most of the barley malt produced (Fischbeck, 2002). In 2000, the main barley malt exporters were France, Belgium, Germany, Canada and Australia whereas the importers were Japan, Brazil and the Russian Federation (FAO, 2004). The United States, China and Germany are the major beer-producing countries, having together produced in 2001 more than one-third of the world total of 132.2 million tonnes of beer (FAO, 2004).

In some countries (Table 8.3) such as Morocco, India, China and Ethiopia, barley is used as an important food crop in daily diets (Bekele et al., 2001; Ceccarelli et al., 1999). Furthermore, some non-alcoholic drinks based on barley and malt are consumed and, for instance, "barley tea" has a longer history of use than green tea in Japan. For production of barley tea, six-rowed barley in Canada, two-rowed barley in Australia and the others are imported.

Table 8.3. **Use of barley for food consumption, 2002**

Country/region	Consumption ('000 tonnes)	Share in total food consumption (%)
Algeria	480	6.7
China (People's Republic of)	661	9.2
Ethiopia	887	12.3
Germany	170	2.4
India	1 108	15.4
Republic of Korea	220	3.1
Morocco	1 071	14.9
Poland	205	2.8
Ukraine	161	2.8
United States	149	2.0
World total	7 207	100.0

Source: FAO (2004).

Processing of barley

The processing of barley for food can be divided into the following categories: *i)* fractionation to produce pure barley starch, and fibre; *ii)* germination to produce barley malt; *iii)* milling and flaking; *iv)* production of non-alcoholic beverages. Barley starch is used in the food industry as a thickener and after hydrolysis as a sweetener as well as in the paper industry as coating material. The insoluble residue from the ethanol production, the distillers' grains as well as the fibre fraction, are used for feed. Barley malt is mainly used for beer production while smaller amounts are used by the whisky distilling industry and by bakeries. Both brewers' and distillers' grains are used as animal feeds as well.

Fractionation process for starch and fibre (Annex 8.A1)

When producing starch, barley grains are first milled in two steps. After removing the hull fraction, the kernels are soaked with enzymes in water. After the separation of fibres and protein, the barley starch is purified in hydrocyclones and dried resulting in a pure barley starch product.

Part of this starch and fermentable sugars are also used as raw materials in alcohol production. After continuous mashing with enzymes and fermentation with yeast, the fermented mash is distilled in a number of columns. The resulting products are both high-quality grain alcohol as well as industrial ethanol. The by-products such as carbon dioxide may be sold to industry and concentrated stillage solubles is sold to farmers as feeding stuffs. Hulls are usually burned in the processor's power plant.

Malting and fermentation

An overview of the typical processing steps from barley to beer is shown in Annex 8.A2. The objective of malting is to promote the production of endogenous enzymes capable of hydrolysing the grain macromolecules to soluble compounds.

The hydrolysis starts during malting and proceeds further during wort production in the brewery or distillery.

Malting process

The malting process consists of three stages: steeping, germination and kilning. During steeping, the barley is washed with water and the moisture content is increased, normally to 43-48%. During germination, the acrospire grows under the hulls and rootlets break through the end of the grain. Rootlets, rich in protein, are mechanically separated from the kilned malt.

Beer and whisky production

Beer production consists of three main stages: wort production, fermentation and downstream processing. To receive wort that contains fermentable sugars and other nutrients for yeast fermentation, the grain components are solubilised and hydrolysed by the enzymes produced during germination. After mashing, insoluble particles are separated and the mash cake is washed with hot water. The sweet wort is sterilised by boiling before the addition of yeast.

Malt whisky production is analogous to that of beer. The fermented mixture is distilled and the distillate is stored for several years. In grain whisky production, high enzyme malt is used mainly as enzyme source and the main raw material is cooked maize and unmalted wheat or barley. The whole mash is cooled and fermented, which allows the continuous the action of malt enzymes during fermentation. After fermentation, the entire mixture is distilled and the residue, distillers' grains, is dried and used as feeding stuff.

Malt syrups and extracts

Sweet wort can also be concentrated. The resulting malt syrup is used for baking, making candy and other food purposes. Malt extracts are prepared by vacuum concentration of the wort to obtain extracts and syrups of different colours, solid contents and enzyme activity. Depending on the drying temperatures, malt extracts with different diastatic (enzymatic) acivities may be produced.

Dry milling and flaking

In contrast to wheat, barley has a multi-cellular aleurone layer with thick cell walls. The endosperm cell walls are also thick and consist mainly of ß-glucan, so the elasticity of barley endosperm is different from that of wheat. This implies difficulties in milling barley as the flour has a low ash and fibre content compared to wheat. Barley flour is much fluffier and less dense than wheat flour. It is mainly ground by roller milling.

The flow diagram of the milling process is shown in Annex 8.A3. The initial steps of milling are to clean and moisturise the barley grain. This is achieved using separators to remove stones, sticks and other foreign material. After separating, the barley goes through an aspirator where airflow removes light impurities such as dust and straws. The next step is moisturising with intensive dampener to move the grain into the most favourable condition for subsequent grinding. Grain and fresh water are delivered together into the machine. The grain/water mixture enters a special high-speed rotor causing a uniform and intensive blend. After moisturising, the grain is ground by rollers and sieved. The fine flour is separated and coarse particles are ground for the second time. This procedure is repeated up to five times to get barley flour with different ash contents and to remove the hulls. Depending on the ash content, the flour shorts (the rejected dark fraction of the flour) may be used as feed.

Barley flours are used to bake special "flat" barley breads especially if darker barley flours are used or to bake mixed breads with wheat. Because the taste of barley is quite strong, flours with lower ash content are used. To produce barley kernels and flakes, barley grains are cleaned and dehulled. After dehulling, kernels are either pin milled to crushed kernels or cut to produce flakes. The cut barley kernels are moisturised by steam injection and roller flaked, after which the flakes are dried before packaging. The steps of barley flaking and the pin milling process are shown in Annex 8.A4.

Barley kernel, crushed barley and pearled barley are used as pot barley, to make porridge, pie fillings and so on. It can be cooked as an alternative to rice, pasta or potatoes, or added to stews. Barley flakes are used for porridge and gruel or as an ingredient in muesli or breakfast cereals. Barley kernels can also be used to produce "corn flakes" type of toasted barley flakes where barley kernels are pressure cooked, flaked and toasted, or they are produced by extrusion.

Production of non-alcoholic barley beverages

Barley grains are selected to remove foreign objects. After the selection, naked barley is steamed to gelatinise the starch and dried. The processing enhances the flavour of naked barley tea. The other barleys are applied directly to roasting. Roasting is repeated two or three times at 200-280°C until heat reaches the grain centre. The degree of roasting can be controlled by the amount of grain or the strength of heat source. According to Briggs (1978), roasting causes heat-dextrination of starch, a decline in hemicelluloses, caramelisation of sugars and formation of melanoidins from the interaction between reducing sugars and amino acids. The consequence is development of dark colour and flavour as well as an increase in acidity and solubility of the product. Roasted barley is cooled and sieved for packaging. In order to prepare barley tea bags, the product is milled and packaged. Other than barley tea grain and barley tea bags, barley tea condensate and packed barley tea are available.

Typical criteria used to determine barley quality

Barley quality criteria vary depending on its use. The most important quality parameters for different uses are discussed below.

Germination

For malting purposes, the most important quality parameter is a uniform germination. Normally, barley is not suitable for malting immediately after harvesting due to dormant grains. Dormancy means that grains are viable but not all of them are ready to germinate. Dormancy is common after a cool and damp season, but occurs less after a hot and dry harvesting season. Some dormancy is needed to avoid pre-germination on the field. Storage conditions affect the length of dormancy (Palmer, 1989; Riis et al., 1989). At least 96% of grains must germinate (Briggs et al., 1981). Uniform start of germination leads to homogenous modification of the endosperm. Homogeneity means in this case the synchronous germination of individual grains.

Moisture content

The moisture content is considered one of the most important quality criteria of malting barley. Wet barley respires more rapidly than dry barley, which may lead to a rise in temperature. High temperature and humidity may then activate the growth of bacteria and fungi, and lead to germination losses and production of mycotoxins. Safe storage conditions are a moisture level of 10-12% and a temperature of 15°C (Briggs et al., 1981). To avoid spoilage, immediate drying of barley after harvesting is needed.

Protein and starch contents

Low protein content is preferred for malting barley, preferably between 8.0-10.5% dry matter. In general, the lower the protein content, the higher the starch content, and thus the higher the sugar content for the final malt. Proteins are partly degraded in malting and mashing to amino acids and soluble peptides, which are needed as yeast nutrients and to produce good foam of beer. A high protein content of the barley may retard water up-take during steeping and result in a high-soluble protein content in wort, which may lead to a problem of haze formation in beer. Low protein content is also preferred for barley starch production to have high yields. For feed use, higher protein content is desired.

Whole and minimally processed grain is fed to farm animals primarily as an energy source. The most important consideration in evaluating barley for its energy value is its test weight. Higher test weights mean that the kernels have a higher starch and lower fibre content.

Analogous to malting barley, barley aimed for starch production should have a high starch and low protein content. The average starch content in barley grains is 60%. In the industrial starch production process, barley starch is fractionated – according to its specific weight – into two categories: starch with a large granule size and that with a small granule size. The large granule size starch is used in different modifications for food and fine paper and the small granule size starch for ethanol production. The best starch varieties used by industry are mostly composed of large granules. A loose internal grain structure is an important characteristic, allowing easy separation of starch granules from other components.

Grain structure and size

Barley cell walls encapsulate starch granules embedded in a protein matrix. With thin cell walls and loose packing of endosperm, the large mealy grains allow a rapid water up-take and uniform distribution of water and enzymes synthesised during germination. On the contrary, due to thick cell walls and tightly packed endosperms, small steely grains retard mass transfer in the endosperm. Large, plumb kernels are desired for malting. The fraction above the 2.5 mm sieve is normally used for malting and the rest is included in the feed fraction. A larger uniform grain size is desired because it enables homogenous water up-take and modification.

For feed use, barley grain is considered to have a poorer nutritive value than wheat or maize because of its higher fibre and consequently lower starch content. The barley hull has approximately 13% fibre, and dehulling is not practical for feed uses because the hull is fused to the seed by a cementing substance produced by the caryopsis.

Enzyme potential

Enzyme activity in barley is low or the enzymes exist in bound form. The major aim of malting is to produce or release bound enzymes to be active during germination and later in wort production. Numerous enzymes are found in malt. The enzyme spectra needed for different uses, for example, beer or whisky production, varies. The major enzyme groups include starch-, protein- and cell wall-degrading enzymes. The enzyme potential of barley can only be predicted after germination.

Comparative analyses

This chapter suggests parameters that barley developers and breeders should measure when undertaking comparative analyses of new varieties of barley. Measurement data from the new variety should ideally be compared to those obtained from the near isogenic non-modified line grown under identical conditions. A developer can also compare values obtained from new varieties with data on other barley varieties or with literature values of conventional counterparts presented in this chapter. Critical components include key nutrients and key toxicants for the food source in question. Key nutrients are those components in a particular product, which may have a substantial impact in the overall diet. These may be major constituents (fats, proteins and carbohydrates) or minor compounds (vitamins and minerals). Key toxicants are those toxicologically significant compounds known to be inherently present in the species, i.e. compounds whose toxic potency and level may impact on human and animal health. Similarly, the levels of known anti-nutrients and allergens should be considered. As part of the comparative approach, selected secondary plant metabolites, for which characteristic levels in the species are known, are analysed as further indicators of the absence of unintended effects of the genetic modification on the metabolism.

The final grain composition and quality are influenced by prevailing environmental conditions (Duffus and Cochrane, 1993). Barley composition is known to vary quite markedly from one area to another, as well as from year to year within any given area. For effective comparison it is therefore important that the new variety and its comparator (that is, the control) are grown at the same site(s) (preferably in adjacent plots) and at the same time.

Traditional characteristics screened by barley developers

Phenotypic characteristics provide important information related to the suitability of new varieties for commercial distribution. The selection of new varieties may depend on parental data. Plant breeders developing new varieties of barley evaluate many parameters at different stages in the developmental process. In the early stages of growth, breeders evaluate stand count and seedling vigour. As plants mature, insect resistance and resistance to fungal diseases, for example mildew, net blotch, scald, barley stripe, rusts, smuts and head blight, viral diseases and nematode diseases, are evaluated. At near maturity or maturity, heading, maturation, lodging, shedding and pre-matured germination are evaluated. The matured plant is measured for plant height, ear height, number of shoots, ears and seeds, and yield. The harvested grain is measured for yield, moisture, test weight, shape, size, visual quality, component's contents, malting and milling quality, and palatability.

Nutrients in barley and barley products

Barley grain

Whole barley grain is mostly used for feeding animals. For food purposes, barley is mainly used as dehulled grain or high-fibre content products. Food produced from barley is a good source for many nutrients such as protein, fibre, minerals and B vitamins.

The fibre content of barley is high and rich in β-glucan that is mainly soluble. Fibre-rich cereals such as barley are beneficial for balancing the human diet in a manner that is of no relevance for animals. Low-digestible carbohydrates, especially β-glucan and resistant starch, have a positive impact on lowering post-prandial blood glucose levels. Further, β-glucan has been reported to reduce the blood cholesterol level. Barley products are thought to be good for diabetics, obese and overweight people and for those who have a high blood cholesterol level (Kahlon and Chow, 1997). The β-glucan from barley is also known to stabilise digestion processes in young farm animals, especially in piglets (Bolduan and Jung, 1985). However, due to its viscosity-enhancing property, β-glucan causes undesirable effects in the digestive tract, especially of young avians. But with increasing age of the birds, the anti-nutritive effect decreases (Jeroch et al., 1993). The β-glucan levels are shown in Table 8.6.

Although barley has a relatively high protein content, it does not have the same baking characteristics as wheat gluten. Therefore, typical barley bread has low bread volumes. Barley flour is primarily used in combination with other flours to make multigrain breads.

The composition of barley is presented in Table 8.4, the proximate composition in Table 8.5 and the chemical composition in Table 8.6. The starchy endosperm consists of food reserves in the form of highly digestible carbohydrates (mainly starch), whereas the bran contains high levels of fibre and comparatively more minerals and fat than the endosperm.

Table 8.4. **Composition of barley grain**

Fraction	% kernel (by weight)	Key nutrients
Hulls (husks)	9-14	Cellulose, lignin, silica, pentosan, phenolic compounds
Seed coat	5.5-6.5	Cellulose, lipid
Aleurone layer	11-13	Lipid, protein, β-glucan, arabinoxylan, minerals, vitamins
Embryo	2.5-4.0	Lipid, storage protein, cellulose, sugars, minerals, vitamins
Endosperm	65-68	Starch, protein, β-glucan, arabinoxylan

Sources: Compiled from Briggs (1978); Palmer (1989).

Table 8.5. **Proximate composition of barley grain**

Parameter	% of dry matter
Protein	7.6-14.4
Fat	1.3-2.8
Crude fibre	4.0-8.0
Acid detergent fibre (ADF)	2.4-9.9
Neutral detergent fibre (NDF)	13.8-30.8
N-free extract	62.0-81.4
Ash	2.0-5.0

Sources: Compiled from Briggs (1978); Aherne (1990); Hunt (1995); Bull and Bradshaw (1995); Novus International (1996); NRC (1998; values converted from 89% dry matter to 100% dry matter); Anderson and Schroeder (1999); Lardy and Bauer (1999); USDA (2001).

Carbohydrates

Carbohydrates constitute the bulk of the total dry matter of the barley grain (Table 8.6). Most of the carbohydrate in barley is starch which serves as energy source during germination. Over 96% of the total grain cellulose is present in the hulls (husks) (Duffus and Cochrane, 1993). Mono- and di-saccharides (sucrose, glucose, fructose and maltose) are present in lesser amounts, but their concentration is twice as high as in other cereals. Of the non-starch polysaccharide fraction, the content of arabinoxylan (total 6.7% of which 0.4% is water soluble; Stölken et al., 1996) and β-glucan (4.6%; Stölken et al., 1996) is of relevance when barley is fed to young monogastrics, due to the negative effects on digestion. It is noteworthy that contrary to this, the low-digestible carbohydrates, especially β-glucan and resistant starch, have a positive impact on human health due to their role in lowering post-prandial blood glucose levels and in reducing the blood cholesterol level.

Table 8.6. **Chemical composition of barley grain**

Component	% of dry matter
Carbohydrates	78-83
Starch	63-65
Sucrose	1-2
Other sugars	1
Water-soluble polysaccharides	1-1.5
Alkali-soluble polysaccharides	8-10
Cellulose	4-5
β-glucan	1-4
Lipids	2-3
Protein	10-12
Albumins and globulins	3.5
Prolamins (hordeins)	3-4
Glutelins (hordenins)	3-4
Nucleic acids	0.2-0.3
Minerals	2
Other	5-6

Sources: MacGregor and Fincher (1993); Lyons (1978; β-glucan); Marins de Sa and Palmer (2001; β-glucan).

Proteins

The proteins of barley can be divided into four solubility groups: albumins (water-soluble); globulins (soluble in dilute saline); prolamins (soluble in alcohol/water mixtures); and glutelins (soluble only in dilute acid or alkali). Prolamins, called hordeins in barley, are the major storage proteins and account for 35-50% of the total nitrogen in the grain. The albumins, globulins, glutelins consist predominantly of structural and metabolic proteins (Kreis and Shewry, 1992).

The protein content of barley grains varies considerably. The precise composition depends on the growth conditions and on the rate and timing of nitrogen fertilisation (Duffus and Cochrane, 1993). For this reason it is important that an appropriate comparator is used for the comparative analysis. The typical protein fractions are listed in Table 8.6.

In general, protein content and protein quality of barley grain are not sufficient for high-performing monogastric farm animals. Consequently, their diets have to be supplemented with other protein sources. The low content of essential amino acids (e.g. lysine and methionine) in barley proteins is a direct consequence of the high content of hordeins that are relatively low in these amino acids. The amino acid composition of crude protein in barley grain fractions is listed in Table 8.7.

Hordeins have been reported to interfere with the brewing process; the amount of extract that ultimately can be derived from malt is inversely related to the protein (hordein) content of the original grain.

Table 8.7. **Amino acid composition of barley and its fractions**

g amino acid/100 g crude protein

Amino acid	Barley	Bran	Flour
Alanine	4.4-4.6	4.1-5.0	3.9-4.4
Arginine	4.2-6.2	4.6-5.7	4.6-5.5
Aspartic acid	6.8-7.4	6.4-8.6	5.7-7.1
Cystine	1.0-1.79	0.3-2.3	1.4-2.1
Glutamic acid	21.9-26.1	20.6-26.6	23.3-28.5
Glycine	4.2-5.1	3.9-5.0	3.4-4.3
Histidine	1.9-3.3	1.4-2.2	2.2-2.4
Isoleucine	3.1-3.9	3.4-3.7	3.5-3.7
Leucine	5.4-7.1	6.6-7.5	6.6-7.0
Lysine	3.1-4.2	3.3-5.0	3.4-4.1
Methionine	1.4-3.2	1.7-2.3	1.6-2.7
Phenylalanine	4.2-5.4	5.1-5.4	5.0-5.5
Proline	11.4-12.4	9.9-11.9	10.1-12.8
Serine	3.7-5.4	4.4-4.7	4.0-4.4
Threonine	3.0-3.7	3.2-3.8	3.0-3.6
Tyrosine	1.9-2.8	2.5-3.3	2.9-3.2
Valine	3.9-5.3	4.7-6.1	5.2-5.4

Sources: Compiled from Bhatty (1993); Briggs (1978); Harrold (1999); Bull and Bradshaw (1995); NRC (1998); Ensminger et al. (1990). Values taken from the last four references were calculated from dry matter basis to percentage of protein, based on reported protein levels.

Vitamins

The vitamin content of barley grains varies widely. Ungerminated barley does not contain vitamins A, C and D, although the carotenoids and sterols that are present may act as precursors for vitamins A and D, respectively (Briggs, 1978). Vitamin E, a mixture of tocopherols, occurs in barley oil.

Barley is unique among cereals in having all eight naturally occurring tocopherols. Tocolpherols are found exclusively in germ tissue (embryo, scutellum) and tocotrienols in the starchy endosperm and aleurone (Morrison, 1993). The tocol derivatives of barley are presented in Table 8.8.

Barley also contains B vitamins. These vitamins are mainly present in the embryo and the aleurone layer (Palmer, 1989). Typical ranges of B vitamin and folate concentrations in barley are shown in Table 8.9.

Table 8.8. **Tocol derivatives of barley**

Tocopherols	mg/kg barley	Tocotrienols	mg/kg barley
α-tocopherol	2.0-11.7	tocotrienol α-T-3	11.0-49.3
β-tocopherol	0.4-4.0	tocotrienol β-T-3	2.7-14.3
γ-tocopherol	0.3-12.9	tocotrienol γ-T-3	2.0-14.0
δ-tocopherol	0.1-0.9	tocotrienol δ-T-3	0.7-3.9

Sources: Compiled from Newman and Newman (1992); Morrison (1993).

Table 8.9. **B vitamin and folate content in barley grain, barley flour and malt flour**

Vitamin	Barley grain (µg/g)	Barley flour (µg/g)	Malt flour (µg/g)
Thiamine (vitamin B1)	1.2-16	3.7-4.0	3.1
Riboflavin (vitamin B2)	0.8-3.7	1.0-1.1	3.1
Niacin	46-147	55-63	56
Pantothenic acid	3.7-4.4	1.5	5.8
Pyridoxine (vitamin B6)	2.7-11.5	1.0-4.0	6.6
Folates	0.19-0.3	0.08-0.19	3.8

Sources: Compiled from Briggs (1978); USDA (2001); Fineli (2001).

Minerals

The major constituents of the mineral fraction of barley are magnesium, phosphorus, potassium, calcium and sodium. The average mineral content varies significantly, and this appears to be due to a number of factors, including the variety in question, the growing and soil conditions, and fertilizer application. Major constituents based on a compilation of worldwide data are given in Table 8.10.

Table 8.10. **Macroelements in barley grain (86-89% dry matter)**

	Ranges g/kg dry matter
Calcium (Ca)	0.4-0.7
Magnesium (Mg)	0.9-1.5
Phosphorus (P)	2.3-4.2
Sodium (Na)	0.2-2.7
Potassium (K)	3.0-5.9

Sources: Novus International (1996); NRC (1982; values converted from percentage dry matter to g/kg dry matter); NRC (1998; values converted from percentage as fed to g/kg dry matter using dry matter value reported).

A high portion of phosphorus in barley grain is bound to the phytate complex (51-66%), making much of the phosphorous unavailable to monogastric animals. Yet barley contains more phosphorous than common cereal grains and the phosphorous bioavailability of barley is higher than that in other grains (Harrold, 1999). The amounts of copper, iron, manganese and zinc present in barley grain may vary to a large extent due to growing conditions and this has to be taken into account when diets for farm animals are formulated (Novus International, 1996). As with vitamins, these minerals are mainly concentrated in the embryo and the aleurone layer (Duffus and Cochrane, 1993).

Lipids

In the mature barley grain, the lipid content is approximately 3%. Lipids constitute only a small part of the dry matter in most barley tissues yet they comprise significant reserves in the embryo and the aleurone layer of the grain. They are essential for the functional integrity of the cells. The composition and distribution of lipids in the different parts of barley grain are presented in Tables 8.11A and 8.11B. The total fat content, analysed as ether extract, is presented in Tables 8.5 and 8.6.

The majority of the lipids in barley are acyl lipids containing the fatty acids commonly found in higher plants, that is, myristic acid (14:0), palmitic acid (16:0), stearic acid (18:0), oleic acid (18:1, n-9), linoleic acid (18:2, n-6) and linolenic acid (18:3, n-3). The typical relative fatty acid composition of barley fat is presented in Table 8.12.

Barley contains approximately 0.8 mg/g sterols. Barley sterols include stigmasterol, β-sitosterol, campesterol and cholesterol (Piironen et al., 2000). These may occur in the free form, as glycosides, esterified with fatty acids or as acylated glycosides. Of the sterols, β-sitosterol is the primary sterol, comprising about 60% of the total sterols in barley. Campesterol is the next most abundant sterol found in barley (Piironen et al., 2000).

Table 8.11. **Composition and distribution of lipids in the principal parts of barley grain**

A. Composition

Compartment	Lipid class (weight %)		
	Nonpolar lipid (NL)	Glycolipids (GL)	Phospholipids (PL)
Whole grain	65-75	6-26	9-20
Embryo	76-90	6	18
Bran-endosperm	64-68	13	23
Aleurone	82		
Coleorhiza	74	4	22
Coleoptile	67	6	27
Scutellum	88	3	8
Hull	76	18	6

B. Distribution

Tissue	Tissue in grain (weight %)	Lipid in tissue (weight %)	Lipid as fraction of total lipid (%)
Whole grain	100	2-4.2	100
Embryo	3-6	19.6-24.0	17.9-37
Endosperm	88-97	1-3	63-72
Hull	6.8	2.4	5.0

Source: Adapted from Morrison (1993).

Table 8.12. **Fatty acid composition of total lipids in barley, malt and various parts of barley grain**

Anatomical part	Fatty acids (% of total fatty acids as detected)						
	14:0	16:0	18:0	18:1	18:2	18:3	Other
Barley	<1	20-28	1-2	10-21	44-60	4-9	
Malt	<1-2	17-31	1-3	5-12	48-65	7-11	
Embryo (dissected)	Trace	19-24	1	14-18	51-56	7-8	
Aleurone (fraction)	<1	20-21	1	12-15	55-58	6-7	14:1, 16:1 (1)
Endosperm (fraction)	<1	19-23	1-3	9-23	51-62	2-5	
Hulls (husks)	1-6	20-40	1-5	10-20	43-50	7-16	16:1 (1-4)
Bran	1	21	1	13	57	6	16:1 (trace)

Sources: Adapted from Morrison (1993); NRC (1994; values only for barley grain calculated from percentage as fed to percentage of fatty acids by totalling the fatty acids and dividing each by the total).

Whole plant

In addition to the production of grains, immature barley plants are used for forage, pasture or hay. The typical constituents in barley silage and straw are shown in Table 8.13. However, depending on the harvesting stage, the crop quality and composition may differ considerably. For example, when harvested at the dough stage the dry matter content may be between 28-42% and the crude ash content 6.2-7.7% of dry matter (Jaakkola et al., 2001, 2003).

Table 8.13. **Chemical composition of barley whole plant silage (18.5-39% dry matter) and straw (86-91% dry matter) from barley**

	Whole plant silage (g/kg of dry matter)	Straw (g/kg of dry matter)
Proximates		
Ash	75-188	64-75
Crude protein	67-120	38-44
Crude fat	29-39	17-19
Crude fibre	356-68	420-438
Acid detergent fibre (ADF)	345	590
Neutral detergent fibre (NDF)	563-568	725
Minerals		
Calcium	4.8-6.0	3.0
Phosphorus	3.0-3.3	0.7-0.8
Magnesium	1.4-1.8	0.9-2.3
Potassium	24.3-29.5	23.7
Sodium	1.3-1.5	1.4-3.7

Sources: Jeroch et al. (1993); NRC (1982, 2000, 2001).

Processing by-products

By-products from the dry milling of barley and from the beer and malt industry have long been employed as ingredients in animal feeds. According to the processing technology and the extraction rate, various classifications of barley by-products from dry milling are possible. Maltsters' pellets, brewers' grains and brewers' yeast are by-products from the brewing industry. Brewers' grain is a bulky by-product including spent grain and hops. The principal by-products from the brewing and milling processes (see figures in Annexes 8.A2 and 8.A3) and their mean chemical composition are listed in Table 8.14.

Table 8.14. **Crude nutrients in processing by-products**

g/kg dry matter

	High-grade feed[1]	Bran	Low-grade feed[2]	Hulls	Malsters' pellets (sprouts and hulls)	Brewers' grains	Brewers' yeast
Crude ash	43	5	78	71	66	49	85
Crude protein	138	137	111	85	296	244	530
Crude fat	34	38	37	44	10	79	20
Crude fibre	81	140	209	276	164	177	15

Notes: 1. By-product obtained from processing of screened and dehulled barley into pearl barley. 2. By-product obtained from processing of screened barley into pearl barley.

Sources: Kling and Woehlbier (1983); Jeroch et al. (1993).

Anti-nutrients and other compounds in barley and barley products

Anti-nutrients

The content of common anti-nutrients in cereals, including barley, is considered to be low when compared with legumes such as faba beans, peas and lupines.

Protease and amylase inhibitors

Protease inhibitors, especially trypsin inhibitors, may decrease the digestibility and nutritional value of ingested protein and retard growth when sufficient amounts are present in the diet. Amylase inhibitors may affect the digestibility of starch (Aherne, 1990). Both protease and amylase inhibitors have been identified in barley (Palmer, 1989). However, they do not appear to be responsible for any serious anti-nutritional activity in humans (Klopfenstein, 2000), probably because both inhibitor types tend to be heat labile.

Amylase inhibitor accumulates in barley grain during grain development (Duffus and Cochrane, 1993). Chymotrypsin inhibitors are present in the starchy endosperm and the aleurone layer (Kreis and Shewry, 1992).

Lectins

Lectins, sometimes called phytohemagglutenins, are glycoproteins that bind to certain carbohydrate groups on cell surfaces, such as intestinal epithelial cells, where they cause lesions and severe disruption and abnormal development of the microvilli. Although more commonly associated with legumes, cereal grains including barley are also known to contain lectins. However, their potential for physiological significance is unknown (Liener, 1989).

Phytic acid

Phytic acid (myo-inositol hexaphosphate) chelates minerals such as iron, zinc, phosphorus, calcium, potassium and magnesium. The bioavailability of these minerals can thus be reduced by the presence of phytic acid in monogastric animals, although in humans phytic acid does not seem to have a major effect on potassium, phosphorus or magnesium assimilation. Ruminants, on the other hand, are more readily able to utilise phytate-complexed phosphorus because they have abundant amounts of microbial phytase which degrades phytate in the rumen (Harland, 1993). Bull and Bradshaw (1995) report phytic acid levels ranging from 0.70-0.76% for barley grain.

Hordeins

Barley, along with other gluten-containing cereals such as wheat and rye, is also associated with a condition known as gluten-sensitive enteropathy (also called coeliac disease), which affects genetically predisposed individuals (FAO, 2004). Gluten is a complex of two major storage proteins in cereals, namely prolamin (hordeins in barley, gliadins in wheat) and glutelin (hordenins in barley, glutenins in wheat). The sensitivity response is triggered by the prolamin fraction of the cereal storage proteins that are hordeins in barley (gliadins in wheat).

Other compounds

Barley also contains a number of other constituents, some of which, at higher intakes, have been suggested to have a role in protection against diseases (Thompson, 1994). These include simple phenolic acids, lignans and the flavonoids.

Ferulic, vanillic, *o*- and p-coumaric, syringic, *p*-hydroxybenzoic, sinapic and chlorogenic acids occur free in barley. Water soluble esters of *p*-hydroxybenzoic, protocatechuic, ferulic, vanillic, p-coumaric, syringic, caffeic, sinapic and isoferulic acids have been detected as have glycosides of several of these and of gentisic, chlorogenic and dihydroxybenzoic acids (Briggs, 1978). Phenolic acids, principally ferulic but also p-coumaric acid, are covalently associated with arabinoxylans and constitute approximately 0.05% of cell walls in the starchy endosperm and 1.2% of aleurone walls. The insoluble, bound p-coumaric acid of barley grain is concentrated on the outer grain layers (MacGregor and Fincher, 1993). Bacterial enzymes in the human colon slowly and partially degrade the aleurone cell walls. This degradation results in the release of feruloylated oligosaccharides, which can then be further degraded to release ferulic acid. The phenolic acids are good antioxidants (Rice-Evans et al., 1997).

The flavonoids are a large group of phenolic compounds that occur widely in plants, and many of them have good antioxidant properties. Barley contains a range of flavonoids. Catechin, epicatechin, anthocyanins and proanthocyanins also occur in barley grains (Briggs, 1978).

Barley also contains phytoestrogenic compounds, that is, isoflavones and lignans. Minor amounts of isoflavones are present in barley (Murphy and Hendrich, 2002). Lignans are phenolic dimers, which are predominantly present in the bran. Lignans are converted by fermentation in the large intestine to mammalian lignans (Thompson, 1994). The plant lignan secoisolariciresinol occuring in barley is converted by intestinal microbes into enterodiol and enterolactone (Murphy and Hendrich, 2002).

Food use

Identification of key barley products consumed by humans

Some 140 million tonnes of barley are produced annually worldwide (FAO, 2004). In industrialised countries, the consumption of barley as food has lost most of its earlier importance in human nutrition (Fischbeck, 2002). The strong taste and "gummy" mouth feeling of whole barley kernels is limiting its food use. The major products are whole and crushed or pearled barley kernels, flours and flakes.

The predominant food product of barley is malt that is primarily used in the brewing industry. Barley malts, malt extracts and syrups are used in small amounts in food products to give better flavour and colour, for example in breakfast cereals and baked goods. The largest use is in fermented bakery products. Malt extract is a source of soluble sugars, protein and amylase in the dough and promotes the activity of yeast resulting in good bread texture and bigger loaf volume, good flavour and colour to the finished baked products. Further applications of malt products are for non-fermented bakery products, for example crackers, cookies and muffins. Malted barley rich in enzymes is also used for bakery products as a source of amylases to compensate the low α-amylase activity in bread wheat flours.

Although most of barley starch is used for manufacturing fine-quality papers, it also serves as good raw material for the food industry, where it is used as sweetener and binder. In the brewing industry, barley starch is used, together with barley malt, in the production of beer. Starch fermentation products are also distilled to pure grain alcohol for vodka-type products as well as industrial ethanol that is sold mainly to the pharmaceutical industry.

Modest quantities of non-alcoholic drinks based on barley and malt are consumed in various parts of the world. Both barley and malt are roasted and hot water infusion of the whole or ground products are consumed. Examples of such beverages are "malt coffee" or "barley tea". "Barley water" is made by soaking pot or pearled barley. There are also various malted beverages available, often in the form of "malty milk", in which malt extract is blended with milk. The mixture is dried and sold as soluble powder (Briggs, 1978). "Barley tea" is consumed in many Asian countries. Six-rowed barley, two-rowed barley and six-rowed naked barley are processed to barley tea in Japan. Furthermore, there are some minor food products such as barley germ-oil used as a food supplement or barley sprouts that are occasionally consumed. Both are of limited importance in the human diet.

Identification of key constituents and suggested analysis for food use

The suggested key constituents to be analysed for human uses of barley are shown in Table 8.15. As all food products are derived from the whole grain, it is considered sufficient, in most circumstances, to analyse key constituents of kernels only. In the production of malts, the seeds undergo a germination process activating the formation of a number of enzymes that, in turn, have a role in mobilising the seed reserves and enhancing the brewing process. The major enzymes produced during germination are starch-, protein- and cell wall-degrading enzymes. Therefore, depending on the nature and purpose of the specific modification, additional analyses of different fractions may also be useful.

It is not yet clear to what extent lectins, trypsin inhibitors and amylase inhibitors may be significant anti-nutrients in barley. However, it would not be desirable for their levels to be increased. The literature is not abundant with reference values and therefore the suggestion that these constituents be measured should remain optional.

Table 8.15. **Suggested constituents to be analysed in barley for food use**

	Whole grain	Flour	Malt
Proximates[1]	X	X	X
Starch	X	X	X
β-glucan	X	X	X
Amino acids	X		
α-tocopherol	X		
B vitamins	X	X	

Note: 1. Proximates include protein, fat, crude fibre, ash and nitrogen-free extracts (sugars, starch, soluble fraction of hemicellulose).

Feed use

Identification of key barley products consumed by animals

The key barley products used in animal feed can be divided into three categories: *i)* whole and minimally processed grain; *ii)* whole plant forages; *iii)* by-products of processing.

Whole and minimally processed grain

Whole and minimally processed grain is fed to farm animals primarily as an energy source and also to supply protein, vitamins and minerals. Barley is the most widely cultivated animal-feed cereal throughout Europe. As to the high digestibility it can be used most effectively in pig feeding, but it is also a valuable component in concentrates for ruminants and poultry (Kling and Woehlbier, 1983).

Most barley for animal feed is derived from winter varieties, which are somewhat higher in crude fibre and correspondingly lower in their energetic feeding value as compared to the summer varieties, which are predominantly used for brewing. However, in years in which growing conditions affect the quality of brewers' barley adversely, significant quantities are then used for animal feed (De Boer and Bickel, 1988). There are no known feed restrictions for barley grain in animal diets (Hoffmann, 1997).

The most important consideration in evaluating barley for its energy value is its test weight (that is, weight per bushel). Barley batches with higher test weights are higher in starch and lower in fibre. Some have observed that the two-row varieties of barley have a tendency to be higher in starch and lower in fibre, though compositional analysis has not revealed any major differences (Boyles et al., 2002).

Barley is a valuable grain for "finishing" beef cattle in the United States. Most barley is subjected to processing to break or alter the hard shell so that the barley is more amenable to animal digestion. Common processing techniques include grinding, cold rolling, moist rolling (16% moisture), tempering (soaking 24 hours in water) and rolling, steam rolling and steam flaking. With processing, its energy value is slightly less than maize because of the higher fibre, but it has more protein (Boyles et al., 2002).

Barley is used in swine diets, especially in those geographic regions where maize cannot be economically produced. In these regions, it competes with wheat as a feed, though it is considered to have a poorer nutritive value because of its higher fibre content. The barley hull has approximately 13% fibre, and dehulling is not practical for feed uses because the hull is fused to the seed by a cementing substance produced by the caryopsis. Dehulled varieties have been developed, but even though the composition looks attractive from its higher amino acid content, especially in respect of lysine, variable results have been obtained in feeding trials. It has been postulated that β-glucans, structurally similar to cellulose, are part of the cause of variable animal performance observed. However, the exact role of these substances has not been confirmed through research.

Whole plant

In the last decades, whole plant silage is becoming a more important feed for ruminants as well as for other species (Jeroch et al., 1993). For this production type, winter and summer varieties are used, sometimes sown in combination with a fast-growing grass variety. In addition, the straw, as a by-product of the harvested grains, can be used as low-quality forage for ruminants. The whole plant silages are high in fibre and low in protein and may be used in extensive cattle production, to provide some nutrients.

By-products of processing

The amount of milling by-products used in animal feed is probably a function of the demand for flour, grouts and pearl barley. These barley products are predominantly used in nutritional products for human consumption (Becker and Nehring, 1967). By-products of the dry milling of barley have long been employed as ingredients in animal feeds. Generally millers remove 80-83% of the kernel for flour and the rest goes into the production of livestock feeds (Becker and Nehring, 1967).

In milling by-products resulting from pearl barley production, the residues amount to 50-60% of raw barley. The individual by-products have largely lost their identity during the milling process. The by-products from the various production steps are combined in a single product (generally termed "barley feed") that is sold to the feeding industry. Individual by-products are not generally marketed (Becker and Nehring, 1967).

The predominant criterion for the feeding value of the milling by-products is the fibre content, as the digestibility of total nutrients is negatively affected by this fraction. Accordingly, low-grade barley feed and hulls are poor quality feeding stuffs for monogastric farm animals (Kling and Woehlbier, 1983).

By-products from brewing, such as brewers' grain, is also known as valuable feeding stuffs. As their moisture content is very high, they are mostly fed fresh to cattle and dairy cows. After drying they may also be used as constituents of concentrates for poultry (Jeroch et al., 1993). Depending on economic value, the various brewers' by-products are sold separately or as mixers with grains. Consequently, brewers' grains vary considerably in their chemical composition.

Brewers' grain is a bulky by-product of the beer or malt industry and the product includes spent grain and hops. It is a good source of by-pass protein for dairy cattle but is ow in calcium and phosphorus. Intake is limited to 20-25% of the grain mixture dry matter and 15-25% of the total ration dry matter. It has a short storage life of two to five days in summer and five to seven days in winter. Because of its bulkiness and cost, distribution is usually limited to a distance of 167-333 kilometres from the brewery (Amaral-Phillips and Hemken, 2002).

Identification of key constituents and suggested analyses for feed use

The composition of grain, the by-products of processing and the whole plant appear to be representative of all the products that could be fed to animals. The nutritional and compositional parameters of barley, which are of importance for animal feed use, are shown in Table 8.16. Analysing either whole grain or by-products of processing will provide equivalent information on these parameters.

It is not yet clear to what extent lectins, trypsin inhibitors and amylase inhibitors may be significant anti-nutrients of barley. However, it would not be desirable for their levels to be increased. Because the literature is not abundant with reference values, the suggestion that these constituents were measured should remain optional.

Table 8.16. **Suggested constituents to be analysed in barley for feed use**

	Whole grain	Processing by-products	Whole plant
Proximates	X	X	X
Amino acids	X	X	
Phytic acid	X		
β-glucan	X	X	

The key analysis for animal feeds is the proximate analysis. Feeds are typically evaluated in terms of six components: dry matter; crude ash (mineral matter); crude protein (N x 6.25); crude fat (ether extract); crude fibre (composed of cellulose, hemicellulose and lignin); and nitrogen-free extracts (starch, sugars, soluble fraction of hemicellulose). For proximate analysis of animal feeds, acid-detergent fibre (ADF) and neutral detergent fibre (NDF) are preferred to crude fibre analysis. They give an improved indication of the digestibility and the energetic feeding value of the feed, which is particularly important for feed evaluation.

Annex 8.A1
Barley fractionation for starch and ethanol production

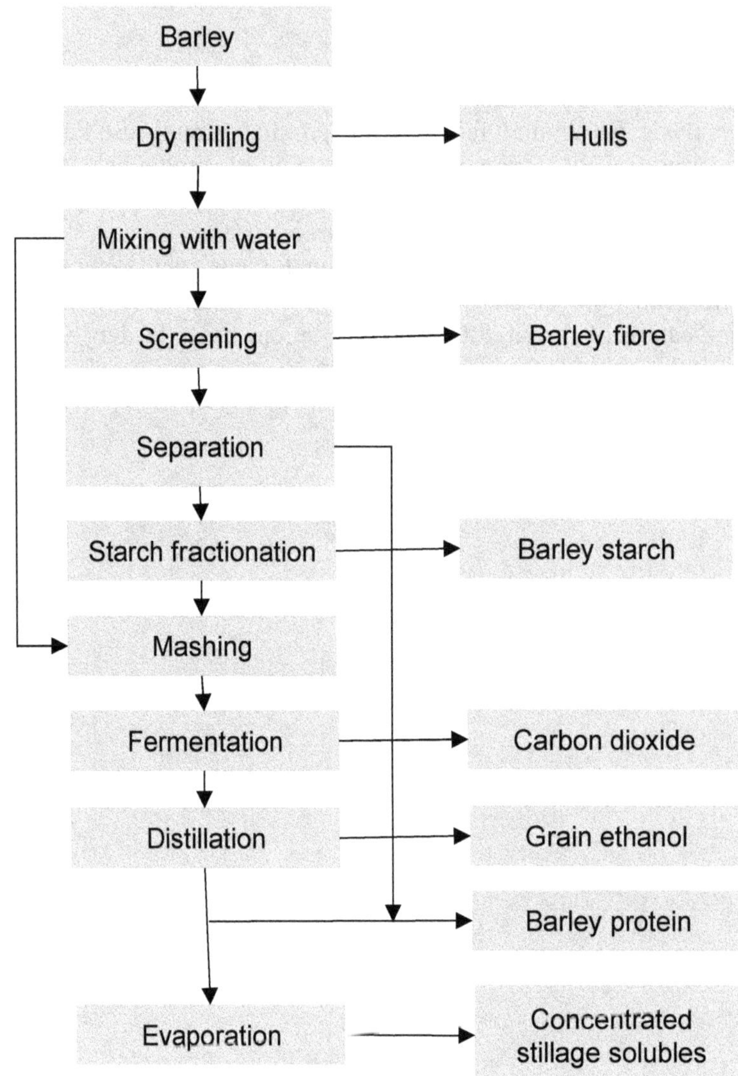

Annex 8.A2
Overview of malt, beer and malt syrup production steps

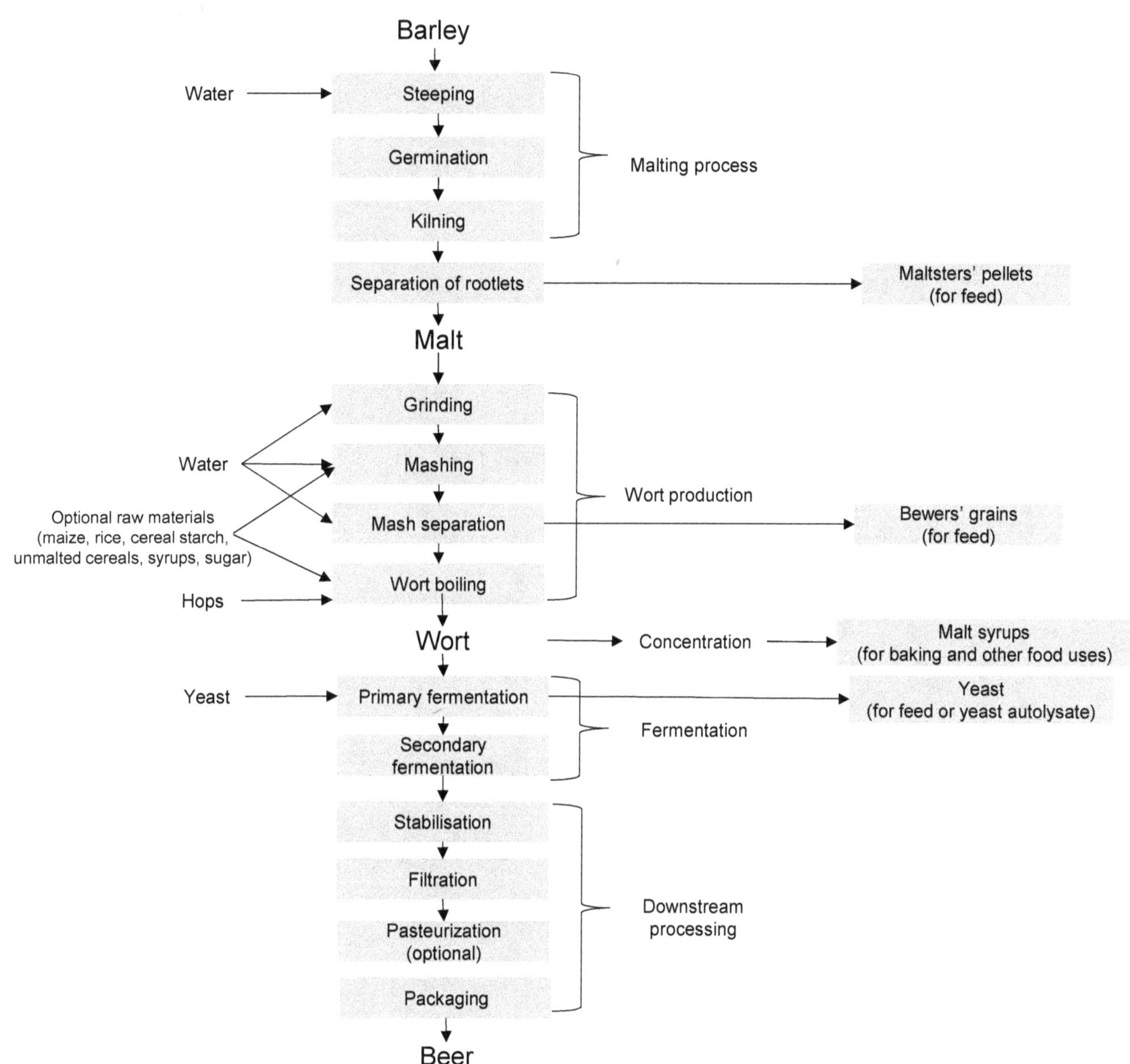

Annex 8.A3
Barley flour milling

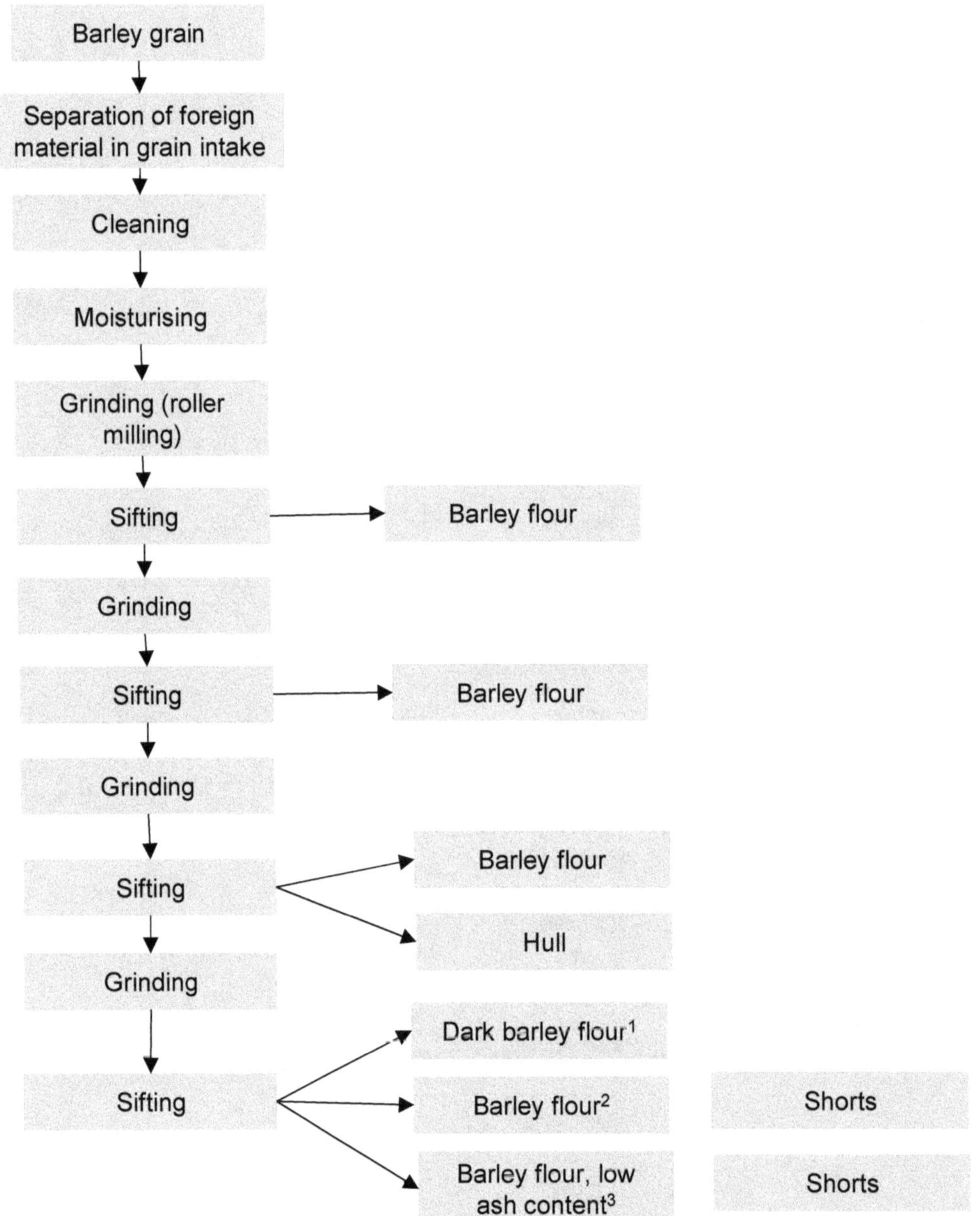

Notes: 1. When milling dark barley flour (ash content 1.2-1.3%), no shorts is taken. 2. When milling barley flour (ash content 1.0-1.2%), some shorts is taken for feed use. 3. When milling low ash content barley flour (ash content below 0.9%), more shorts is taken for feed use.

Annex 8.A4
Barley flaking process

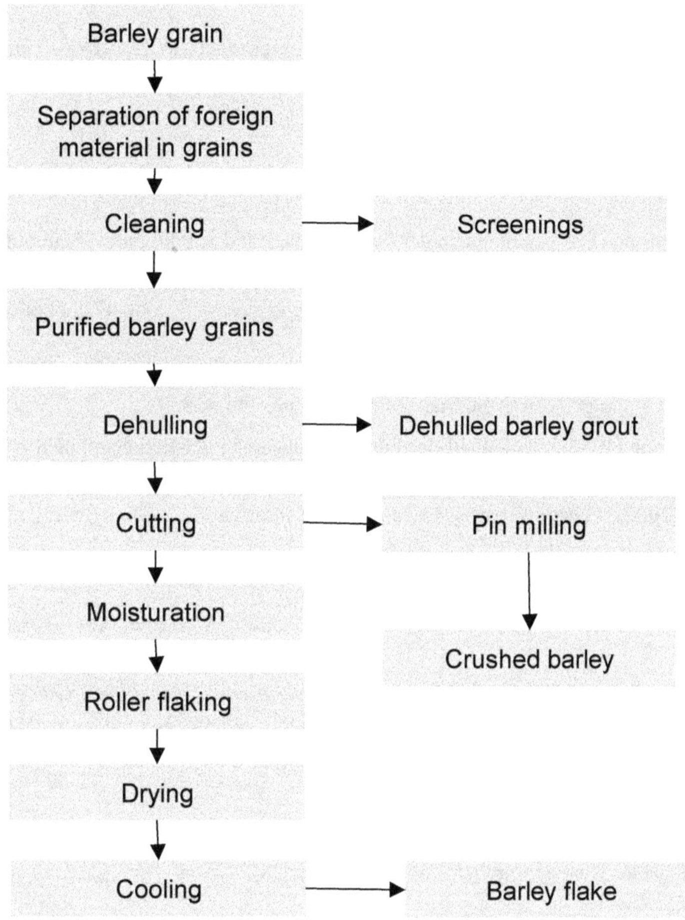

References

Aherne, F. (1990), "Barley: Hulles", *Nontraditional Feed Sources for Use in Swine Production*, Butterworths, London.

Amaral-Philips, D.M. and R.W. Hemken (2002), "Using by-products to feed dairy cattle", University of Kentucky, Cooperative Extension Service, College of Agriculture.

Anderson, V. and J.W. Schroeder (1999), *Feeding Barley to Dairy Cattle*, EB-72, North Dakota State University, Fargo, North Dakota, http://library.ndsu.edu/tools/dspace/load/?file=/repository/bitstream/handle/10365/9318/eb72_2010.pdf?sequence=1.

Becker, M. and K. Nehring (1967), Handbuch der Futtermittel, Band 3, Paul Parey, Hamburg-Berlin.

Bekele, B. et al. (2001), "Occurrence and distribution of barley yellow dwarf virus (BYDV) isolates in central Ethiopia", *International Journal of Pest Management*, Vol. 47, Issue 2, pp. 115-119, http://dx.doi.org/10.1080/09670870151130570.

Bhatty, R.S. (1993), "Non-malting uses of barley", in: MacGregor, A.W. and R.S. Bhatty (eds.), *Barley: Chemistry and Technology*, AACC, St. Paul, Minnesota, pp. 355-418.

Bolduan, G. and H. Jung (1985), "Zur Wahl der Futtergetreideart in der Ferkelaufzucht", *Tierzucht*, Vol. 39, pp. 555-557.

Boyles, S.L. et al. (2002), "Feeding barley to cattle. Beef information", Ohio State University Extension, available at: http://beef.osu.edu/library/barley.html.

Briggs, D.E. (1978), *Barley*, Chapman & Hall, London.

Briggs, D.E. et al. (1981), *Malting and Brewing Science, Volume 1: Malt and Sweet Wort*, Chapman & Hall, London.

Bull, R.C. and L. Bradshaw (1995), *Barley: General Considerations. A Nutritional Guide to Feeding Pacific Northwest Barley to Ruminants*, Bull, R.C. (ed.), University of Idaho, College of Agriculture, Moscow, Idaho, EXP 776, available at: http://www.ibrarian.net/navon/paper/A_Nutritional_Guide_to_Feeding_Pacific_Northwest_.pdf?paperid=3201068.

Ceccarelli, S. et al. (1999), "The Icarda strategy for global barley improvement", ICAEDA, available at: www.researchgate.net/publication/237704968_The_ICARDA_Strategy_for_Global_Barley_Improvement.

De Boer, F. and H. Bickel (1988), *Livestock Feed Resources and Feed Evaluation in Europe*, Elsevier.

Duffus, C.M. and M.P. Cochrane (1993), "Formation of the barley grain – Morphology, physiology and biochemistry", in: MacGregor, A.W. and R.S. Bhatty (eds.), *Barley: Chemistry and Technology*, AACC, St. Paul, Minnesota, pp. 31-72.

Ellis, R.P. (2002), "Wild barley as a source of genes for crop improvement", in Slafer, G.A. et al. (eds), *Barley Science, Recent Advanges from Molecular Biology to Agronomy of Yield and Quality*, Food Products Press, Binghampton, New York, pp. 65-83.

Ensminger, M.E. et al. (1990), *Feeds & Nutrition*, The Ensminger Publishing Co., Clovis, California.

FAO (2004), *FAOSTAT Data*, http://faostat.fao.org (accessed February 2004).

FAO (2000), *Report of a Joint FAO/WHO Expert Consultation on Foods Derived from Biotechnology*, World Health Organization, Geneva, 29 May-2 June 2000, http://www.fao.org/fileadmin/templates/agns/pdf/topics/ec_june2000_en.pdf (accessed 11 February 2015).

FAO (1996), *Report of a Joint FAO/WHO Expert Consultation on Biotechnology and Food Safety*, Food and Agriculture Organization, Rome, Italy, 20 September- 4 October 1996 http://www.fao.org/ag/agn/food/pdf/biotechnology.pdf (accessed 11 February 2015)

Fineli (2001), *Finnish Food Composition Database*, National Public Health Institute, Nutrition Unit, Helsinki, www.ktl.fi/fineli.

Fischbeck, G. (2002), "Contribution of barley to agriculture: A brief overview", in Slafer, G.A.J.L et al. (eds.), *Barley Science, Recent Advantages from Molecular Biology to Agronomy of Yield and Quality*, Food Products Press, Binghampton, New York, pp. 1-14.

Harlan, J.R. (1995), "Barley. *Hordeum vulgare, (Gramineae-Triticinae)*", in: Smart, J. and N.W. Simmonds (eds.), *Evolution of Crop Plants*, Longman Scientific and Technical, Harlow, Essex, England.

Harland, B.L. (1993), "Phytate contents of foods", in: Spiller, G.A. (ed.), *CRC Handbook of Dietary Fibre in Human Nutrition*, CRC Press, Boca Raton, Florida, pp. 617-623.

Harrold, R.L. (1999), *Feeding Barley to Swine and Poultry*, EB-73, North Dakota State University, Extension Service, Fargo, North Dakota.

Hoffmann, M. (1997), Futterspezifische Restriktionen für Getreide und Eiweißfuttermittel in der Schweine- und Rinderfütterung, Sächs, Landeskontrollverband, Lichtenwalde.

Hunt, C.W. (1995), "Feeding value of barley grain for beef and dairy cattle", in: Bull, R.C. (ed.), *A Nutritional Guide to Feeding Pacific Northwest Barley to Ruminants*, Univeristy of Idaho, College of Agriculture, EXP 776.

Jaakkola, S. et al. (2003), "Concentrate supplementation of dairy cow diet based on whole crop barley and wheat silage", in: Niemeläinen, O. and M. Topi-Hulmi (eds.), *Proceedings of the NJF's 22nd Congress "Nordic Agriculture in Global Perspective"*, 1-4 July 2003, Turku, Finland, MTT Agrifood Research Finland, p. 10.

Jaakkola, S. et al. (2001), "Whole-crop barley silage for dairy cows", in: *Production and Utilization of Silage, with Emphasis on New Techniques*, NJF Seminar No. 326, Lillehammer, 27-28 September, pp. 69-74.

Jeroch, H. et al. (1993), Futtermittelkunde, Gustav-Fischer, Jena-Stuttgart.

Kahlon, T.S. and F.I. Chow (1997), "Hypocholesterolemic effects of oat, rice and barley dietary fibers and fractions", *Cereal Foods World*, Vol. 42, No. 2, pp. 86-92.

Kling, M. and W. Woehlbier (1983), *Handelsfuttermittel*, Ulmer-Stuttgart.

Klopfenstein, C.F. (2000), "Nutritional quality of cereal-based foods", in: Kulp, K. and J.G. Ponte, Jr. (eds.), *Handbook of Cereal Science and Technology*, Second edition, Marcel Dekker, Inc., New York, pp. 705-723.

Kreis, M. and P.R. Shewry (1992), "The control of protein synthesis in developing barley seeds", in: Shewry, P.R. (ed.), *Barley: Genetics, Biochemistry, Molecular Biology and Biotechnology*, C.A.B. International, Wallingford, United Kingdom, pp. 319-334.

Lardy, G. and M. Bauer (1999), "Feeding barley to beef cattle", North Dakota State University, EB-70, Fargo, North Dakota.

Liener, J.E. (1989), "The nutritional significance of lectins", in: Kinsella, J.E. and W.G. Soucie (eds.), *Food Proteins*, American Oil Chemists' Society, Champaigne, Illinois, pp. 329-353.

Lyons, T.P. (1978), "Beta-glucan measurement and control", *Technical Quarterly of the Master Brewing Association of the Americas*, Vol. 15, No. 2, April/May/June, pp. 102-105.

MacGregor, A.W. and G.B. Fincher (1993), "Carbohydrates of the barley grain", in: MacGregor, A.W. and R.S. Bhatty (eds.), *Barley: Chemistry and Technology*, AACC, St. Paul, Minnesota, pp. 73-130.

Marins de Sa, R. and G.H. Palmer (2001), "The relationship between single grain analyses and malt modification", *ICBD Research Newsletter*, Vol. 2, pp. 3-5, December.

Morrison, W.R. (1993), "Barley lipids", in: MacGregor, A.W. and R.S. Bhatty (eds.), *Barley: Chemistry and Technology*, AACC, St. Paul, Minnesota, pp. 199-246.

Murphy, P.A. and S. Hendrich (2002), "Phytoestrogens in foods", in: Taylor, S.L. (ed.). *Advances in Food and Nutrition Research*, Vol. 41, Academic Press, London, pp. 196-246.

Newman, C.W. and R.K. Newman (1992), "Nutritional aspects of barley seed structure and composition", in: Shewry, P.R. (ed.), *Barley: Genetics, Biochemistry, Molecular Biology and Biotechnology*, C.A.B. International, Wallingford, United Kingdom, pp. 351-368.

Novus International (1996), *Raw Material Compendium: A Compilation of Worldwide Data Sources*, 2nd edition, Brussels.

NRC (2001), *Nutrient Requirements of Dairy Cattle*, Seventh Revised Edition, National Research Council, National Academies Press, Washington, DC, available at: www.nap.edu/catalog/9825/nutrient-requirements-of-dairy-cattle-seventh-revised-edition-2001.

NRC (2000), *Nutrient Requirements of Beef Cattle*, Seventh Revised Edition: Update 2000, National Academies Press, Washington, DC, available at: www.nap.edu/catalog/9791/nutrient-requirements-of-beef-cattle-seventh-revised-edition-update-2000.

NRC (1998), *Nutrient Requirements of Swine*, Tenth Revised Edition, National Research Council, National Academies Press, Washington, DC, available at: www.nap.edu/catalog/6016/nutrient-requirements-of-swine-10th-revised-edition.

NRC (1994), *Nutrient Requirements of Poultry*, Ninth Revised Edition, National Research Council, National Academies Press, Washington, DC, available at: http://books.nap.edu/catalog/2114.html.

NRC (1982), *United States – Canadian Tables of Feed Composition: Nutritional Data for United States and Canadian Feeds*, Third Revision, National Academies Press, Washington, DC, available at: www.nap.edu/catalog/1713/united-states-canadian-tables-of-feed-composition-nutritional-data-for.

OECD (2000), "Report of the Task Force for the Safety of Novel Foods and Feeds", prepared for the G8 Summit held in Okinawa, Japan on 21-23 July 2000, C(2000)86/ADD1, OECD, Paris, available at: www.oecd.org/chemicalsafety/biotrack/REPORT-OF-THE-TASK-FORCE-FOR-THE-SAFETY-OF-NOVEL.pdf (accessed 28 Jan. 2015).

OECD (1997), "Report of the OECD workshop on toxicological and nutritional testing of novel foods", SG/ICGB(1998)1, OECD, Paris, final version available at: www.oecd.org/officialdocuments/publicdisplaydocumentpdf/?doclanguage=en&cote=sg/icgb(98)1/final.

OECD (1993), *Safety Evaluation of Foods Derived by Modern Biotechnology: Concepts and Principles*, OECD, Paris, available at: www.oecd.org/science/biotrack/41036698.pdf.

Palmer, G.H. (1989), "Cereals in malting and brewing", in: Palmer, G.H. (ed.), *Cereal Science and Technology*, Aberdeen University Press, Aberdeen, United Kingdom, pp. 61-242.

Piironen, V. et al. (2000), "Plant sterols: Biosynthesis, biological function and their importance to human nutrition", *Journal of the Science of Food and Agriculture*, Vol. 80, Issue 7, pp. 939-966, http://dx.doi.org/10.1002/(SICI)1097-0010(20000515)80:7<939::AID-JSFA644>3.0.CO;2-C.

Rice-Evans, C.A. et al. (1997), "Antioxidant properties of phenolic compounds", *Trends in Plant Science*, Vol. 2, Issue 4, April, pp. 152-159, http://dx.doi.org/10.1016/S1360-1385(97)01018-2.

Riis, P. et al. (1989), "Controlled, rapid and safe removal of dormancy in malting barley", *Proceedings of the European Brewing Convention 22nd Congress*, IRL Press, Oxford, United Kingdom, pp. 195-202.

Stölken, B. et al. (1996), Beta Glucane und Pentosane in Getreidesortimenten aus Mecklenburg-Vorpommern, Kongressband 1996, VDLUFA – Darmstadt, *Schriftenreihe*, Vol. 44, pp. 241-244.

Thompson, L.U. (1994), "Antioxidants and hormone-mediated health benefits of whole grains", *Critical Reviews in Food Science Nutrition*, Vol. 34, pp. 473-497.

USDA (2001), "Food group 20, Cereal grains and pasta", *USDA Nutrient Database for Standard Reference*, Release 14, United States Department of Agriculture, available at: www.nal.usda.gov/fnic/foodcomp.

WHO (1991), *Strategies for Assessing the Safety of Foods Produced by Biotechnology*, Report of a Joint FAO/WHO Consultation, World Health Organization of the United Nations, Geneva, out of print.

Chapter 9

Alfalfa (*Medicago sativa*) and other temperate forage legumes

*This chapter, prepared by the OECD Task Force for the Safety of Novel Foods and Feeds with Canada and the United Kingdom as lead countries, deals with the composition of alfalfa (*Medicago sativa*) and other temperate forage legumes which are important in animal feeds. It contains elements that can be used in a comparative approach as part of a safety assessment of feeds (and foods) derived from new varieties. Background is given on alfalfa (lucerne) production, processing and characteristics screened by breeders. Then alfalfa nutrients, anti-nutrients and secondary metabolites are detailed, followed by key products and compositional parameters suggested for analysis of new varieties for feed use and for food use. It also provides information on other forage legumes (clovers, trefoil, sainfoin, vetch, other species) and their key components.*

Introduction

Forage legumes are an essential component of agricultural systems in temperate regions of the world. The benefits of forage legumes include providing top-quality animal feed, suitable ground cover and a valuable source of nitrogen. The nitrogen-fixing ability of the legume occurs through inoculation with rhizobia. The root nodules contain the rhizobia, which have a symbiotic relationship with the legume, allowing for the fixation of nitrogen for the plant. In return, the legumes supply the rhizobium bacteria with a source of fixed carbon derived from the photosynthetic process. This allows the legumes to survive and grow with little or no nitrogen added to the soil. When legumes are used as cover crops, they contribute large amounts of nitrogen to the soil for uptake by the subsequent crop.

Leguminosae is one of the largest plant families in the world. The genera *Trifolium* and *Medicago* are prominent in sustainable farming systems within temperate regions. In Canada, for example, more than 26 million hectares are devoted annually for livestock grazing and forage production. Of this, 4 million hectares are tame or seeded pasture and 6.5 million hectares are cultivated tame hay and fodder crops.

Legumes are favoured by ruminants, whether for grazing or as well-preserved silage or hay. The lower content of structural fibre and the higher protein content of legumes when compared to grasses results in an improved voluntary intake and digestion process as well as a more efficient absorption of nutrients (Ulyatt et al., 1977, Beever and Thorp, 1996). By feeding legumes, animal production response is also improved mainly due to the high concentration of protein and minerals within legumes. Legumes are generally grown in combination with grasses to reduce the persistent, high-viscosity foam (bloating hazard) that occurs with low-fibre, high-protein legume species (Howarth et al., 1991; Popp et al., 2000). Although there is contradictory evidence (Majak et al., 1980, Clark and Reid, 1974), saponins have also been implicated in bloat (Klita et al., 1996). The presence of condensed tannins in legume forages disrupts the foam and prevents bloat (Tanner et al., 1995; Lees, 1992).

This chapter will review alfalfa (*Medicago sativa*), the most common forage legume grown in the temperate regions, and will introduce the other prominent legumes.

Alfalfa (*Medicago sativa* l.)

Production

Alfalfa, also known as lucerne, is a herbaceous perennial legume that grows throughout the world in a variety of climates. It is widely distributed in temperate zones including the United States, southern Canada, Europe, the People's Republic of China, southern Latin America and South Africa. More than 33 million hectares of alfalfa are cultivated throughout the world. This was one of the first forages to be domesticated and with its high yield potential, it soon became a popular choice for livestock feeding.

Alfalfa breeders recognise three types of cultivated alfalfa (lucerne) as members of a single species, *M. sativa*. The three subspecies are ssp *medicago* (purple alfalfa), ssp *falcata* (yellow alfalfa) and spp *varia* (variegated alfalfa). Ssp *varia* is a probable hybrid between ssp *medicago* and ssp *falcata*. Common purple alfalfa is a high yielding, early maturing yet less hardy species. Yellow alfalfa has a lower yield but a higher level of hardiness than common purple alfalfa. Cultivars used in drier, cooler regions of Canada have a higher proportion of *M. falcata* germplasm, conferring winter dormancy and hence winter hardiness and a higher tolerance to grazing than those bred from mainly ssp *medicago* germplasm (Frame et al., 1998).

Alfalfa is believed to have originated in the Islamic Republic of Iran; however, related plants are found throughout central Asia and Siberia. Its cultivation around Lake Lucerne in Switzerland is thought to have resulted in the crop taking the name, lucerne. Until the early 1900s, alfalfa was not grown successfully in the northern hemisphere due to the lack of cold hardiness. Wendelin Grimm introduced the hybrid variegated lucerne from Germany which formed the basis of cultivars capable of surviving the cold winters in the northern United States and Canada (Frame et al., 1998). Alfalfa is the world's most important forage crop (Michaud et al., 1988).

A wide range of soil and climatic conditions are suitable for alfalfa production, however, well-drained soil with a neutral pH and good fertility produce an optimum forage. This long-lived perennial is more drought-tolerant than most other temperate forage legumes, including birdsfoot trefoil and red clover (Peterson et al., 1992), which become dormant under severe drought conditions. Alfalfa also has a tolerance for alkaline soils and a high salt content. However, it is intolerant of acidic soils with a pH below 6, poor drainage or water logging (Sheaffer et al., 1988).

Unlike other forage legumes, alfalfa is generally grown in monoculture, although it can be mixed with other legumes and/or grasses. A grass/alfalfa sward may reduce weed invasion, provide a more balanced nutrient composition for successful ensiling, or the grass may utilise transferable nitrogen from alfalfa (Chamblee and Collins, 1988). However, mixtures may not always improve dry matter yields relative to alfalfa monoculture.

Alfalfa stands decline in yielding ability with age under irrigation and "optimum" management conditions (Hayman and McBride, 1984). Progressive annual decline in alfalfa dry matter yields has been attributed to many factors: competition from companion grasses, weed pressure, injury or loss of plants from pests and/or diseases, winter damage, poor drainage, or management factors such as uncontrolled grazing, over-frequent cutting or inadequate fertilization.

Alfalfa has a poor persistence if continuously stocked; sufficient regrowth between defoliations is critical to ensure stand survival. Alfalfa is best used in a rotational grazing system.

One of the important functions of alfalfa is its ability to fix nitrogen from the atmosphere and enhance the nitrogen balance of the soil, which the plant utilises in turn. This eliminates the need for nitrogen fertilizer. *Rhizobium meliloti* is one of the main bacterial groups that infects and induces nitrogen-fixing nodules on the roots of alfalfa plants. Estimates of nitrogen fixation by alfalfa vary widely but are generally higher than for other temperate forage legumes (Frame et al., 1998). Soil mineral nitrogen or fertilizer nitrogen imposes a restriction on nitrogen fixation. The deficiency of certain minerals such as potassium, calcium or magnesium or excessive soil acidity may also limit nitrogen fixation (Frame et al., 1998).

Processing

In addition to its use in grazing systems, alfalfa is primarily used for hay, silage, artificially dried forage and pelleted meal. Cutting alfalfa at the 10% bloom stage and then at five to seven week intervals was shown to maximise dry matter production, provide forage of reasonable nutritive value and help to maintain sward longevity in Minnesota (Sheaffer et al., 1988). In certain regions, under ideal conditions, six to nine cuttings per year may be achieved, whereas in other regions two to five cuttings may be the maximum. To ensure plant survival through the winter, the last harvest should be early enough to allow plants to build up carbohydrate and nitrogen reserves before cessation of growth, but not allow a heavy canopy to develop prior to winter. Studies have demonstrated that three cuts per year yielded more than four cuts, although the forage was of lower nutritive value (Brink and Marten, 1989).

Before processing, alfalfa is cut and allowed to dry to varying moisture content in the field. Hay should be dried to approximately 85% dry matter. The optimum dry matter content for chopped silage is 30% when stored in bunker silos, 35% in concrete tower silos and 45% in oxygen limiting silos. Composition of alfalfa hay compared with silage is shown in Table 9.1. To produce dehydrated alfalfa, a regular supply of forage with a high protein content is required, as well as cutting before the growth reaches the bud stage. Proper dehydration of alfalfa can increase the utilisation of forage protein by ruminants.

Alfalfa meal or alfalfa leaf meal is hay that has been dried (either naturally or artificially) and ground. Alfalfa leaf meal is of better quality, and contains not more than 18% crude fibre. Alfalfa meal includes stem fractions and therefore higher fibre. Alfalfa leaf meal and alfalfa meal are good sources of carotene. When processing alfalfa, it is important to retain the nutritious leaf fraction as much as possible during handling.

Table 9.1. **Quality of alfalfa hay or silage made from the same second cut crop**

Component	Silage	Hay
Dry matter gm/kg	413	850
Neutral detergent fibre gm/kg dry matter	354	352
Acid detergent fibre gm/kg dry matter	265	257
Crude protein gm/kg dry matter	212	197
Nonprotein nitrogen gm/100 gm of total N	49.4	7.7

Source: Adapted from Broderick (1995).

Within the alfalfa plant, the leaves have a higher concentration of nutrients than the stems, with the exception of potassium. Magnesium concentrations decline with crop maturity. In late-cut hay, magnesium may be restricted to levels well below the minimum for animal requirements if soil potassium levels are high due to preferential uptake of potassium by the plants (Frame et al., 1998). Alfalfa has a low concentration of sodium, therefore salt supplementation of cattle and sheep on alfalfa pasture has been beneficial to their health and production (Jagusch, 1982).

Traditional characteristics screened by alfalfa developers

For registration/public release of new varieties of alfalfa in Canada and the United States, only phenotypic characteristics are required to be considered. The main indicators of alfalfa quality for livestock feeding include the proximates, acid detergent fibre, neutral detergent fibre, lignin and minerals (summarised in Forage Genetics Inc., 2003, Table 2). Published values of these components vary widely in the literature, depending on geographical location, environmental conditions, variety, time of harvest and storage conditions. Therefore, it is important to make comparisons only with appropriate comparators, e.g. near isogenic lines, reference cultivars or commercial varieties grown at the same time under similar conditions and locations.

Table 9.2. **Constituents typically monitored in forages for livestock feeding**

Constituent	Importance
Moisture	Feeding value
Proximates: – Protein – Fat – Ash	Nutrition/feeding value
Acid detergent fibre	Digestibility
Neutral detergent fibre	Digestibility
Lignin	Digestibility/anti-quality factor
Minerals: – Calcium (Ca) – Copper (Cu) – Iron (Fe) – Magnesium (Mg) – Manganese (Mn) – Phosphorus (P) – Potassium (K) – Sodium (Na) – Zinc (Zn)	Nutrition

Source: Forage Genetics (2003), personal communication.

Nutrients in alfalfa

Tables 9.3-9.6 summarise proximate, amino acid, fatty acid and mineral composition of alfalfa from a variety of databases.

Table 9.3. **Proximate, lignin, acid detergent fibre (ADF) and neutral detergent fibre (NDF) composition of late vegetative/early bloom alfalfa**

	NRC 1971	NRC 1982	Ensminger et al.	NRC 1996	Monsanto	Range
Dry matter	90.1	23.0	91.0	19.0	17.9-29.2	17.9-91.0
Crude protein	19.7	19.0	17.9	25.0	15.3-25.8	15.3-25.8
Crude fat	2.2	3.1	2.6	2.9	1.3-3.2	1.3-3.2
Crude fibre	29.8	25.0	25.8			25.0-25.8
Neutral detergent fibre (NDF)		40.0	36.8	39.3	26.5-35.7	26.5-40.0
Acid detergent fibre (ADF)		31.0	29.0		23.1-33.4	23.1-33.4
Lignin	7.7	7.0	5.8	7.9	3.9-9.7	3.9-9.7
Ash	8.7	9.5	8.4	9.2	8.8-15.3	8.4-15.3

Note: Data except for dry matter presented on a percentage dry matter basis.

Sources: NRC (1971, 1982, 1996); Ensminger et al. (1990); Monsanto (2003).

Anti-nutrients and secondary metabolites in alfalfa

Bloat potential in ruminants

Legumes are unusual in that the very characteristics that make them valuable as ruminant feed (a high content of readily digestible protein and carbohydrate), can predispose animals to bloating, a potentially serious condition that can result in death.

The etiology of bloat and plant and animal risk factors are reviewed in Clark and Reid (1974), Colvin and Backus (1988), Howarth et al. (1991) and Popp et al. (2000).

Table 9.4. **Amino acid composition of alfalfa**

	Hay NRC[1]	**Hay** NRC[2]	**Hay** literature[3]	**Hay** Monsanto[4]	**Hay** range	**Silage** range[5]
Alanine			0.70	0.79-1.59	0.70-1.59	0.69-0.94
Arginine	1.14	1.18	0.62	0.71-1.54	0.62-1.54	0.27-0.51
Aspartic acid			1.40	1.75-3.52	1.40-3.52	1.83-1.95
Cysteine		0.32	0.20	0.18-0.35	0.18-0.35	
Glutamic acid			1.20	1.52-3.03	1.20-3.03	1.27-1.48
Glycine	1.03		0.60	0.71-1.47	0.60-1.47	0.67-0.76
Histidine	0.50	0.44	0.28	0.37-0.74	0.28-0.74	0.14-0.28
Isoleucine	0.96	0.97	0.50	0.66-1.26	0.50-1.26	0.55-0.76
Leucine	1.64	1.68	0.90	1.11-2.25	0.90-2.25	0.90-1.23
Lysine	1.27	1.17	0.59	0.99-1.81	0.59-1.81	0.32-0.74
Methionine	0.36	0.36	0.18	0.24-0.48	0.18-0.48	0.06-0.21
Phenylalanine	1.07	1.09	0.65	0.72-1.59	0.72-1.59	0.53-0.79
Proline			0.70	0.75-1.34	0.70-1.34	0.89-1.14
Serine	0.97		0.60	0.75-1.36	0.60-0.36	0.57-0.67
Threonine	1.08	1.00	0.60	0.61-1.15	0.60-1.15	0.63-0.72
Tryptophan		0.35		0.16-0.31	0.16-0.35	
Tyrosine	0.74		0.50	0.50-1.16	0.50-1.16	0.25-0.41
Valine	1.22	1.20	0.60	0.79-1.55	0.60-1.55	0.76-0.94

Note: Data presented on a percentage of dry matter basis.

Sources: 1. NRC (1982). 2. NRC (2001). 3. Cunningham et al. (1994); Phuntsok et al. (1998). 4. Monsanto (2003). 5. Christensen (2004a); Phuntsok et al. (1998).

Table 9.5. **Fatty acid composition of alfalfa**

	Hay (gm/100 gm of fatty acids)[1]	**Silage** (gm/100 gm of dry matter)[2]
C12:0	0.70	0.01-0.03
C14:0	2.90	0.01-0.02
C16.0	27.6	0.41-0.47
C16.1	0.20	0.04-0.05
C17:0	2.15	0.01-0.11
C18:0	36.5	0.06-0.07
C18:1	4.11	0.06-0.07
C18:2	0.75	0.34-0.42
C18:3		0.14-0.63
Other	24.90	0.35-0.92
Total	100	2.09-2.10

Sources: 1. Bas et al. (2003). 2. Christensen (2004b).

The condition and its incidence

Primary bloat or frothy bloat (tympanites) is the over-distension of the rumen caused by the accumulation of fermentation gases in a stable protein foam or froth (Tanner et al., 1995), and usually occurs as an outbreak in several animals on pasture that contains high levels of leguminous plants. Primary bloat can also occur in feedlot cattle. When an animal is experiencing pasture bloat, the stable froth is produced in the rumen in a "layer" on top of the ruminal contents (mostly liquid) and prevents the gas bubbles from rising to the top and dispersing their contents. Once the froth has formed and natural eructation is prevented, the rumen motility is initially increased, causing further frothing. Finally, there is a loss of muscle tone and rumen motility. Death is a result of several factors, including the depressive effect of rumen distension on the heart and lungs and absorption of toxins from the rumen.

Table 9.6. **Mineral composition of late vegetation to early bloom alfalfa**

Expressed on dry matter basis

	NRC 1971	NRC 1982	Ensminger et al.	NRC 2000	Preston	NRC 2001	Monsanto	Range
Sodium (Na) g/100 gm	0.15	0.19	0.15	0.12		0.03	0.02-0.21	0.02-0.21
Potassium (K) g/100 gm	2.08	2.09	2.56	2.51	2.50	2.56	1.39-4.31	1.39-4.31
Calcium (Ca) g/100 gm	1.40	1.96	1.63	1.41	1.41	1.56	0.90-1.53	0.90-1.96
Phosphorus (P) g/100 gm	0.21	0.30	0.22	0.22	0.26	0.31	0.22-0.45	0.22-0.45
Magnesium (Mg) g/100 gm	0.30	0.27	0.34	0.34		0.33	0.11-0.45	0.11-0.45
Iron (Fe) mg/100 gm	0.02	0.03	0.02	0.02		0.021	0.02-1.54	0.02-1.54
Sulfur (S) g/100 gm	0.30	0.37	0.30	0.30	0.27	0.33		0.27-0.37
Copper (Cu) mg/kg	13.4	10.0	12.6	12.7		10.0	5.3-10.2	5.3-13.4
Cobalt (Co) mg/kg	0.01	0.13	0.29	0.29		0.65		0.01-0.65
Manganese (Mn) mg/kg	31.5	43.0	36.2	36.0		49.0	34.6-109.5	31.5-109.5
Zinc (Zn) mg/kg		18.0	30.2	30.0	22.0	26.0	18.1-36.0	18.0-36.0
Selenium (Se) mg/kg			0.55	0.55		0.20		0.20-0.55
Chlorine (Cl) g/100 gm	0.38	0.47	0.38	0.34	0.38	0.55		0.34-0.55

Sources: NRC (1971, 1982, 2000, 2001) ; Ensminger et al. (1990); Preston (2003); Monsanto (2003).

The main risk factor in pasture bloat is the rapid ingestion of immature/fast-growing legumes in pre-flowering stages. Alfalfa, red clover and white clover have similar bloat potential. Other forage legumes are considered to be of low risk.

Ingestion of only the most succulent parts of the plant is an important risk factor, in addition to the sward type. Frost and growth of alfalfa at low temperatures have been shown to increase bloat risk by increasing the leaf cell constituents (soluble protein, pectic polysaccharides) implicated in pasture bloat (MacAdam and Whitesides, 1996). Wetness of the pasture has also been suspected to be a risk factor for bloat. It is, however, more likely that the real risk is the fast growth brought on by wet and favourable weather.

Several animal factors contribute to bloat (Mendel and Boda, 1961, Howarth et al., 1991). Young animals are considered more susceptible to acute and severe bloat than older animals, and it is suspected that animals can adapt to eating bloating pastures and are less susceptible after exposure. Fasting has also been shown to predispose animals to pasture bloat, but the mechanism is not established. Since there are individual differences in the ability of cattle to tolerate rumen distension and the presence of contributory factors in any given situation, some animals only suffer sub-clinical or mild bloating. While the toleration of mild bloat allows adaptation to new pastures, sub-clinical and mild bloat have been recognised as causing major losses on clover dominant pastures in the form of reduced feed intake and subsequent lower weight gains (Latimori et al., 1992; Rossi et al., 1997).

Bloat is a common problem in all areas in which temperate legumes are used as ruminant feed and has long been recognised as a major problem in countries like New Zealand, where clover forms an important part of the pastures (Carruthers et al., 1987). Due to its association with clover, bloat has been considered a risk factor on organic farms, where clover often constitutes more than 50% of the sward content. However, research both in the United Kingdom and elsewhere in Europe on organic farms suggests that the incidence of clinical bloat is not higher than on conventional farms (Weller et al., 1996; Frankow-Lindberg and Danielsson, 1997).

Although there is no widely recognised test for bloating potential, selection for a low initial rate of digestion (to four hours) has been successfully used as the criterion in developing a "bloat-reduced" cultivar of alfalfa AC Grazeland Br (Coulman et al., 2000). In addition to the initial rate of digestion, a number of other factors known to influence the bloat potential of forages can be measured, such as leaf venation pattern, fibre content and digestibility, cell wall thickness, ease of nucleation of rumen bacteria, and preferential synthesis of protein and reduction in lipids in chloroplasts (Lees et al., 1982; Howarth et al., 1979; Fay et al., 1981; Lees, 1984; Stifel et al., 1968).

Saponins

Saponins are divided into two groups, including the steroidal saponins, which occur as glycosides in some pasture grasses, and the triterpenoid saponins, which occur in many temperate legumes, particularly alfalfa. Because saponins have a distinct foaming characteristic (Marston et al., 2000), historically they have been considered a primary cause of bloat in animals grazing temperate forages. The development of low saponin varieties of alfalfa that still cause bloat suggests the importance of other factors as the main causal agent(s) (Majak et al., 1980).

Figure 9.1. **The monodesmodosidic medicagenic acid**

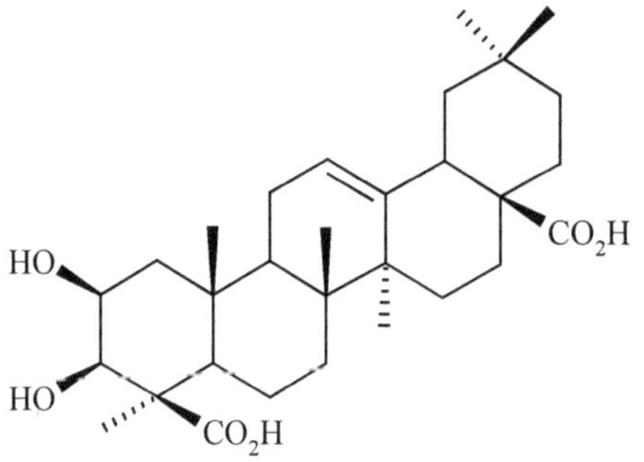

A total of some 24 saponins have been identified in alfalfa (Bialy et al., 1999) but the soyasapogenols, zanhic acid glycosides and medicagenic acid are quantitatively the most important (see Table 9.7; Oleszek et al., 1992; Massiot et al., 1988, 1991). Saponins can have a positive or a negative role in plants. Supplementation with saponins has been shown to decrease ammonia production and protozoal count, and improve growth rates in lambs (Makkar and Becker, 2000). The toxicity of the various saponins to animals differs (Hostettmann and Marston, 1995). Triterpenoid saponins may reduce feed palatability and feed degradation in the rumen and their presence greatly limits the use of alfalfa in some non-ruminant diets (Lu and Jorgensen, 1987; reviewed in Oleszek, 1996). Poultry rations containing 10% alfalfa meal depress chick growth and egg production due to saponins (Birk, 1969; Bondi et al., 1973; Pedersen et al., 1972).

Saponins are highly toxic to fish and amphibians (Cheeke, 1971; Khalil and El Adawy, 1994; Makkar and Becker, 2000), but not to ruminants and swine (Bins and Pedersen, 1964). Symptoms of saponin toxicity, believed largely due to the medicagenic and zanhic acid content in alfalfa, include irritation to mouth and digestive tract, increased membrane permeability and, in acute cases, haemolysis (Oleszek, 1996). Zanhic acid glycosides may also cause production of intestinal gases. Ensiling of alfalfa can reduce the total saponin and medicagenic acid content (Kalac et al., 1996).

Table 9.7. **Crude saponin and medicagenic acid content in various cultivars of alfalfa grown in Mexico**

Cultivar	Crude saponin (g/kg dry matter)	Medicagenic acid (g/kg dry matter)
Sundor	17.7	0.023
Maxidor	11.7	0.027
Valenciana	8.8	0.165
Condor	8.5	0.024
Puebla 76	8.3	0.097
Inia 76	6.8	0.115
NK-819	5.9	0.013
Pierce	4.9	0.031

Source: Data from Pérez et al. (1997).

A number of analytical methods for various saponins have been used with varying success. Biological methods have been used but are dependent on the inhibition of the growth of the fungus, *Trichoderma viride*; these methods measure exclusively medicagenic acid glycosides. A high pressure liquid chromotography method was developed by Oleszek (2004), but the method has not been sufficiently modified to make it a practical routine procedure. There does not appear to be enough reliable data in the literature for meaningful comparisons with database values. It is important that analysis of appropriate comparators be conducted if saponin analysis is to be undertaken.

Condensed tannins

Condensed tannins (proanthocyanidins) derive from the flavonoid biosynthesis pathway and are essentially oligomers of flavan-3-ols of varying size and complexity. The chemistry, biochemistry and molecular regulation of these plant metabolites are reviewed in Marles et al. (2003). They are widespread in the plant kingdom. A universal characteristic of condensed tannins is their ability to bind reversibly or irreversibly to proteins in feed, saliva and microbial cells, with microbial enzymes, and with endogenous proteins or other feed components and to inhibit ruminant microorganism activity (Bae et al., 1993; Hagerman and Robbins, 1993; Jones et al., 1994; Tanner et al., 1994; Molan et al., 2001). The protein-binding capacities among oligomers from different plant species and developmental stages differ with variations in proanthocyanidin and protein structure (Hagerman and Butler, 1981; Butler et al., 1984).

Condensed tannins are also metal chelators and strong antioxidants (Muir, 1997; Stoutjeskijk et al., 2001; Slabbert, 1992). They have the potential to eliminate pasture bloat, improve the efficiency of conversion from plant to animal protein (ruminal bypass protein), reduce greenhouse gases, reduce gastrointestinal parasites and inhibit insect feeding (Waghorn, 1990; Waghorn and Shelton, 1992; Neizen et al., 1995, 1998; Broderick and Albrecht, 1997; Aerts et al., 1999; Muir et al., 1999; McMahon et al., 2000; Butter et al., 2001; McSweeney et al., 2001). Tannins and saponins can act in an additive fashion in the rumen (Makkar et al., 1995).

Condensed tannin levels exceeding 40-50 g kg^{-1} dry matter in forages may reduce protein and matter (DM) digestibility of the forages by ruminants and, consequently, at high concentrations condensed tannins may be regarded as "anti-nutritional" compounds (Barry, 1989). However, at low to moderate levels (20-40 g kg^{-1} dry matter) tannins can increase the quantity of dietary protein, especially essential amino acids, flowing to the small intestine increasing production without any effect on feed intake (Aerts et al., 1999).

Figure 9.2. **A flavan-3-ol (epigallocatechin) monomer and a model proanthocyanidin oligomer showing the mechanism of extension through additional 4-8 and 4-6 inter-flavonoid linkages**

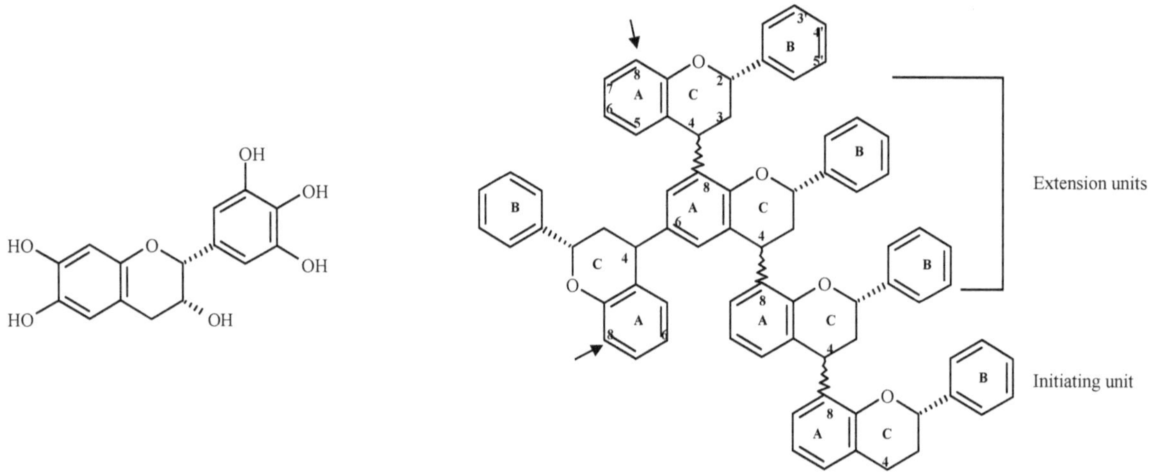

Extension units

Initiating unit

Although having a potential detrimental effect on protein digestibility, generally, legumes that contain condensed tannins in excess of 50 g kg^{-1} DM do not cause bloat (Table 9.8). Dietary condensed tannins may provide a means to beneficially manipulate protein digestion and/or prevent pasture bloat in ruminants.

Research efforts are being directed to genetically modify alfalfa to derepress its anthocyanidin biosynthetic pathway, or to isolate genes encoding steps of this pathway and introduce them into alfalfa and clover from other plant species (reviewed in Marles et al., 2003). A host of condensed tannin biosynthetic and regulatory genes have been discovered to contribute to these strategies (reviewed in Marles et al., 2003). In addition, the *Lc* anthocyanin regulatory gene from maize induces small amounts of condensed tannin in alfalfa forage (Ray et al., 2003), and the forage has a reduced initial rate of digestion and reduced gas production *in vitro* (Wang et al., 2003).

Table 9.8. **Extractable and bound condensed tannin in bloating and bloat-safe temperate legumes measures by the butanol-HCl method**

Forage	Condensed tannin (g kg^{-1} dry matter)		
	Extractable	Bound	Total
Bloat safe			
Big trefoil (*Lotus pedunculatus*)	61	15	77
Birdsfoot trefoil (*Lotus corniculatus*)	36	11	47
Sulla (*Hedysarum coronarium*)	33	12	45
Sainfoin (*Onobrychis vicifolia*)	29		
Potentially bloating			
Red clover (*Trifolium pratense*)	0.4	1.3	1.7
Alfalfa (*Medicago sativa*)	0.0	0.5	0.5

Source: Barry and McNabb (1999).

Oestrogen agonists and antagonists

Adverse effects on reproductive health of farm animals grazing legumes have been recognised since the early 1940s when there was a substantial outbreak of infertility in Australian sheep grazing subterranean clover. This was subsequently shown to result from the presence of a variety of naturally occurring oestrogen mimetics, the so-called "phytoestrogens". Two types of phytoestrogens are now recognised: the coumesterols (coumestrol) related to the coumarins and quantitatively more important in alfalfa, and the isoflavonoids more widely distributed in *Trifolium* spp (Livingston, 1978). These compounds can also be induced in alfalfa with pathogen stress (Latunde-Dada and Lucas, 1985). Levels of coumestrol in alfalfa forage range from 2.99-104.37 ppm (Monsanto, 2003). The structure of the more important isoflavonoids recognised in red clover are shown in Figure Table 9.9. Other isoflavone conjugates have been identified (Klejdus et al., 2001). Phytoestrogen infertility appears to be species specific, and ruminants such as cattle and sheep are more susceptible than other animals (Stob, 1983; Moule et al., 1963; reviewed in Howarth, 1988).

Formononetin and biochainin A are the two isoflavones found in the greatest amounts in forage legumes (Smolenski et al., 1981) and together can reach 15 g kg^{-1} dry matter in some red clover cultivars. Concentrations in white clover are usually substantially lower (0.5 g kg^{-1} dry matter). The major metabolic transformation of the isoflavones occurs in the rumen. Biochainin A is demethylated to genistein and via ring cleavage to 4-ethylphenol and organic acids with the loss of all oestrogenic activity. Formononetin is mainly demethylated to daidzein and then to equol by hydrogenation and ring cleavage (Lundh, 1995). However, unlike the end products of biochainin A metabolism, equol is a more potent oestrogen mimetic than either of its parent compounds.

Figure 9.3. **Isoflavonoid linkages**

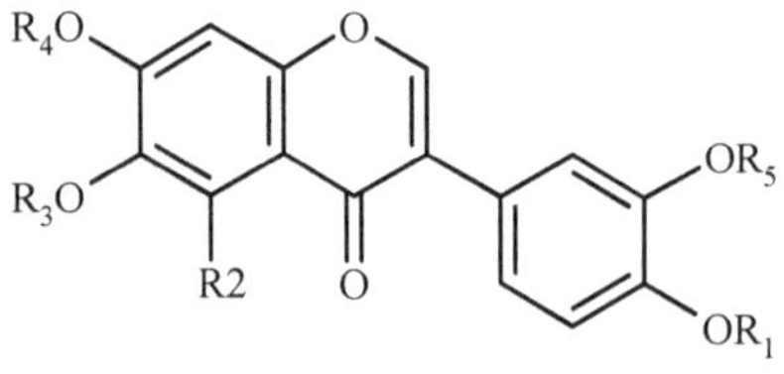

Table 9.9. **Structure of the isoflavonoids identified in red clover**

Compound	R_1	R_2	R_3	R_4	R_5
Daidzein	H	H	H	H	H
Daidzin	H	H	H	Glucose	H
Genistein	H	H	OH	H	H
Genistin	H	H	OH	Glucose	H
Formononetin	CH_3	H	H	H	H
Ononin	CH_3	H	H	Glucose	H
Biochainin A	CH_3	H	OH	H	H
Sissotrin	CH_3	H	OH	Glucose	H
Trifoside	Glucose	H	H	CH_3	H
Calycosin	CH_3	H	H	H	OH
Pectolinarigenin	CH_3	OH	OCH_3	H	H
Pratensein	CH_3	OH	H	H	OH
Pseudobaptigenin	$-CH_2-$	H	H	H	-O-

Source: He et al. (1996).

Coumesterol (Figure 9.4), present as the major phytoestrogen in alfalfa but also occurring in white clover, is not metabolised by the rumen flora and is absorbed in its original form. Coumesterol is known to be approximately 30-fold more effective than genistein in mice and to cause oestrogen-related disorders in animals. Concentrations in healthy plants rarely exceed 30-60 mg kg^{-1} dry matter, but coumesterols can accumulate in plants subject to fungal attack. Significant genetic variation exists in alfalfa for coumesterol (Hanson et al., 1965). In addition, doses showing no effects in the short term may induce hormonal effects if consumed over a longer period.

Figure 9.4. **Coumesterol (coumestrol)**

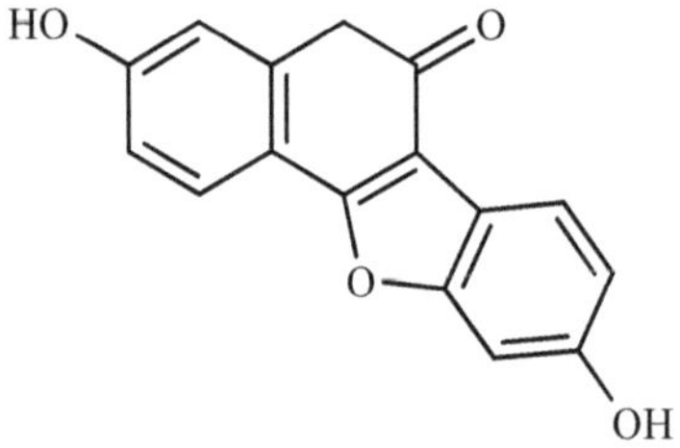

Cyanogenic glycosides

The cyanogenic glycosides are composed of an α-hydrozynitrile type aglycone and a sugar moiety which is usually D-glucose. They are widely distributed in the plant kingdom but, within the temperate forage legumes, are considered a cause for concern only in some cultivars of white clover and *Lotus corniculatus* (Vetter, 2000).

In populations of white clover, plants that produce hydrocyanic acid (HCN) following damage of the leaves, and plants that fail to do so co-exist within the same population. The difference in cyanogenic glucoside content is caused by variation in two genes: *Ac* regulating the production of the cyanogenic glucosides linamarin and lotaustralin, and *Li* regulating the production of the hydrolysing enzyme linamarase. Only plants that contain at least one active allele of each of the genes *Ac* and *Li* are cyanogenic. White clover and *Lotus corniculatus* contain varying amounts of both kinds of metabolites (reviewed in Smolenski et al., 1981).

Figure 9.5. **Linamarin a cyanogenic glucoside from white clover**

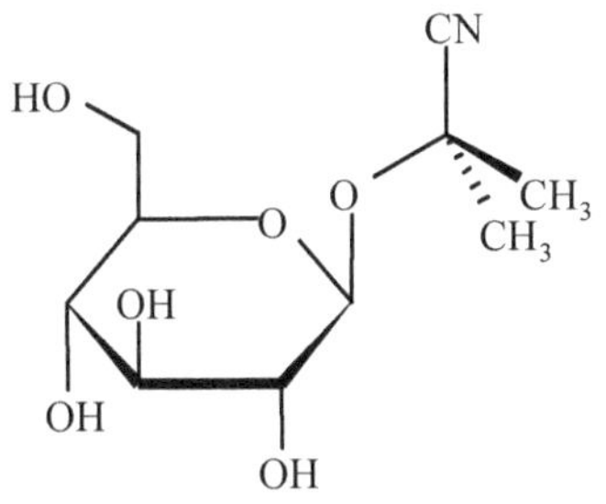

CN
HO
O
O
CH_3
CH_3
OH
OH
OH

Ruminant animals are more susceptible to HCN poisoning than non-ruminants due to fast microbial breakdown of cyanogenic glycosides (Smolenski et al., 1981). Hydrocyanic acid released from linamarin and lotaustralin is further metabolised within the grazing animal to inorganic thiocyanate, which is goitrogenic. North American cultivars of white clover have a notably lower HCN-generating potential than most European cultivars, although there are considerable differences among cultivars bred within countries (Wheeler, 1989). In Switzerland, cultivars with a mean HCN content above 370 mg HCN kg^{-1} dry matter are excluded from the national list. The HCN potential is also greatly affected by environmental factors and is increased by moisture stress, predation, low light intensity, cool grazing conditions and low soil phosphorus supply (Vickery et al., 1986).

Other secondary metabolites

The occurrence of a wide range of secondary metabolites in temperate legumes has been documented in various natural product databases and resources (see, for example, www.ars.usda.gov) (Duke, 1992; Chapman and Hall, 1982-98). This is particularly true for alfalfa and for *Trifolium pratense*, both of which have elicited interest as herbal products with claimed health benefits. However, in most instances, quantitative data is not available and would have to be generated as part of a comparative assessment. Measurement of known secondary metabolites would only be justified if there were reason to suspect some change to the metabolic pathway involved in their generation or if they were of known toxicity. Canavanine is a potentially toxic structural analogue of L-arginine that is a stored by many legumes including alfalfa (Rosenthal and Nkomo, 2000). Under normal conditions, L-canavanine is found in seeds, cotyledons and the emerging shoots and only in very low amounts in older vegetative tissues. However, its cytotoxicity might warrant its inclusion in a comparative analysis.

Feed use of alfalfa and alfalfa products

Alfalfa is rich in protein, vitamins and minerals and is a main component in livestock rations. The natural phenolic defence compounds of alfalfa include a simple alkaloid, saponins, coumestans (which are increased by exposure to the pea aphid), isoflavones (pterocarpans and medicarpan, both of which are induced in response to fungal infection) and flavones (Massiot et al., 1988, 1991; Oleszek et al., 1992; Stochmal et al., 2001; Ray et al., 2003; reviewed in Howarth, 1988).

Other natural components can be found in natural product databases (e.g. Duke, 1992). Alfalfa accumulates only trace amounts of condensed tannins in forage (Goplen et al., 1980; Ray et al., 2003).

Throughout the world, alfalfa is recognised as a premium forage for feeding to dairy cattle and horses. It can also be a valuable feed for beef cattle, sheep and other livestock with high nutrient requirements such as lactating ewes and dairy goats or backgrounded calves.

However, alfalfa is a forage legume with high bloat potential (Howarth et al., 1991; Popp et al., 2000). AC Grazeland Br is the world's first bloat-reduced variety (60-80% bloat-reduced), and was selected for a lower initial rate of digestion (Coulman et al., 2000). Plants of this variety have a thicker cell wall and fast regrowth (Goplen et al., 1993; Najda, 2002).

Alfalfa meal is not a suitable feedstuff for use at high dietary levels by non-ruminant animals except horses, rabbits and gestating sows. The problems associated with alfalfa use by monogastric animals include low protein digestibility, low digestible energy, moderately high fibre, saponins and phenolics content, and low palatability.

The feeding value of alfalfa is largely determined by the stage of growth as the nutritive value decreases as the plant matures. The leaves of the alfalfa plant are abundant in nutrients, including protein, vitamins E and K, calcium, magnesium, potassium and carotene. The dry matter yield of alfalfa increases with advancing maturity but the nutritive value is reduced. Plant maturity results in a decline of the leaf:stem ratio, an increase in lignin content of the stem and leaf loss through leaf shatter.

In ruminants, the utilisation of alfalfa protein is inefficient and causes problems because of rapid turnover in the rumen and a high proportion of protein nitrogen which is lost as ammonia, although ruminal protein degradability declines with plant maturity (Amrane and Michaeletdoreau, 1993). The crude protein content is generally higher in ensiled alfalfa than in hay, due mainly to greater leaf loss in hay making; however, much more of the nitrogen comprises non-protein nitrogen (NPN) in silage (Broderick, 1995).

Compared with grass, alfalfa has higher intake characteristics and a higher animal production response per unit of DM ingested. The potential reasons are the rapid passage of digesta out of the rumen (which stimulates appetite), high concentration of soluble protein (which assists in microbial synthesis in the rumen), the stimulation of cellulose digestion, a low concentration of cell wall in the dry matter, and an adequate supply of minerals and vitamins (Conrad and Klopfenstein, 1988).

Specialty protein extracts of alfalfa are also used in livestock feeding; for example, xanthophyll is sometimes used to impart a yellowish colour to poultry eggs and flesh.

Food uses of alfalfa

The use of sprouted seeds as food has a very long history. Currently, in North America, sprouted mung bean and alfalfa seeds are often available in the fruit and vegetable sections of grocery stores and a wider range of sprouted seeds or seeds for sprouting, including mung beans and alfalfa as well as adzuki bean, Chinese cabbage, clover, lentil, onion and radish are sold in natural and health food stores.

Although only sprouts of alfalfa and clover from the list of forage legumes discussed in this chapter were identified as used for food in a brief search of the Internet, it should not be assumed that other forage legumes might not at some stage be considered for sprouting for human food use. In addition to sprouts, protein extracts of alfalfa (e.g. rubisco) have received attention for possible use in various food applications.

Most people in a North American dietary context would be expected to consume minor quantities of these foods, roughly 60 mL (8-20 g serving, depending on the type of sprout) and only on an occasional basis. Among the small segment of committed users, amounts of 1-2 cups per day may, however, be common.

A decision regarding the importance of assessing the nutrient composition of forage legumes used as sprouted seeds in human diets should be guided by the frequency and quantity of such sprouts in a given country and their contribution to nutrient intake. The fact that they are often promoted as being highly nutritious may also be a consideration in requesting data.

Table 9.10, showing the composition of alfalfa sprouts, is extracted from the *USDA National Nutrient Database for Standard Reference*, Release #16. This database provides data for sprouted alfalfa, kidney beans, mung beans, navy beans, pinto beans, lentils, peas, radish seeds, soybeans and wheat.

A comparison of the nutrient composition of one cup of alfalfa sprouts to recommended intakes of these nutrients suggests that the contribution is minor. A suggested minimum compositional analysis where alfalfa is likely to be sold for food use would be the analysis of fresh forage or sprouted alfalfa seed for the parameters listed in Table 9.11 with the addition of vitamin C, beta-carotene, folate and phytoestrogens to provide a basis for assessment of potential unintended effects with relevance to human food use.

Table 9.10. **Composition of raw sprouted alfalfa seeds**

Nutrient	Unit	Value per 100 g	Value per 240 mL (33 g)	Sample count	Standard error	Value per 100 g dry matter
Water	g	91.140	30.076	10	1.226	
Energy (calculated)	kcal	29.000	9.570	0		327.314
Protein	g	3.990	1.317	10	0.563	45.034
Total lipid (fat)	g	0.690	0.228	10	0.141	7.788
Ash		0.400	0.132	10	0.044	4.515
"Carbohydrate,by difference"	g	3.780	1.247	0		42.664
"Fibre, total dietary"	g	2.500	0.825			28.217
"Sugars, total"	g	0.180	0.059	3	0.012	2.032
Calcium (Ca)	mg	32.000	10.560	10	4.659	361.174
Iron (Fe)	mg	0.960	0.317	10	0.114	10.835
Magnesium (Mg)	mg	27.000	8.910	10	3.978	304.740
Phosphorus (P)	mg	70.000	23.100	10	7.914	790.068
Potassium (K)	mg	79.000	26.070	10	9.79	891.648
Sodium (Na)	mg	6.000	1.980	10	1.094	67.720
Zinc (Z)	mg	0.920	0.304	10	0.273	10.384
Copper (Cu)	mg	0.157	0.052	10	0.017	1.772
Manganese (Mg)	mg	0.188	0.062	10	0.019	2.122
Selenium (Se)	mcg	0.600	0.198	0		6.772
Vitamin C	mg	8.200	2.706	10	0.678	92.551
Thiamin	mg	0.076	0.025	10	0.005	0.858
Riboflavin	mg	0.126	0.042	10	0.017	1.422
Niacin	mg	0.481	0.159	10	0.044	5.429
Pantothenic acid	mg	0.563	0.186	10	0.069	6.354
Vitamin B6	mg	0.034	0.011	10	0.005	0.384
"Folate, total"	mcg	36.000	11.880	10	0.8	406.321
Vitamin B12	mcg	0.000	0.000	0		0.000
Vitamin A (carotenoids)	IU	155.000	51.150	0		1 749.436
Vitamin E	mg	0.020	0.007	0		0.226
Vitamin K	mcg	30.500	10.065	0		344.244
Threonine	g	0.134	0.044	1		1.512
Isoleucine	g	0.143	0.047	1		1.614
Leucine	g	0.267	0.088	1		3.014
Lysine	g	0.214	0.071	1		2.415
Valine	g	0.145	0.048	1		1.637

Source: USDA Agricultural Research Service (2003).

Identification of key products and suggested analysis for new forage varieties

Forage legumes are an essential component of the livestock feed industry. They also provide several environmental benefits including a reduction in soil erosion and the ability to fix nitrogen from the atmosphere. Forages are an excellent source of crude protein, carbohydrates, vitamins, calcium, magnesium, potassium, iron, cobalt and carotene. It is important that these nutrients are considered when evaluating novel legumes.

These plants are introduced into the growing environment as seeds and if the conditions are favourable, growth begins. An important stage of plant growth is the initiation of flowering or inflorescence. It is recommended that the forage legume be cut or grazed at this time, due to its optimum nutritive value and yield. After this point, the proportion of lignin increases and digestibility decreases. Forages are well adapted to the environments in which they grow. Plants have been selected to withstand frost and winter damage, drought, salinity, acidity or alkalinity.

Forage legumes can be processed in a variety of ways including hay, silage, pelleted meal or dehydrated cubes or simply remain as pasture. With the exception of pasture legumes, forages are processed to preserve nutrients and assist with handling of the product. These processes may influence the structure of the plant and the nutritive value for the animal.

The chemical composition of forages varies with physiological age and therefore forage quality analyses are essential. Analyses listed in Table 9.11 should be considered for new varieties. Additional analyses to be further considered, on a crop-by-crop basis, are listed in Table 9.12. When evaluating a novel forage, the compositional analysis should be conducted on material sampled at the late vegetative/early bloom stage of growth, when hay and silage cuts are normally taken.

The risk of bloat and the presence of saponins within forage legumes are the main factors that limit the use of these plants.

Table 9.11. **Suggested minimum compositional parameters to be analysed in hay or fresh forage legumes used for animal feed**

Parameter	Fresh forage/hay
Crude protein	X
Neutral detergent fibre	X
Acid detergent fibre	X
Lignin (ADL or other)	X
Crude fat	X
Ash	X
Minerals (calcium, phosphorus)	X
Amino acids	X

Table 9.12. **Additional compositional parameters to be considered for analysis in hay, silage or fresh forage for legumes used for animal feed, on a crop-by-crop basis**

Parameter	Fresh forage/hay	Silage
Total condensed tannins	X	X
Total saponins	X	X
– Medicagenic acid	X	
– Zanhic acids	X	
Phytoestrogens		
– Coumesterol (and its methyl derivatives)	X	X
– Formononetin	X	X
– Daidzein	X	X
Cyanogenic glycosides		
– Liminarin	X	
– Lotaustralian	X	
Canavanine	X	

Forage legumes other than alfalfa

The information presented in the review of alfalfa in this chapter is largely applicable to all temperate legumes. The remainder of this chapter serves to introduce other important forage legumes used in livestock feeding. Key components to be analysed in new forage varieties are identified in Table 9.11.

*Red clover (***Trifolium pratense** *L.)*

The *Trifolium* species are widely distributed. Within this genus there are 240 species found in most temperate regions. The total area where they can be found in the United States is believed to exceed that of alfalfa (Smith et al., 1985). Alfalfa and clovers collectively meet the legume pasture, hay and silage production requirements of temperate, humid and subhumid regions (Rumbaugh, 1990).

Red clover (*Trifolium pratense* L.) is an important forage legume grown in northern temperate areas of the world, especially Europe and North America. In the 1980s, 7 million hectares of red clover were grown in North America out of a world total of 20 million hectares (Smith et al., 1985). Red clover is thought to have originated in southeastern Europe and Asia Minor.

Physiology

Red clover is adapted to a wide range of soil and environmental conditions, especially poorly drained soils, and is relatively tolerant of lower soil pH and lower soil fertility. A deep tap root allows red clover a high degree of resistance to drought, although not as much tolerance as alfalfa or sainfoin. The optimum temperature for growth is 20-25°C and the optimum pH is 6-7.5. Although generally considered a short-lived perennial, improved US types of red clover are relatively productive for three and sometimes four years.

Carbohydrates are important in red clover for the plant's survival overwinter. The polysaccharide starch is the principal storage carbohydrate, which accumulates in the roots during the growing season and is depleted during winter. Taking more than one autumn harvest reduces carbohydrate accumulation, and therefore reduces yield at the first harvest in the following year.

Red clover grown in monoculture or in combination with grasses is a major hay crop in several regions of the world. In North America, it is grown in the humid northeast and in the Pacific northwest of the United States under irrigation and used as an annual in southeast United States (Taylor and Smith, 1995). For red clover, the yield potential is high; red clover varieties tend to have slightly lower forage yields than alfalfa in the area south of the US-Canada border (Undersander et al, 2002). Red clover plants survive better in severe winters when sown with a grass, rather than as monoculture (Belzile, 1987). It is commonly grown for silage and pasture, and not commonly harvested for dry hay due to its slow drying rate. A common production practice in parts of North America is to harvest the second-cut crop for seed, following a first cut-crop for forage or silage. Silage management for red clover-dominant swards includes a first cut at the early flowering stage and a second cut six to eight weeks later. Traditionally, red clover was regarded as a "difficult" crop for silage making due to low dry matter, low water soluble carbohydrate contents and a high buffering capacity, which slowed the attainment of low pH for good fermentation. A satisfactory silage fermentation is more likely to result from red clover/grass mixtures because of higher dry matter and water soluble carbohydrate concentrations and lower nitrogen contents (Frame et al., 1998). Red clover has high bloat potential (Howarth et al., 1991).

Nitrogen fixation

Rhizobial inoculation is not usually carried out in European countries since most soils contain *R. leguminosarum* bv. *trifolii*. If red clover use is extended to soils without a previous history of clover growth, rhizobial inoculation of the seed is essential.

The total amount of nitrogen fixed by red clover and the contribution it makes to the nitrogen content of the soil can vary. The factors for variation in nitrogen fixation include climatic and soil conditions, presence and efficacy of *Rhizobium*, companion species and stage of plant development. Nitrogen fixation can contribute up to 80% of total nitrogen assimilation in red clover (Heichel et al., 1985). However, the rate of N_2 fixation may be greatly reduced due to drought, accumulation of inorganic nitrogen in soil, soil acidity or plant defoliation (Maag and Nosberger, 1980).

Feed

Red clover has a high nutritive value for ruminants. It can improve the quality of autumn-saved forage for out-wintered livestock where the climate allows this practice. The digestibility of red clover's primary growth declines with advancing maturity in a linear fashion and is related to the declining leaf:stem ratio. The decline in digestibility is associated with increasing lignin content and a reduction in degradability of polysaccharides other than starch (Taylor and Quesenberry, 1996).

The nutrient composition of red clover is shown in Tables 9.13 and 9.14. Compared with grasses, red clover is usually higher in concentrations of pectin, lignin, nitrogen, calcium, magnesium, iron and cobalt (Frame et al., 1998). Alfalfa and red clover have similar nutrient content. One of the main differences is that red clover contains polyphenol oxidases, which are enzymes that play a role in inhibiting plant proteases (protein degrading enzymes) and proteolysis (protein breakdown) in the silo. As a result of the polyphenol oxidase action, red clover protein is not broken down during silage fermentation to the same extent as alfalfa protein. Therefore, red clover has more undegradable protein (bypass protein 25-35%) than alfalfa (15-25%). Additional research has shown that when red clover and alfalfa are of similar fibre content, red clover may be more digestible than alfalfa, providing a more energy-dense forage to the diets of lactating dairy cows (Hoffman and Broderick, 2001). Unfortunately, red clover does not stand up to continuous stocking, but works well in a rotational stocking system. Red clover does not accumulate condensed tannins in forage (Sarkar et al., 1976). As discussed in the anti-nutritional factors of alfalfa section, isoflavonoids are more common in clover species than alfalfa.

*White clover (***Trifolium repens** *L.)*

On a world basis, white clover (*Trifolium repens* L.) is the most important true clover species for grazed swards within the genus *Trifolium*. White clover is used primarily in Western Europe and North America, New Zealand and Australia. There are approximately 15 million hectares of pasture with white clover in Australasia and 5 million hectares in the United States. This legume is thought to have originated in the Mediterranean area (Taylor et al., 1980).

Table 9.13. **Proximate, lignin, acid detergent fibre and neutral detergent fibre composition of red clover (*Trifolium pratense* L.) harvested at early bloom stage**

Expressed on dry matter basis, in %

	Hay NRC 1971	**Hay** NRC 1982	**Hay** Ensminger et al.	**Hay** NRC 1996	**Hay** Hoffman et al.	**Range for Hay**	**Silage** literature range[1]
Dry matter	87.3	89.0	87.0	89.0		0.87-0.89	21.1-53.5
Crude protein	21.4	16.0	21.4	20.8	18.4	16.0-21.4	14.9-22.5
Crude fat	3.9	2.8	3.9	3.0		2.8-3.9	4.3
Crude fibre	20.4	28.8	20.4			20.4-28.8	-
Neutral detergent fibre				48.0	34.9	34.9-48.0	31.7-50.5
Acid detergent fibre					24.4	24.4	24.9-37.0
Lignin		10.0		16.67	4.3	4.3-10.0	4.2-4.3
Ash	9.7	8.5	9.7	7.0		7.0-9.7	1.9-11.5

Note: 1. Dewhurst et al. (2003); Broderick et al. (2001); Coblentz et al. (1998); Hoffman et al. (1997); Hoffman et al. (1993); Al-Mabruk et al. (2004).

Sources: NRC (1971, 1982, 1996); Ensminger et al. (1990); Hoffman et al. (1993); Dewhurst et al. (2003); Broderick et al. (2001); Coblentz et al. (1998); Hoffman et al. (1997); Hoffman et al. (1993); Al-Mabruk et al. (2004).

Table 9.14. **Mineral composition of late vegetation to early bloom red clover (*Trifolium pratense* L.)**

Expressed on dry matter basis, in %

	NRC 1971	NRC 1982	Ensminger et al.	NRC 2001	Range
Sodium (Na), mg/100 g		0.19		0.18	0.18-0.19
Potassium (K), mg/100 g	2.57	1.62	3.24	1.81	1.62-3.24
Calcium (Ca), mg/100 g	1.77	1.53	1.55	1.38	1.38-1.77
Phosphorus (P), mg/100 g	0.31	0.25	0.37	0.24	0.24-0.37
Magnesium (Mg), mg/100 g	0.51	0.43	0.39	0.38	0.38-0.51
Iron (Fe), mg/100 g		0.018	0.073	0.024	0.018-0.073
Sulfur (S), mg/100 g		0.17		0.16	0.16-0.17
Copper (Cu), mg/kg		11.0	21.1	11.0	11.0-21.1
Coablt (Co), mg/kg		0.16	0.23	0.16	0.16-0.23
Manganese (Mn), mg/kg		73.0	86.7	108.0	73.0-108.0
Zinc (Zn), mg/kg		17.0	52.0	17.0	17.0 52.0
C1, mg/100 g		0.32		0.32	0.32

Sources: NRC (1971, 1982, 2001); Ensminger et al. (1990).

Production

White clover is usually grown in association with suitable grass species such as perennial ryegrass, or in the United States, Kentucky bluegrass (*Poa pratensis*). Grass/clover swards may be utilised successfully by a range of grazing systems for both intermittent (rotational), continuous grazing or a blend of both types in the same season. In Atlantic Canada, white clover is allowed to stockpile from late summer for use in late autumn, thus extending the grazing season (Fraser et al., 1993; Kunelius and Narasimhalu, 1993). White clover has high bloat potential (Howarth et al., 1991), likely due to the large amount of foliage.

White clover plays an important role in arable cropping, particularly in sustaining or building up soil fertility, whether as a green manure or as a legume-rich phase within a crop rotation (Barney, 1987; Ten Holte and Van Keulen, 1989). It also has a role, when undersown in arable crops such as corn (maize), in protecting the soil from erosion and minimising damage from harvesting operations (Lampkin, 1990). In monoculture or in combination with grass, white clover acts as a protective ground cover or soil-stabilization plant (Parente and Frame, 1993). There is increased interest in the use of white clover as an understorey to supply the nitrogen requirements of a cereal crop.

Physiology

White clover is capable of spreading and establishing itself in suitable niche situations in grazed pastures. It can tolerate severe defoliation better than other types of legumes, is more persistent and has the ability to colonise bare spaces (Burdon, 1983). White clover is adapted to a wide range of soils but it does not thrive in poorly drained soils (McAdam, 1983), shallow drought-prone soils (Foulds, 1978, Thomas, 1984) or saturated, unamended peat (Burdon, 1983). Unlike red clover and alfalfa, white clover has a continual generation of new leaves.

Nitrogen fixation

For nitrogen fixation, rhizobial populations of the strain *Rhizobium leguminosarum* bv. *trifolii* infect the roots of white clover and are highest in soils in which *Trifolium* species have been or are currently prevalent. Otherwise, white clover needs to be inoculated with effective and competitive strains of Rhizobia (Newbould et al., 1982).

Feed

The nutrient composition of white clover is shown in Tables 9.15 and 9.16. The digestibility of white clover is higher than that of other temperate forage legumes. White clover is almost always grown in association with grasses; approximately 10-20% white clover allows for optimal animal productivity (Curll, 1982; Stewart, 1984). Dry matter intake by a variety of livestock has been shown to be higher for white clover than for grass, regardless of feed form (fresh, dried, hay or silage) (Thomson, 1984). The physical, chemical and plant anatomical features all contribute to the superior intake quality of white clover. Sheep spend less time masticating white clover, and the weight per bite is heavier due to a greater bulk density (Edwards et al., 1995). Heifers spend a longer time grazing and ruminating on grass than clover (Orr et al., 1996). The rate of particle degradation in the rumen is faster with white clover than with ryegrass (Moseley and Jones, 1984, Ulyatt et al., 1986) and there is enhanced ruminal digestion with the legume (Beever and Thorp, 1996).

In addition to a faster rate of intake for white clover than for grass at comparable digestibility levels, ingested nutrients in white clover may be utilised more efficiently (Beever et al., 1985) and more efficient use made of metabolizable energy for animal production (Rattray and Joyce, 1974).

White clover does not accumulate condensed tannin in forage, but accumulates these polymers in flowers (Sarkar et al., 1976; Foo et al., 1982). Some white clover cultivars can contain cyanogenic glycosides.

*Alsike clover (*Trifolium hybridum *L.)*

Alsike clover (*Trifolium hybridum* L.) is grown in temperate and subartic areas of Europe, Asia, North and South America and some regions of Australasia. This legume tends to yield and grow better in cooler climates. It is thought to have originated in northern Europe. This short-lived perennial has similar persistence to red clover.

Table 9.15. **Proximate, lignin, acid detergent fibre and neutral detergent fibre composition of white clover (*Trifolium repens* L.) harvested at late vegetative/early bloom stage**

Expressed on a dry matter basis, in %

	Hay NRC 1971	Hay NRC 1982	Hay Ensminger et al.	Hay NRC 1996	Range for Hay	Silage Dewhurst et al.
Dry matter	17.7	90.0	89.0	89.0	17.7-90.0	24.2
Crude protein	28.2	22.0	22.4	22.4	22.0-28.2	26.1
Crude fat	3.3	2.7	2.7	2.7	2.7-.3	
Crude fibre	15.7	21.2	20.8		15.7-21.2	
Neutral detergent fibre			36.0	36.0	36.0	26.9
Acid detergent fibre		32.0	32.0		32.0	27.4
Lignin		7.0	6.6	7.0	6.6-7.0	
Ash	11.9	10.1	9.4	9.4	9.4-11.9	10.0

Sources: NRC (1971, 1982, 1996; Ensminger et al. (1990); Dewhurst et al. (2003).

Table 9.16. **Mineral composition of late vegetation to early bloom white clover (*Trifolium repens* L.)**

Expressed on dry matter basis, in %

	NRC 1971	NRC 1982	Ensminger et al.	NRC 2001	Range
Sodium (Na), g/100 g	0.39	0.13	0.13	0.13	0.13-0.39
Potassium (K), g/100 g	2.13	2.62	2.44	2.44	2.13-2.44
Calcium (Ca), g/100 g	1.40	1.35	1.45	1.45	1.35-1.45
Phosphorus (P), g/100 g	0.51	0.31	0.34	0.33	0.31-0.51
Magnesium (Mg), g/100 g	0.45	0.48	0.47	0.47	0.45-0.48
Iron (Fe), g/100 g	0.034	0.041	0.047	0.047	0.034-0.047
Sulfur (S), g/100 g	0.33	0.21	0.21	0.21	0.21-0.43
Copper (Cu), mg/kg		10.0	9.40	9.41	9.40-10.0
Coablt (Co), mg/kg		0.16	0.16	0.16	0.16
Manganese (Mn), mg/kg	307.2	95.0	123.1	123.0	95.0-307.2
Zinc (Zn), mg/kg		17.0	17.0	17.9	17.0-17.9
C1, g/100 g	0.61	0.30	0.30	0.30	0.30-0.61

Sources: NRC (1971, 1982, 2001); Ensminger et al. (1990)..

Production

This legume grows best in cool temperate conditions, but is adaptable to wet, infertile or acid soils that are unsuitable for red clover or alfalfa (Townsend, 1995). However, it is intolerant of drought or salinity. Alsike clover tends to be very tolerant of cold and frost and therefore allows for its establishment and growth in cooler climate areas. The majority of the world's alsike clover seed is produced in North America, including in Alberta, Idaho and Oregon.

This forage legume is usually grown in combination with grasses and other legumes. In North America, it is recommended to grow alsike in a mixture with red clover and a grass such as timothy (Townsend, 1995). The agronomic and management requirements of alsike clover are similar to those of red clover, and the forage causes bloat (Howarth et al., 1991).

Feed

As with other legumes, alsike clover is rich in protein and minerals, although it declines in digestibility as the plant matures. It is very palatable for livestock and continues to bloom throughout the season. It is used for pasture and hay, although the high moisture content makes it difficult to dry for hay production. The regrowth after taking a cut of hay is excellent for use in a fall grazing system. It is important to note that hay or pasture containing more than 5% alsike clover is not recommended for horses; it is associated with alsike clover poisoning characterised by liver damage and photosensitization in horses. The causal toxin is not known, and may originate with an associated fungus rather than the clover itself (Knight and Walter, 2003). The composition of alsike clover is shown in Tables 9.17 and 9.18.

Table 9.17. **Proximate analysis of alsike clover (*Trifolium hybridum* L.) harvested at late vegetative/early bloom stage**

Expressed on a dry matter basis, in %

	NRC 1971	Ensminger et al.	NRC 1982	Range
Dry matter	87.4	88.0	19.0	19.0-88.0
Crude protein	14.2	14.2	24.1	14.2-24.1
Crude fat	2.7	2.8	3.2	2.7-3.2
Crude fibre	30.1	29.9	17.5	17.5-30.1
Ash	8.7	8.7	12.8	8.7-12.8

Sources: NRC (1971, 1982); Ensminger et al. (1990).

Table 9.18. **Mineral composition of late vegetation to early bloom alsike clover (*Trifolium hybridum* L.)**

Expressed on a dry matter basis, in %

	NRC 1971	Ensminger et al.	NRC 1982	Range
Sodium (Na), g/100 g	0.46	0.46	0.46	0.46
Potassium (K), g/100 g	2.74	2.22	2.62	2.22-2.74
Calcium (Ca), g/100 g	1.29	1.30	1.32	1.29-1.32
Phosphorus (P), g/100 g	0.26	0.25	0.28	0.25-0.28
Magnesium (Mg), g/100 g	0.32	0.45	0.31	0.31-0.45
Iron (Fe), g/100 g	0.045	0.026	0.046	0.026-0.046
Sulfur (S), g/100 g	0.21	0.19	0.17	0.17-0.21
C1, g/100 g	0.78	0.78	0.77	0.77-0.78
Copper (Cu), mg/kg	6.0	6.0	6.0	6.0
Manganese (Mn), mg/kg	117.0	69.0	117.0	69.0-117.0

Sources: NRC (1971, 1982); Ensminger et al. (1990).

*Subterranean clover (*Trifolium subterraneum *L.)*

Subterranean clover (*Trifolium subterraneum* L., also known as subclover) is a winter annual that is very important in the drylands of Australia. This legume is thought to have originated in the Mediterranean region and was developed for pastoral use and soil improvement, especially in Australia, where it is used in rotation with cereal cropping. It is also used in the northwest United States, southern Europe, Latin America and New Zealand to a lesser degree. It is adapted to regions with hot dry summers and moist winters with mild temperatures (6-14°C) and abundant rainfall.

Production

Subterranean clover germinates rapidly in the moist autumn, grows during winter and spring, flowering and seeding in late winter/early spring and then survives the dry summer as a dormant seed. This efficient system is designed to escape the damaging summer drought. This legume grows best when soil fertility levels are relatively high, especially with high phosphorus and sulfur, regardless, it is valued for its ability to grow in less fertile, acidic soils (Frame et al., 1998).

Along with its use in grazing, this legume is used for erosion control, hydro-seeding road side banks and as a green manure or weed smothering cover in horticultural and orchard situations (Caporali et al., 1993).

Nitrogen fixation

If a pasture is being renewed by sowing with subclover or in a mixture with grass, inoculation with a rhizobium strain is advisable unless there has been a long history of satisfactory subclover growth. Using the correct strain of *Rhizobium leguminosarum* bv. *trifolii* has a positive impact on the establishment and performance of subclover.

Feed

Subclover is outstanding among annual forage legumes for its tolerance to grazing (Caporali et al., 1993). An annual seed crop is essential for subclover persistence in pasture. Therefore, it is important that the sward's potential to produce a seed crop is not jeopardised by overgrazing. In common with other legume species, subclover is rich in crude protein compared to grasses. The protein concentration declines steadily with advancing plant maturity, as does the digestibility.

Grazed subclover in irrigated swards has high digestibility and nitrogen content and low NDF, ADF and lignin content (Frame et al., 1998). Effective rumen-degradable protein in the leaf can be so low that microbial protein synthesis in the rumen is limited, adversely affecting animal production (Mulholland et al., 1996). Table 9.19 shows proximate composition of subterranean clover.

Table 9.19. **Proximate analysis of subterranean clover (*Trifolium subterraneum* L.) harvested at early bloom stage**

Reported on dry matter basis, in %

	NRC
Dry matter	90
Crude protein	30.5
Fat	3.7
Crude fat	10.1
Ash	11.1

Source: Adapted from NRC (1971).

Birdsfoot trefoil (Lotus corniculatus L.*) and greater lotus (*Lotus spp.*)*

The species within the *Lotus* genus are referred to as pioneer legumes because they are suitable for developing pastures on acidic, infertile soils in cool, moist areas of the world (Frame et al., 1998). Both perennials and annuals are components of this genus. There are a large number of species of *Lotus* (Zandstra and Grant, 1968; USDA, 2003). Three examples used for forage include birdsfoot trefoil (*Lotus corniculatus* L.), marsh birdsfoot trefoil (big trefoil or lotus) (*Lotus uliginosus* Schkuhr. also called *L. pedunculatus*) and narrow-leaf birdsfoot trefoil (*Lotus tenuis*).

Birdsfoot trefoil was not introduced to North America until the early 1900s; however, it was very common in Europe, Africa and Asia. The majority of species are found in the Mediterranean region and this is thought to be their area of origin. Approximately 1.2 million hectares are grown in northeastern North America on acidic, infertile and low-input management systems (Beuselinck and Grant, 1995). Greater lotus can be found in Britain, France and Germany as well as the northwestern United States.

Production

Birdsfoot trefoil is suited to clay soils which are too wet or too acidic for alfalfa. Birdsfoot trefoil is drought tolerant, even more so than alfalfa (Peterson et al., 1992). It also persists in poorly drained soil more than alfalfa or red clover (Barta, 1986) and is highly tolerant of saline soils (Schachtman and Kelman, 1991). Narrow-leaf birdsfoot trefoil is adapted to poorly drained soils and sown in central Europe and the northern United States, especially on saline and alkaline soils. The *Lotus* species are slow to become popular due to their slow establishment, slow growth rate and poor competitive ability (McKersie et al., 1981). The greater lotus species is not winter hardy. Birdsfoot trefoil is very winter hardy once established, although less than alfalfa, but it does not survive in harsh Canadian prairie conditions. Unlike alfalfa, which has a significant period of flower-free growth, lotus plants have a short non-flowering period.

Birdsfoot trefoil is very useful on marginal land, and is a non-bloating legume (Howarth et al., 1991) due to the presence of forage condensed tannins (Foo et al., 1982; Sarkar et al., 1976). Big trefoil also contains tannins (Foo et al., 1982). A number of reports from different areas of the world confirm the use of lotus species, especially birdsfoot trefoil, for pasture renovation in a variety of situations, ranging from lowland grazing to alpine pastures (Frame et al., 1998). If this legume is used for a combination of hay and pasture, the hay crop should be taken at the early bloom stage and the subsequent regrowth grazed at the first flower. Weed control is very important, especially in the establishment year, since birdsfoot trefoil is not competitive in a weedy stand (Beuselink and Grant, 1995). This legume produces less forage with hay yields of 25-30% less than alfalfa. It is recommended that birdsfoot trefoil be used only in areas that are not suitable for alfalfa production due to soil acidity, poor drainage or low fertility.

Feed

There is little information available on the chemical composition of Lotus trefoil forage, but birdsfoot trefoil nutritive value is similar to that of alfalfa (Marten and Jordan, 1979). The composition of birdsfoot trefoil is shown in Tables 9.20 and 9.21. The lignin content of birdsfoot trefoil is lower than in other legumes such as white clover, red clover or alfalfa. The *Lotus* species contain varying amounts of floral and forage condensed tannins (Sarkar et al., 1976; Foo et al., 1982; Muir et al., 1999; Muir, unpublished), as well as varying amounts of flavonols (Harney and Grant, 1964, 1965) and cyanogenic glycosides (Grant and Sidhu, 1967). *Lotus uliginosis* has a moderate condensed tannin content ranging from 40-245 $mg.g^{-1}$ dry weight (Lees et al., 1994; Muir et al., 1999). *Lotus corniculatus* produces small to moderate amounts (Muir et al., 1999). Some *Lotus corniculatus* plants have very high levels of cyanogenic glycosides (Zandstra and Grant, 1968).

The more upright types of birdsfoot trefoil are suited to hay and silage production with a possibility of two to three cuts per season. This legume is of major importance for hay, silage and grazing in the northern United States and eastern Canada (Beuselinck and Grant, 1995). Birdsfoot trefoil is highly palatable to livestock, even though it accumulates condensed tannins. Therefore, these pastures are best used in a rotational stocking system (Van Keuren and Davis, 1968, Van Keuren et al., 1969). Early spring grazing or continuous stocking will weaken and eliminate a stand of birdsfoot trefoil. Birdsfoot trefoil is persistent, and will last for several years if managed properly.

Table 9.20. **Proximate, lignin, acid detergent fibre and neutral detergent fibre composition of birdsfoot trefoil (*Lotus corniculatus* L.) harvested at late vegetative/early bloom stage**

Expressed on a dry matter basis, in %

	NR 1971[1]	NRC 1982	Ensminger et al.	NRC 1996	Hoffman et al.	Range
Dry matter	89.0	92.0	91.0	91.0	100	89.0-100
Crude protein	16.0	16.3	15.3	15.9	17.0	15.3-16.3
Crude fat	2.2	2.5	2.1	2.1		2.1-2.5
Crude fibre	29.6	30.7	32.3			29.6-32.3
Neutral detergent fibre			47.0	47.5	44.4	44.4-47.5
Acid detergent fibre		36.0	36.0		35.8	35.8-36.0
Lignin		9.0		9.1	9.8	9.1-9.8
Ash	7.6	7.0	7.4	7.4		7.0-7.6

Sources: NRC (1971, 1982, 1996); Ensminger et al. (1990); Hoffman et al. (1993).

Table 9.21. **Mineral composition of early bloom birdsfoot trefoil (*Lotus corniculatus* L.)**

Expressed on dry matter basis, in %

	NRC	Ensminger et al.	Range
Sodium (Na), g/100 g	0.07	0.07	0.07
Potassium (K), g/100 g	1.92	1.92	1.92
Calcium (Ca), g/100 g	1.70	1.7	1.70
Phosphorus (P), g/100 g	0.27	0.23	0.23-0.27
Magnesium (Mg), g/100 g	0.51	0.51	0.51
Iron (Fe), g/100 g	0.023	0.023	0.023
Sulfur (S), g/100 g	0.25	0.25	0.25
Copper (Cu), mg/kg	9.0	9.3	9.0-9.3
Cobalt (Co), mg/kg	0.11	0.11	0.11
Manganese (Mn), mg/kg	29.0	28.7	28.7-29.0
Zinc (Zn), mg/kg		77.2	77.2

Sources: NRC (1971); Ensminger et al. (1990).

*Sainfoin (*Onobrychis viciifolia Scop.*)*

Sainfoin (*Onobrychis viciifolia scop.*) is also known as St. Foin, cock's head, esparcet, holy clover or holy grass. In French, *sainfoin*, is interpreted to mean "healthy hay", which is probably referring to its non-bloat characteristics. This perennial legume is indigenous to temperate western Asia and southern Europe. It can be found on dry calcareous soils of the western United States and Canada (Miller and Hoveland, 1995), although its lack of genetic variability has prevented it from becoming agriculturally important in either country.

Production

Sainfoin grows well on calcareous soils having a pH of 6 or higher, which tend to be too dry or too barren for clover or alfalfa. It is even more drought-resistant than alfalfa; however, it yields less (Rogers, 1976). Sainfoin yields best on deep, well-drained soils, and will not withstand wet soils or high water tables. It is somewhat intolerant of saline soils and tends to grow well on soils that are low in phosphorus. Sainfoin requires soil rich in lime and can withstand cold temperatures. It is not as winter hardy as the locally recommended cultivars of alfalfa, and tends to be very susceptible to invasion from weeds because of its slow growth during the establishment year.

Grown in monoculture or in combination with grasses such as fescue or cocksfoot, this legume competes poorly with creeping, rooted grasses. The stage of growth at the time of cutting determines the quality of the hay or silage, cutting at mid-flowering for hay and early flowering for silage. Growth after the first harvest is nutritious and preferred by livestock. However, overgrazing should be avoided since re-growth will be limited, especially if grazing is intensive. Sainfoin is very palatable and is grazed by livestock in preference to alfalfa. Forage dry-matter yields of sainfoin are about 20% lower under dryland conditions compared with alfalfa, and may be 30% or more lower in irrigated areas.

Unlike alfalfa, sainfoin does not drop its lower leaves; stems remain succulent as the plant matures so that quality does not decrease as rapidly. Unfortunately, use of sainfoin has been limited by the cost and availability of seed. Seed supplies have been inadequate, primarily because reliance on native insect pollinators provides inconsistent seed yields. Also, with the increase in cheap sources of N fertilizer, this legume's popularity has declined.

This legume is recommended only for short-term rotations in pure stands or for planting in grass legume mixtures (along with alfalfa, birdsfoot trefoil, meadow bromegrass or orchard grass) that persist after sainfoin declines. The seeding of sainfoin with a noncompetitive grass may help to boost yields and reduce weed pressure. The advantages of sainfoin for pasture use include excellent quality and palatability that give superior animal performance without the danger of bloat (Howarth et al., 1991). Addition of 10-20% sainfoin to an alfalfa diet also suppressed most of the bloating in steers (McMahon et al., 1999).

Nitrogen fixation

The nitrogen fixation abilities in sainfoin are poor in comparison with alfalfa and clover. For good establishment and growth, sainfoin must be inoculated with a special rhizobium prior to planting. Nitrogen-fixing bacteria may be short lived or ineffective so that nitrogen fertilisation may be required for this legume.

Feed

Sainfoin is rich in protein similar to other legumes; however, its digestibility is limiting. It has lower crude protein and digestibility than alfalfa (Karnezos et al., 1994). Sainfoin is rich in minerals compared to grasses, but its calcium and sodium contents are much lower than in other forage legumes (Spedding and Diekmahns, 1972). Sainfoin is higher in carbohydrates than alfalfa and lower in crude protein, fibre and ash. Sainfoin forage and flowers contain moderate levels of condensed tannin ranging from 27-75 mg.g-1 dry weight (Koupai-Abyazani et al., 1993; Marais et al., 2000). Substantially lowered beef production costs occur when cattle are raised in alfalfa mixed pastures that include sainfoin as a source of condensed tannin (Popp et al., 2000.).

*Cicer milkvetch (*Astragalus cicer*)*

Cicer milkvetch (*Astragalus cicer*) is a long-lived perennial that is native to the European continent. This legume is grown in a wide variety of environments since it performs well on poor, infertile soil. Cicer milkvetch is grown in Canada on a small scale.

Production

This pasture legume is a hardy forage plant with deep roots and a creeping growth habit. It is tolerant to drought, slight acidity and alkalinity, but is intolerant to waterlogged soils. Cicer milkvetch is more accepting of late spring and early frosts than alfalfa.

Two years are required after establishment to produce any hay or pasture. Cicer milkvetch tolerates grazing and grows well throughout the season. An advantage to this legume is its bloat-safe property, which occurs because of its reticulate leaf veins and epidermal thickness (Howarth et al., 1979; Lees et al., 1982). Yields for this legume are comparable to alfalfa in a longer growing season area. Due to its slow spring growth and slow recovery after harvest, it may only be harvested two or three times per season. It is competitive in combination with grasses and therefore requires an equally competitive grass if the legume is to be equally maintained. These grasses include creeping foxtail, meadow bromegrass, orchard grass and tall fescue.

Feed

The protein content of cicer milkvetch equals or exceeds that of other legumes. This high protein level is due to the leaf:stem ratio – which is 40% higher than alfalfa – as well as its ability to hold its leaves during the drying and baling processes. The moisture content, when harvested, is on average 4-8% higher than alfalfa or sainfoin. This results in an extended drying time that is approximately three days longer than other legumes. It is especially well suited for use in a pasture environment and resists damage from overgrazing. Cicer milkvetch tends to be readily consumed by all classes of livestock, either in the form of hay or pasture. No cases of bloat have been reported for cicer milkvetch.

*Sweet clover (*Melilotus officinalis*)*

Sweet clover (*Melilotus officinalis*) is a hardy, drought-tolerant biennial that has adapted to a wide range of soils. This legume is tolerant of alkalinity but not acidity (Gorz and Smith, 1978). The yellow type of sweet clover is more drought-tolerant, shorter in stature and earlier maturing than the white type of sweet clover (*Melilotus alba*).

Production

This legume is used in Canada and the United States cornbelt, in areas with alkaline soils, for both hay and pasture, as well as an aid for erosion control with its deep root system. Sweet clover should be cut prior to the bud stage for good quality hay. For grazing, regrowth will occur if a 30-centimetre stubble is maintained.

Feed

As with other clover species, this legume has the potential to cause problems with bloat, although the potential is not as high as alfalfa and clovers (Howarth et al., 1991). Sweet clover produces coumarins, a sweet-smelling phenolic that develops into dicoumarol under sub-optimal hay-curing conditions (wet, mouldy). Dicoumarol is an anti-coagulant that causes livestock death from internal bleeding (sweet clover disease). Low-coumarin varieties have been developed (Goplen, 1971, 1981).

Serradella (*Ornithopus spp*)

Serradella is a summer annual which is native to south-western Europe. This legume is a winter or cool season annual when it is grown in mild regions, such as southern Australasia. There are two species, pink or French serradella (*Ornithopus sativus* Brot.) which is cultivated for forage in some parts of Europe, Australia, high altitudes in Kenya and South Africa; and Yellow serradella (*Ornithopus compressus* L.) which occurs widely in natural pastures in countries surrounding the Mediterranean on non-calcareous soils.

Production

It grows on all soil types on which subterranean clover is grown but also on sandy, gritty soils where clover cannot grow (Gladstones and McKeown, 1977). The yellow type of serradella is confined to areas that receive at least 500 mm of rainfall per year.

Feed

Similar to most legumes, the crude protein and digestibility decline with advancing plant maturity, although the rates of decline are slower than for alfalfa or red clover (Iglesias and Lloveras, 2000).

Once established, serradella can be grazed in systems similar to those for subclover with similar stocking rates but it can also be cut for silage (Taylor and Hughes, 1978). The dry matter yields for this legume are quite variable and are dependent on several factors. The pink serradella variety tends to have a high nutritive value (Gladstones and Barrett-Lennard, 1964). In north-western Spain, pink serradella, planted in early fall, can be used alongside corn in a double-cropping system (Iglesias and Lloveras, 1998). Serradella is used as an understorey for grazing in agroforestry situations in New Zealand due to its nitrogen-fixing ability. It can also be used as an understorey in vineyards, growing while the vines are dormant, controlling weeds and supplying nitrogen to the vines (Lloveras, 1987).

*Sulla (***Hedysarum coronarium** *L.)*

Sulla (*Hedysarum coronarium* L.) is also known as Italian sainfoin, French honeysuckle or sweet vetch. This short-lived perennial is thought to have originated in the western Mediterranean region and North Africa (Duke, 1981).

Production

Sulla is the main legume in southern Italy with approximately 250 000 hectares used for grazing and hay (Martiniello and Ciola, 1994). It has been evaluated for use in North America but occupies few acres of commercial production (Allen and Allen, 1981). Sulla is mainly sown alone, but can be grown with a cereal or in a mixture with other legumes and on soils with a pH greater than 6-6.5.

Feed

The forage is of high nutritive value (especially the leaflets), and therefore it is important that sulla is cut prior to the onset of flowers for an optimal hay product. With respect to grazing this legume is best utilised in a rotational grazing system. *Hedysarum* species contain floral and forage condensed tannins, which eliminate the risk of bloat (Skadhauge et al., 1997).

References

Aerts, J.R. et al. (1999), "Polyphenols and agriculture: Beneficial effects of proanthocyanidins in forages", *Agriculture, Ecosystems & Environment*, Vol. 75, Issues 1-2, August, pp. 1-12, http://dx.doi.org/10.1016/S0167-8809(99)00062-6.

Allen, O.N. and E.K. Allen (1981), *The Leguminosae, A Source Book of Characteristics, Uses and Nodulation*, University of Wisconsin Press, Madison, Wisconsin.

Al-Mabruk, R.M. et al. (2004), "Effects of silage species and supplemental vitamin E on the oxidative stability of milk", *Journal of Dairy Science*, Vol. 87, No. 2, February, pp. 406-412.

Amrane, R. and B. Michaeletdorcau (1993), "Effect of maturity stage of Italian ryegrass and lucerne on ruminal nitrogen degradability", *Annales de Zootechnie*, Vol. 42, No. 1, pp. 31-37, http://dx.doi.org/10.1051/animres:19930103.

Bae, H.E. et al. (1993), "Effects of condensed tannins on endoglucanase activity and filter paper digestion by Fibrobacter succinogenes S85", *Applied and Environmental Microbiology*, Vol. 59, No. 7, pp. 2 132-2 138.

Barney, P.A. (1987), "The use of *Trifolium repens, Trifolium subterraneum* and *Medicago lupulina* as overwintering leguminous green manures", *Biological Agriculture & Horticulture*, Vol. 4, Issue 3, pp. 225-234.

Barry, T.N. (1989), "Condensed tannins: Their role in ruminant protein and carbohydrate digestion and possible effects upon the rumen ecosystem", in: Nolan, J.V. et al. (1988), *The Roles of Protozoa and Fungi in Ruminant Digestion*, Penambul Books, Armidale, NSW, Australia, pp. 153-169.

Barry, TN. and W.C. McNabb (1999), "The implications of condensed tannins on the nutritive value of temperate forages fed to ruminants", *British Journal of Nutrition*, Vol. 81, pp. 263-272.

Barta, A.L. (1986), "Metabolic response of *Medicago sativa* L. and *Lotus corniculatus* L. roots to anoxia", *Plant, Cell & Environment*, Vol. 9, Issue 2, pp. 127-131, http://dx.doi.org/10.1111/j.1365-3040.1986.tb01575.x.

Bas, P. et al. (2003), "Fatty acid composition of mixed-rumen bacteria: Effect of concentration and type of forage", *Journal of Dairy Science*, Vol. 86, No. 9, September, pp. 2 940-2 948.

Beever, D.E. and C. Thorp (1996), "Advances in the understanding of factors influencing the nutritive value of legumes", in: Younie, D. (ed.), *Legumes in Sustainable Farming Systems*, Occasional Symposium No. 30, British Grassland Society, Reading, pp. 194-207.

Beever, D.E. et al. (1985), "The digestion of fresh perennial ryegrass (*Lolium perenne* L. cv. Melle) and white clover (*Trifolium repens* L. cv. Blanca) by growing cattle fed indoors", *British Journal of Nutrition*, Vol. 54, No. 3, November, pp. 763-775.

Belzile, L. (1987), "Effect of companion timothy on winter survival of red clover", *Canadian Journal of Plant Science*, Vol. 67, pp. 1 101-1 103.

Beuselinck, P.R. and W.F. Grant (1995), "Birdsfoot trefoil", in: Barnes, R.F. et al. (eds.), *Forages*, 5th edition, Vol. 1: An Introduction to Grassland Agriculture, Iowa State University Press, Ames, Iowa, pp. 237-248.

Bialy, Z. et al. (1999), "Saponins in alfalfa (*Medicago sativa* L.) root and their structural elucidation", *Journal of Agricultural and Food Chemistry*, Vol. 47, No. 8, August, pp. 3 185-3 192.

Bins, W. and M.W. Pedersen (1964), "Effecting of feeding high saponin hay to four-month-old Holstein-Friesian bulls", *Proceedings of the 7th Conference Rumen Function*, Chicago, Illinois.

Birk, Y. (1969), "Saponins", in: Liener, E.I. (ed.), *Toxic Constituents of Plant Foodstuffs*, Academic Press, New York, pp. 169-210.

Bondi, A. et al. (1973), "Forage saponins", in: Butler, G.W. and R.W. Bailey (eds.), *Chemistry and Biochemistry of Herbage*, Academic Press, New York, pp. 511-528.

Brink, G.E. and G.C. Marten (1989), "Harvest management of alfalfa – nutrient yield vs. forage quality and relationship to persistence", *Journal of Production Agriculture*, Vol. 2, No. 1, pp. 32-36, http://dx.doi.org/10.2134/jpa1989.0032.

Broderick, G.A. (1995), "Performance of lactating dairy cows fed either alfalfa silage or alfalfa hay as the sole forage", *Journal of Dairy Science*, Vol. 78, No. 2, February, pp. 320-329.

Broderick, G.A. and K.A. Albrecht (1997), "Crop quality and utilization. Ruminal in vitro degradation of protein in tannin-free and tannin-containing forage legume species", *Crop Science*, Vol. 37, No. 6, pp. 1 884-1 891, http://dx.doi.org/10.2135/cropsci1997.0011183X003700060037x.

Broderick, G.A. et al. (2001), "Production of lactating dairy cows fed alfalfa or red clover silage at equal dry matter or crude protein contents in the diet", *Journal of Dairy Science*, Vol. 84, No. 7, July, pp. 1 728.

Burdon, J.J. (1983), "Biological flora of the British Isles: *Trifolium repens*", *Journal of Ecology*, Vol. 71, No. 1, pp. 307-330.

Butler, L.G. et al. (1984), "Interaction of proteins with sorghum tannin: Mechanism, specificity and significance", *Journal of the American Oil Chemists' Society*, Vol. 61, No. 5, May, pp. 916-920, available at: http://link.springer.com/article/10.1007%2FBF02542166#page-1.

Butter, N.L. et al. (2001), "Effect of dietary condensed tannins on gastrointestinal nematodes", *Journal of Agricultural Science*, Vol. 137, No. 4, July, pp. 461-469, http://dx.doi.org/10.1017/S0021859601001605.

Caporali, F. et al. (1993), "Prospects for more sustainable cropping systems in central Italy based on subterranean clover as a cover crop", in: Baker, M.J. (ed.), *Proceedings of the XVII International Grassland Congress, New Zealand and Australia*, Vol. III, New Zealand Grassland Association, Palmerston North, pp. 2 197-2 198.

Carruthers, V.R. et al. (1987), *The 1986 MAF Bloat Survey. Proceedings of the 4th Seminar of the Dairy Cattle Society of the New Zealand Veterinary Association held at Rotorua, New Zealand*, 18-20 November 1987, Massey University, Palmerston North, New Zealand, pp. 11-16.

Chamblee, D.S. and M. Collins (1988), "Relationships with other species in a mixture", in: Hanson, A.A. et al. (eds.), *Alfalfa and Alfalfa Improvement*, Agronomy Monograph No. 29, ASA/CSSA/SSSA, Madison, Wisconsin, pp. 439-461.

Chapman and Hall (1982-98), The Combined Chemical Dictionary on CD-ROM. HeadFAST/CD Software © Head Software Int. PST. Release 2.2.

Cheeke, P.R. (1971), "Nutritional and physiological implications of saponins. A review", *Canadian Journal of Animal Science*, No. 51, No. 3, pp. 621-632, http://dx.doi.org/10.4141/cjas71-082.

Christensen, R.A. (2004a), Unpublished data from the following experiments: Christensen, R.A. et al. (1994), "Infusion of four long-chain fatty acid mixtures into the abomasum of lactating dairy cows", *Journal of Dairy Science*, Vol. 77, No. 4, pp. 1 052-1 069; Overton, T.R. et al. (1996), "Effects of dietary fat with or without nicotinic acid on nutrient flow to the duodenum of dairy cows", *Journal of Dairy Science*, Vol. 79, No. 8, pp. 1 410-1 424; and Christensen, R.A. and J.H. Clark, personal communication.

Christensen, R.A. (2004b), unpublished data from the following experiments: Christensen, R.A. et al. (1998), "Fatty acid flow to the duodenum and in milk from cows fed diets that contained fat and nicotinic acid", *Journal of Dairy Science*, Vol. 81, No. 4, pp. 1 078; Christensen, R.A. et al. (1994), "Effects of amount of protein and ruminally protected amino acids in the diet of dairy cows fed supplemental fat", *Journal of Dairy Science*, Vol. 77, No. 6, pp. 1 618-1 629; and Christensen, R.A. et al. (1994), "Infusion of four long-chain fatty acid mixtures into the abomasum of lactating dairy cows", *Journal of Dairy Science*, Vol. 77, No. 4, pp. 1 052-1 069.

Clark, R.T.J. and C.S.W. Reid (1974), "Foamy bloat of cattle. A review", *Journal of Dairy Science*, Vol. 57, No. 7, July, pp. 753-785.

Coblentz, W.K. et al. (1998), "In situ dry matter, nitrogen, and fiber degradation of alfalfa, red clover, and eastern gamagrass at four maturities", *Journal of Dairy Science*, Vol. 81, Issue 1, January, pp. 150-161, http://dx.Doi.org/10.3168/jds.S0022-0302(98)75562-6.

Colvin, H.W. and R.C. Backus (1988), "Bloat in sheep (Ovis aries). Mini review", *Comparative Biochemistry and Physiology*, Vol. 91, No. 4, pp. 635-644.

Conrad, H.R. and T.J. Klopfenstein (1988), "Role in livestock feeding – greenchop, silage, hay and dehy", in: Hanson, A.A. et al. (eds.), *Alfalfa and Alfalfa Improvement*, Agronomy Monograph No. 29, ASA/CSSA/SSSA, Madison, Wisconsin, pp. 539-566.

Coulman, B. et al. (2000), "A review of the development of a bloat-reduced alfalfa cultivar", *Canadian Journal of Plant Science*, Vol. 80, No. 3, pp. 487-491.

Cunningham, K.D. et al. (1994), "Flows of nitrogen and amino acids in dairy cows fed diets containing supplemental feather meal and blood meal", *Journal of Dairy Science*, Vol. 77, Issue 12, December, pp. 3 666-3 675, http://dx.doi.org/10.3168/jds.S0022-0302(94)77311-2.

Curll, M.L. (1982), "The grass and clover content of pastures grazed by sheep", *Herbage Abstracts*, Vol. 52, No. 9, pp. 403-411.

Dewhurst, R.J. et al. (2003), "Comparison of grass and legume silages for milk production. 1. Production responses with different levels of concentrate", *Journal of Dairy Science*, Vol. 86, No. 8, August, pp. 2 598-2 611.

Duke, J.A. (1992), *Handbook of Phytochemical Constituents of GRAS Herbs and Other Economic Plants*, CRC Press, Boca Raton, Florida.

Duke, J.A. (1981), *Handbook of Legumes of World Economic Importance*, Plenum Press, New York and London.

Edwards, G.R. et al. (1995), "Relationship between vegetation state and bite dimensions of sheep grazing contrasting plant species and its implications for intake rate and diet selection", *Grass and Forage Science*, Vol. 50, Issue 4, December, pp. 378-388, http://dx.doi.org/10.1111/j.1365-2494.1995.tb02332.x.

Ensminger, M.E. et al. (1990), *Feeds & Nutrition*, 2nd Edition, Ensminger Publishing Co., Clovis, California.

Fay, J.P. et al. (1981), "A scanning electron microscopy study of the invasion of leaflets of a bloat-safe and a bloat-casuing legume by rumen microorganisms", *Canadian Journal of Microbiology*, Vol. 27, No. 4, April, pp. 390-399.

Foo, L.Y. et al. (1982), "Proanthocyanidin polymers of fodder legumes", *Phytochemistry*, Vol. 21, Issue 4, pp. 933-935, http://dx.doi.org/10.1016/0031-9422(82)80096-4.

Foulds, W. (1978), "Response to soil moisture supply in three leguminous species", *New Phytologist*, Vol. 80, Issue 3, pp. 535-545.

Frame, J. et al. (1998), *Temperate Forage Legumes*, CAB International, Wallingford, United Kingdom.

Frankow-Lindberg, B.E. and D.A. Danielsson (1997), "Energy output and animal production from grazed grass/clover pastures in Sweden", *Biological Agriculture and Horticulture*, Vol. 14, Issue 4, pp. 279-290, http://dx.doi.org/10.1080/01448765.1997.9755164.

Fraser, J. et al. (1993), "Effects of autumn harvest date on the performance of white clover/grass mixtures in Nova Scotia", *Journal of Agricultural Science*, Vol. 121, Issue 3, December, pp. 315-321, http://dx.doi.org/10.1017/S002185960008549X.

Gladstones, J.S. and R.A. Barrett-Lennard (1964), "Serradella – A promising pasture legume in Western Australia", *Journal of Australian Institute of Agricultural Science*, Vol. 30, pp. 258-262.

Gladstones, J.S. and N.R. McKeown (1977), "Serradella – A pasture legume for sandy soils", Bulletin 4030, Western Australia Department of Agriculture, Perth, Australia.

Goplen, B.P. (1981), "Norgold – A low coumarin yellow blossom sweet clover", *Canadian Journal of Plant Science*, Vol. 61, No. 4, pp. 1 019-1 021, http://dx.doi.org/10.4141/cjps81-154.

Goplen, B.P. (1971), "Polara, a low coumarin cultivar of sweet clover", *Canadian Journal of Plant Science*, Vol. 51, pp. 241-251.

Goplen, B.P. et al. (1993), "Selection of alfalfa for a lower initial rate of digestion and corresponding changes in epidermal and mesophyll cell wall thickness", *Canadian Journal of Plant Science*, Vol. 73, January, pp. 1 111-1 122, available at: http://pubs.aic.ca/doi/pdf/10.4141/cjps93-014.

Goplen, B.P. et al. (1980), "A search for condensed tannins in annual and perennial species of Medicago, Trigonella, and Onobrychis", *Crop Science*, Vol. 20, No. 6, November, pp. 801-804, http://dx.doi.org/10.2135/cropsci1980.0011183X002000060031x.

Gorz, H.J. and W.K. Smith (1978), "Sweetclover", in: Heath, M.E. et al. (eds.), *Forages: The Science of Grassland Agriculture*, 3rd edition, pp. 159-166.

Grant, W.F. and B.S. Sidhu (1967), "Basic chromosome number, cyanogenic glycoside variation, and geographic distribution of *Lotus* species", *Canadian Journal of Botany*, Vol. 45, No. 5, pp. 639-647, http://dx.doi.org/10.1139/b67-070.

Hagerman, A.E. and L.G. Butler (1981), "The specificity of proanthocyanidin-protein interactions", *Journal of Biological Chemistry*, Vol. 256, No. 9, May, pp. 4 494-4 497.

Hagerman, A.E. and C.T. Robbins (1993), "Specificity of tannin-binding salivary proteins relative to diet selection by mammals", *Canadian Journal of Zoology*, Vol. 71, No. 3, pp. 628-633, http://dx.doi.org/10.1139/z93-085.

Hanson, C.H. et al. (1965), "Variation in coumestrol content of alfalfa as related to location, variety, cutting year, stage of growth, and disease", *UDSA Technical Bulletin*, No. 1 333, US Government Printing Office, Washington, DC.

Harney, P.M. and W.F. Grant (1965), "A polygonal presentation ofchromatographic investigations on the phenolic content of certain species of *Lotus*", *Canadian Journal of Genetics and Cytology*, Vol. 7, No. 1, pp. 40-51, http://dx.doi.org/10.1139/g65-007.

Harney, P.M. and W.F. Grant (1964), "A chromatographic study of the phenolics of species of Lotus closely related to L. corniculatus and their taxonomic significance", *American Journal of Botany*, Vol. 51, No. 6, July, pp. 621-627, available at: www.jstor.org/stable/2439989.

Hayman, J.M. and S.D. McBride (1984), "The response of pasture and lucerne to irrigation", *Technical Report 17*, Winchmore Irrigation Research Station, Ministry of Agriculture and Fisheries, p.79 (cited in Douglas, 1986).

He, X.-G. et al. (1996), "Analysis of flavonoids from red clover by liquid chromatography-electrospray mass spectrometry", *Journal of Chromatography*, Vol. 755, Issue 1, November, pp. 127-132, http://dx.doi.org/10.1016/S0021-9673(96)00578-X.

Heichel, G.H. et al. (1985), "Dinitrogen fixation and N and dry matter distribution during 4 year stands of birdsfoot-trefoil and red clover", *Crop Science*, Vol. 25, No. 1, January, pp. 101-105, http://dx.doi.org/10.2135/cropsci1985.0011183X002500010026x.

Hoffman, P. and G. Broderick (2001), *Red Clover Forages for Lactating Dairy Cows. Focus on Forage*, University of Wisconsin.

Hoffman, P.C. et al. (1997), "Performance of lactating dairy cows fed red clover or alfalfa silage", *Journal of Dairy Science*, Vol. 80, No. 12, December, pp. 3 308-3 315.

Hoffman, P.C. et al. (1993), "In situ dry matter, protein, and fiber degradation of perennial forages", *Journal of Dairy Science*, Vol. 76, No. 9, January, pp. 2 632-2 643.

Hostettmann, K. and A. Marston (1995), *Saponins*, Cambridge University Press, Cambridge, Massachusetts.

Howarth, R.E. (1988), "Antiquality factors and non-nutritive chemical components", in: *Alfalfa and Alfalfa Improvement*, Agronomy monograph, No. 29, pp. 493-514.

Howarth, R.E. et al. (1991), *Bloat in Cattle*, Agriculture Canada Publication 1858/E, Communications Branch, Agriculture and Agri-Food Canada, Ottawa, Ontario.

Howarth, R.E. et al. (1979), "Digestion of bloat-causing and bloat-safe legumes", *Annales de Recherches Veterinaires*, Vol. 10, pp. 332-334.

Iglesias, I. and J. Lloveras (2000), "Forage production and quality of serradella in mild winter areas in north-west Spain", *New Zealand Journal of Agricultural Research*, Vol. 43, pp. 35-40.

Iglesias, I. and J. Lloveras (1998), "Annual cool-season legumes for forage production in mild winter areas", *Grass and Forage Science*, Vol. 53, No. 4, December, pp. 318-325, http://dx.doi.org/10.1046/j.1365-2494.1998.00135.x.

Jagusch, K.T. (1982), "Nutrition of ruminants grazing Lucerne", in: Wynn-Williams, R.B. (ed.), *Lucerne for the 1980s*, Special Publication No. 1, Agronomy Society of New Zealand, pp. 73-78.

Jones, G.A. et al. (1994), "Effects of sainfoin (*Onyobrychis viciifolia Scop.*) condensed tannins on growth and proteolysis by four strains of ruminal bacterial", *Applied and Environmental Microbiology*, Vol. 60, No. 4, April, pp. 1 374-1 378.

Kalac, P. et al. (1996), "Changes in saponin content and composition during ensilage of alfalfa (*Medicago sativa* L)", *Food Chemistry*, Vol. 56, Issue 4, August, pp. 377-380, http://dx.doi.org/10.1016/0308-8146(95)00185-9.

Karnezos, T.P. et al. (1994), "Spring lamb production on alfalfa, sainfoin and wheatgrass pastures", *Agronomy Journal*, Vol. 86, No. 3, May, pp. 497-502, http://dx.doi.org/10.2134/agronj1994.00021962008600030008x.

Khalil, A.H. and T.A. El Adawy (1994), "Isolation, identification and toxicity of saponin from different legumes", *Food Chemistry*, Vol. 50, Issue 2, pp. 197-201, http://dx.doi.org/10.1016/0308-8146(94)90120-1.

Klejdus, B. et al. (2001), "Identification of isoflavone conjugates in red clover (*Trifolium pratense*) by liquid chromatographic-mass spectrometry after two-dimensional solid-phase extraction", *Analytical Chimica Acta*, Vol. 450, Issues 1-2, December, pp. 81-97, http://dx.doi.org/10.1016/S0003-2670(01)01370-8.

Klita, P.T. et al. (1996), "Effects of alfalfa root saponins on digestive function in sheep", *Journal of Animal Science*, Vol. 74, No. 5, May, pp. 1 144-1 156.

Knight, A.P. and R.G. Walter (2003), "Plants affecting the skin and liver", in: Knight, A.P. and R.G. Wlater (eds.), *A Guide to Plant Poisoning of Animals in North America*, Teton NewMedia, Jackson, Wyoming, published online by International Veterinary Information Service, New York.

Koupai-Abyazani, M.R. et al. (1993), "The proanthocyanidin polymers in some species of *Onobrychis*", *Phytochemistry*, Vol. 34, Issue 1, August, pp. 113-117, http://dx.doi.org/10.1016/S0031-9422(00)90791-X.

Kunelius, H.T. and P.R. Narasimhalu (1993), "Effect of autumn harvest date on herbage yield and composition of grasses and white clover", *Field Crops Research*, Vol. 31, Issues 3-4, January, pp. 341-349, http://dx.doi.org/10.1016/0378-4290(93)90072-U.

Lampkin, N. (1990), *Organic Farming*, Farming Press Books, Ipswich.

Latimori, N.J. et al. (1992), "Treatment of pastures with paraquat in the control of bloat: Effect on weight gain and residues in animal tissues", *Revista Argentina de Produccion Animal*, Vol. 12, pp. 217-222.

Latunde-Dada, A.O. and J.A. Lucas (1985), "Involvement of the phytoalexin medicarpin in the differential response of callus linesof Lucerne (Medicago sativa) to infection by Verticillium albo-atrum", *Physiological Plant Pathology*, Vol. 26, Issue 1, January, pp. 31-42, http://dx.doi.org/10.1016/0048-4059(85)90028-1.

Lees, G.L. (1992), "Condensed tannins in some forage legumes: Their role in prevention of ruminant pasture bloat", in: Hemingway, R.W. and P.E. Laks (eds.), *Polant Polyphenols 1*, pp. 915-935.

Lees, G.L. (1984), "Cuticle and cell wall thickness: Relations to mechanical strength of whole leaves and isolated cells from some forage legumes", *Crop Science*, Vol. 24, No. 6, November, pp. 1 077-1 081, http://dx.doi.org/10.2135/cropsci1984.0011183X002400060016x.

Lees, G.L. et al. (1994), "Effect of high temperature on condensed tannin accumulation in leaf tissues of big trefoil (Lotus uliginosis Schkuhr)", *Journal of the Science of Food and Agriculture*, Vol. 65, pp. 415-421.

Lees, G.L. et al. (1982), "Morphological characteristics of leaves from some legume forages: Relation to digestion and mechanical strength", *Canadian Journal of Botany*, Vol. 60, No. 10, pp. 2 126-2 132, http://dx.doi.org/10.1139/b82-261.

Livingston, A.L. (1978), "Forage plant estrogens", *Journal of Toxicology and Environmental Health*, Vol. 4, No. 2-3, March-May, pp. 301-324.

Lloveras, J. (1987), "Traditional cropping systems in northwestern Spain (Galicia)", *Agricultural Systems*, Vol. 23, Issue 4, pp. 259-275, http://dx.doi.org/10.1016/0308-521X(87)90047-3.

Lu, C.D. and N.A. Jorgensen (1987), "Alfalfa saponins affect site and extent of nutrient digestion in ruminants", *The Journal of Nutrition*, Vol. 117, pp. 919-927, available at: http://jn.nutrition.org/content/117/5/919.full.pdf.

Lundh, T. (1995), "Metabolism of estrogenic isoflavones in domestic animals", *Proceedings of the Society for Experimental Biology and Medecine*, Vol. 208, No. 1, January, pp. 33-39.

Maag, H.P. and J. Nosberger (1980), "Photosynthetic rate, chlorophyll content and dry matter production of *Trifolium pratense* L. as influenced by nitrogen nutrition", *Angewandte Botanik*, Vol. 54, No. 3/4, pp. 187-194.

MacAdam, J.W. and R.E. Whitesides (1996), "Growth at low temperatures increases alfalfa leaf cell constituents related to pasture bloat", *Crop Science*, Vol. 36, No. 2, pp. 378-382, http://dx.doi.org/10.2135/cropsci1996.0011183X003600020027x.

Majak, W. et al. (1980), "Relationships between ruminant bloat and the composition of alfalfa herbage. II. Saponins", *Canadian Journal of Animal Science*, Vol. 60, September, pp. 699-708, available at: http://pubs.aic.ca/doi/pdf/10.4141/cjas80-081.

Makkar, H.P.S. and K. Becker (2000), "Beneficial effects of saponins on animal production", in: Oleszek, W. and A. Marston (eds.), *Saponins in Food, Feedstuffs and Medicinal Plants*, Kluwer Academic Publishers, Boston, Massachusetts, pp. 281-286.

Makkar, H.P.S. et al. (1995), "In vitro effects of and interactions between tannins and saponins and fate of tannins in the rumen", *Journal of the Science of Food and Agriculture*, Vol. 69, Issue 4, December, pp. 481-493, http://dx.doi.org/10.1002/jsfa.2740690413.

Marais, J.P.J. et al. (2000), "Polyphenols, condensed tannins, and other natural products in Onyobrychis viciifolia (Sainfoin)", *Journal of Agricultural and Food Chemistry*, Vol. 48, No. 8, August, pp. 3 440-3 447.

Marles, M.A.S. et al. (2003), "New perspectives on proanthocyanidin biochemistry and molecular regulation", *Phytochemistry*, Vol. 64, Issue 2, September, pp. 367-383, http://dx.doi.org/10.1016/S0031-9422(03)00377-7.

Marston, A. et al. (2000), "Analysis and isolation of saponins from plant material", in: Oleszek, W. and A. Marston (eds.), *Saponins in Food, Feedstuffs and Medicinal Plants*, Kluwer Academic Publishers, Boston, Massachusetts, pp. 1-12.

Marten, G.C. and R.M. Jordan (1979), "Substitution value of birdsfoot trefoil for alfalfa-grass in pasture systems", *Agronomy Journal*, Vol. 71, No. 1, January, pp. 55-59, http://dx.doi.org/10.2134/agronj1979.00021962007100010013x.

Martiniello, P. and A. Ciola (1994), "The effect of agronomic factors on seed and forage production in perennial legumes sainfoin (*Onobrychis viciifolia* Scop.) and French honeysuckle (*Hedysarum coronarium* L.)", *Grass and Forage Science*, Vol. 49, Issue 2, June, pp. 121-129, http://dx.doi.org/10.1111/j.1365-2494.1994.tb01984.x.

Massiot, G. et al. (1991), "Saponins from the aerial part of alfalfa (Medicago sativa)", *Journal of Agricultural and Food Chemistry*, Vol. 39, No. 1, January, pp. 78-82, http://dx.doi.org/10.1021/jf00001a014.

Massiot, G. et al. (1988), "Reinvestigation of the sapogenins and prosapogenins from alfalfa (Medicago sativa)", *Journal of Agricultural and Food Chemistry*, Vol. 36, No. 5, September, pp. 902-909, http://dx.doi.org/10.1021/jf00083a005.

McAdam, J.H. (1983), "Characteristics of grassland on hill farms in N. Ireland – Physical features, botanical composition and productivity", report, Queens University of Belfast/Department of Agriculture for Northern Ireland, Belfast.

McKersie, B.D. et al. (1981), "Effect of seed size on germination, seedling vigour, electrolyte leakage, and establishment of birdsfoot trefoil (*Lotus corniculatus* L.)", *Canadian Journal of Plant Science*, Vol. 61, No. 2, pp. 337-343, http://dx.doi.org/10.4141/cjps81-048.

McMahon, L.R. et al. (2000), "A review of the effects of forage condensed tannins on rumimal fermentation and bloat in grazing cattle", *Canadian Journal of Plant Science*, Vol. 80, No. 3, pp. 469-485, http://dx.doi.org/10.4141/P99-050.

McMahon, L.R. et al. (1999), "Effects of sainfoin on *in vitro* digestion of fresh alfalfa and bloat in steers", *Canadian Journal of Animal Science*, Vol. 79, pp. 203-212, available at: http://pubs.aic.ca/doi/pdf/10.4141/A98-074.

McSweeney, C.S. et al. (2001), "Microbial interactions with tannins: Nutritional consequences for ruminants", *Animal Feed Science and Technology*, Vol. 91, Issues 1-2, May, pp. 83-93, http://dx.doi.org/10.1016/S0377-8401(01)00232-2.

Mendel, V.E. and J.M. Boda (1961), "Physiological studies of the rumen with emphasis on the animal factors associated with bloat", *Journal of Dairy Science*, Vol. 44, No. 10, October, pp. 1 881-1 898, http://dx.doi.org/10.3168/jds.S0022-0302(61)89980.

Michaud, R. et al. (1988), "World distribution and historical development", in: Hanson, A. et al. (eds.), *Alfalfa and Alfalfa Improvement*, American Society of Agronomy, Madison, Wisconsin, pp. 259-291.

Miller, D.A. and C.S. Hoveland (1995), "Other temperate legumes", in: Barnes, R.F. et al. (eds.), *Forages, 5th edition, Vol. 1, An Introduction to Grassland Agriculture*, Iowa State University Press, Ames, Iowa, pp. 273-281.

Molan, A.L. et al. (2001), "The effect of condensed tannins from Lotus pedunculatus and Lotus corniculatus on the growth of protteolytic rumen bacteria in vitro and their possible mode of action", *Canadian Journal of Microbiology*, Vol. 47, No. 7, July, pp. 626-633.

Monsanto (2003), *Safety, Compositional and Nutritional Assessment of Roundup Ready Alfalfa Events J101 and J103*, U.S. FDA/CFSAN. BNF84.

Moseley, G. and J.R. Jones (1984), "The physical digestion of perennial ryegrass (*Lolium perenne*) and white clover (*Trifolium repens*) in the foregut of sheep", *British Journal of Nutrition*, Vol. 52, No. 2, September, pp. 381-390.

Moule, G.R. et al. (1963), "The significance of oestrogens in pasture plants in relation to animal production", *Animal Breeding Abstracts*, Vol. 31, pp. 139-157.

Muir, A.D. (1997), "Antioxidative activity of condensed tannins", in: Shahidi, F. (ed.), *Natural Antioxidants: Chemistry, Health Effects, and Applications*, AOCS Press, Champaign, Illinois, pp. 204-212.

Muir, A.D. et al. (1999), "The effects of condensed tanninsin the diets of major crop insects", in: Gross, G.G. et al. (eds.), *Plant Polyphenols 2. Chemistry, Biology, Pharmacology, Ecology*, Plenum Press, New York, pp. 867-881.

Mulholland, J.G. et al. (1996), "Nutritive value of subterranean clover in a temperate environment", *Australian Journal of Experimental Agriculture*, Vol. 36, pp. 803-814.

Najda, H. (2002), *Standard Alfalfa*, AC Grazeland Br, Last revision, 12 December.

Neizen, J.H. et al. (1998), "Production, faecal egg counts and worm burdens of ewe lambs which grazed six contrasting forages", *Veterinary Parasitology*, Vol. 80, No. 1, December, pp. 15-27.

Neizen, K.E. et al. (1995), "Growth and gastrointestinal nematode parasitism in lambs grazing either lucerne (*Medicago sativa*) or sulla (*Hedysarum coronarium*) which contains condensed tannins", *Journal of Agricultural Science*, Vol. 125, Issue 2, October, pp. 281-285, http://dx.doi.org/10.1017/S0021859600084422.

Newbould, P. et al. (1982), "The effect of *Rhizobium* inoculation on white clover in improved hill soils in the United Kingdom", *Journal of Agricultural Science*, Vol. 99, Issue 3, December, pp 591-610, http://dx.doi.org/10.1017/S0021859600031282.

NRC (2001), *Nutrient Requirements of Dairy Cattle*, Seventh Revised Edition, National Research Council, National Academies Press, Washington, DC, available at: www.nap.edu/catalog/9825/nutrient-requirements-of-dairy-cattle-seventh-revised-edition-2001.

NRC (2000), *Nutrient Requirements of Beef Cattle*, Seventh Revised Edition: Update 2000, National Research Council, National Academies Press, Washington, DC, available at: www.nap.edu/catalog/9791/nutrient-requirements-of-beef-cattle-seventh-revised-edition-update-2000.

NRC (1996), *Nutrient Requirements of Beef Cattle*, 7th edition, National Research Council, National Academies Press, Washington, DC.

NRC (1982), *United States – Canadian Tables of Feed Composition: Nutritional Data for United States and Canadian Feeds*, Third Revision, National Research Council, National Academies Press, Washington, DC, available at: www.nap.edu/catalog/1713/united-states-canadian-tables-of-feed-composition-nutritional-data-for.

NRC (1971), *Atlas of Nutritional Data on United States and Canadian Feeds*, National Academies Press, Washington, DC.

Oleszek, W. (2004), Personal communication to W.D. Price, US Food and Drug Administration, Rockville, Maryland.

Oleszek, W. (1996), "Alfalfa saponins: Structure, biological activity, and chemotaxonomy", in: Waller, G.R. and K. Yamasaki (eds.), *Saponins Used in Food and Agriculture*, Plenum Press, New York, pp. 155-170.

Oleszek, W. et al. (1992), "Zanhic acid tridesmoside and other dominant saponins from alfalfa (*Medicago sativa* L) aerial parts", *Journal of Agricultural and Food Chemistry*, Vol. 40, pp. 191-196.

Orr, R.J. et al. (1996), "Grazing behaviour and herbage intake rate by Friesian dairy heifers grazing ryegrass or white clover", in: Younie, D. (ed.), *Legumes in Sustainable Farming Systems*, Occasional Symposium No. 30, British Grassland Society, Reading, pp. 221-224.

Parente, G. and J. Frame (1993), "Alternative uses of white clover", *FAO/REUR Technical Series*, No. 29, pp. 30-36.

Pedersen, M.W. et al. (1972), "Growth response of chicks and rats fed alfalfa with saponin content modified by selection", *Poultry Science*, Vol. 51, No. 2, March, pp. 458-464.

Pérez, N. et al. (1997), "Medicagenic acid content in foliage of ten varieties of alfalfa (*Medicago sativa* L) cultivated in Mexico", *Journal of the Science of Food and Agriculture*, Vol. 75, pp. 401-404.

Peterson, P.R. et al. (1992), "Drought effects on perennial forage legume yields and quality", *Agronomy Journal*, Vol. 84, No. 5, September, pp. 774-779, http://dx.doi.org/10.2134/agronj1992.00021962008400050003x.

Phuntsok, T. et al. (1998), "Biogenic amines in silage, apparent postruminal passage, and the relationship between biogenic amines and digestive function and intake by steers", *Journal of Dairy Science*, Vol. 81, No. 8, August, pp. 2 193-2 203.

Popp, J.D. et al. (2000), "Enhancing pasture productivity with alfalfa: A review", *Canadian Journal of Plant Science*, Vol. 80, No. 3, pp. 513-519, http://dx.doi.org/10.4141/P99-049.

Preston, R.L. (2003), *Beef*, March, pp. 40-48.

Rattray, P.V. and J.P. Joyce (1974), "Nutritive value of white clover and perennial ryegrass for young sheep. IV. Utilisation of dietary energy", *New Zealand Journal of Agricultural Research*, Vol. 17, No. 4, pp. 401-406, http://dx.doi.org/10.1080/00288233.1974.10421024.

Ray, H. et al. (2003), "Expression of anthocyanins and proanthocyanidins after transformation of alfalfa with maize Lc", *Plant Physiology*, Vol. 132, No. 3, July, pp. 1 448-1 463.

Rogers, H.H. (1976), "Forage legumes", in: *Plant Breeding Institute Annual Report, 1975*, Plant Breeding Institute, Cambridge, pp. 22-57.

Rosenthal, G.A. and P. Nkomo (2000), "The natural abundance of L-canavanine, an active anticancer agent, in alfalfa, *Medicago sativa* (L.)", *Pharmaceutical Biology*, Vol. 38, No. 1, pp. 1-6, http://dx.doi.org/10.1076/1388-0209(200001)3811-BFT001.

Rossi, D.M. et al. (1997), "Effect of monensin on weight gain and prevention of bloat in steers feeding on lucerne pasture", *Archivos de Medicina Veterinaria*, Vol. 29, pp. 279-282.

Rumbaugh, M.D. (1990), "Special purpose forage legumes", in: Janick, J. and J.E. Simon (eds.), *Advances in New Crops*, Timber Press, Portland, Oregon.

Sarkar, S.K. et al. (1976), "Condensed tannins in herbaceous legumes", *Crop Science*, Vol. 16, No. 4, July, pp. 543-546, http://dx.doi.org/10.2135/cropsci1976.0011183X001600040027x.

Schachtman, D.P. and W.M. Kelman (1991), "Potential of *Lotus* germplasm for the development of salt, aluminum and manganese tolerant pasture plants", *Australian Journal of Agricultural Research*, Vol. 42, pp. 139-149.

Sheaffer, C.C. et al. (1988), "Cutting schedules and stands", in: Hanson, A.A. et al. (eds.), *Alfalfa and Alfalfa Improvement*, Agronomy Monograph No. 29, ASA/CSSA/SSSA, Madison, Wisconsin, pp. 411-437.

Skadhauge, B. et al. (1997), "Leucocyanidin reductase and accumulation of proanthocyanidins in developing legume tissues", *American Journal of Botany*, Vol. 84, No. 4, April, pp. 494-503.

Slabbert, N. (1992), "Complexation of condensed tannins with metal ions", in: Hemingway, R.W. and P.E. Laks (eds.), *Plant Polyphenols. Synthesis, Properties and Significance*, Plenum Press, New York, pp. 421-436.

Smith, R.R. et al. (1985), "Red clover", in: Taylor, N.L. (ed.), *Clover Science and Technology*, ASA/CSSA/SSSA, Madison, Wisconsin, pp. 457-470.

Smolenski, S.J. et al. (1981), "Toxic constituents of legume forage plants", *Economic Botany*, Vol. 35, No. 3, July-September, pp. 321-355, available at: www.jstor.org/stable/4254302.

Spedding, C.R.W. and E.C. Diekmahns (1972), "Grasses and legumes", *British Agriculture Bulletin*, No. 49, Commonwealth Bureau of Pastures and Field Crops, Farnham Royal.

Stewart, T.A. (1984), "Utilising white clover in grass based animal production systems", in: Thomson, D.J. (ed.), *Forage Legumes*, Occasional Symposium No. 16, British Grasslands Society, Hurley, pp. 93-103.

Stifel, F.B. et al. (1968), "Chemical and ultrastructural relationships between alfalfa leaf chloroplasts and bloat", *Phytochemistry*, Vol. 7, Issue 3, March, pp. 335-364, http://dx.doi.org/10.1016/S0031-9422(00)90873-2.

Stob, M. (1983), "Naturally occurring food toxicants extorgens", in: Rechcigl, M., Jr. (ed.), *CRC Handbook of Naturally Occurring Food Toxicants*, CRC Press, Boca Raton, Florida, pp. 81-100.

Stochmal, A. et al. (2001), "Acylated apigenin glycosides from alfalfa (*Medicago sativa L.*) var. Artal", *Phytochemistry*, Vol. 57, No. 8, pp. 1 223-1 226.

Stoutjeskijk, P.A. et al. (2001), "Possible involvement of condensed tannins in aluminium tolerance of Lotus pedunculatus", *Australian Journal of Plant Physiology*, Vol. 28, pp. 1 063-1 074.

Tanner, G.J. et al. (1995), "Proanthocyanidins (condensed tannins) destabilise plant protein foams in a dose dependent manner", *Australian Journal of Agricultural Research*, Vol. 46, pp. 1 101-1 109.

Tanner, G.J. et al. (1994), "Proanthocyanidins inhibit hydrolysis of leaf proteins by rumen microflora *in vitro*", *British Journal of Nutrition*, Vol. 71, No. 6, June, pp. 947-958.

Taylor, A.O. and K.A. Hughes (1978), "Conservation-based forage crop systems for major or complete replacement of pasture", *Proceedings of the Agronomy Society of New Zealand*, Vol. 8, pp. 161-166.

Taylor, N.L. and K.H. Quesenberry (1996), *Red Clover Science*, Kluwer Academic Publishers, Dordrecht, Netherlands.

Taylor, N.L. and R.R. Smith (1995), "Red clover", in: Barnes, R.F. et al. (eds.), *Forages, Vol. 1, An Introduction to Grassland Agriculture*, 5th edition, Iowa State University Press, Ames, Iowa, pp. 217-226.

Taylor, N.L. et al. (1980), "Genetic system relationships in the genus *Trifolium*", *Economic Botany*, Vol. 36, Issue 4, October/December, pp. 431-441.

Ten Holte, L. and H. Van Keulen (1989), "Effects of white clover and red clover as a green crop on growth, yield and nitrogen of sugar beet and potatoes", *Developments in Plant and Soil Sciences*, Vol. 37, pp. 16-24.

Thomas, H. (1984), "Effects of drought on growth and competitive ability of perennial ryegrass and white clover", *Journal of Applied Ecology*, Vol. 21, No. 2, August, pp. 591-602, http://dx.doi.org/10.2307/2403431.

Thomson, D.J. (1984), "The nutritive value of white clover", in: Thomson, D.J. (ed.), *Forage Legumes*, Occasional Symposium No. 16, British Grassland Society, Hurley, pp. 78-92.

Townsend, C.E. (1995), "Registration of C-33, C-34 and C-35 genetic root stocks of diploid alsike clover", *Crop Science*, Vol. 35, pp. 1 519.

Ulyatt, M.J. et al. (1986), "Contribution of chewing during eating and rumination to the clearance of digesta from the reticulorumen", in: Milligan, L.P. et al. (eds.), *Control of Digestion and Metabolism in Ruminants*, Proceedings of the VIth International Symposium on Ruminant Physiology, Banff, Canada, Prentice Hall, Englewood Cliffs, New Jersey, pp. 498-515.

Ulyatt, M.J. et al. (1977), "The nutritive value of legumes", *Proceedings of the New Zealand Grassland Association*, Vol. 38, pp. 107-118, available at: www.grassland.org.nz/publications/nzgrassland_publication_1420.pdf.

Undersander, D.J. et al. (2002), *Forage Variety Update for Wisconsin*, UWEX A 1525 (revised).

USDA Agricultural Research Service (2003), *USDA National Nutrient Database for Standard Reference*, Release 16, Nutrient Data Laboratory, available at: http://ndb.nal.usda.gov.

USDA (2003), *USDA Plants Database*, http://plants.usda.gov.

Van Keuren, R.W. and R.R. Davis (1968), "Persistence of birdsfoot trefoil, *Lotus corniculatus* L., as influenced by plant growth habit and grazing management", *Agronomy Journal*, Vol. 60, No. 1, January, pp. 92-95, http://dx.doi.org/10.2134/agronj1968.00021962006000010030x.

Van Keuren, R.W. et al. (1969), "Effect of grazing mangement on the animal production from birdsfoot trefoil pastures", *Agronomy Journal*, Vol. 61, No. 3, May, pp. 422-425, http://dx.doi.org/10.2134/agronj1969.00021962006100030026x.

Vetter, J. (2000), "Plant cyanogenic glycosides", *Toxicon*, Vol. 38, pp. 11-36.

Vickery, P.J. et al. (1986), "Factors affecting the hydrogen cyanide potential of white clover (*Trifolium repens* L)", *Australian Journal of Agricultural Research*, Vol. 38, No. 1, January, pp. 1 053-1 059.

Waghorn, G.C. (1990), "Beneficial effects of low concentrations of condensed tannins in forages fed to ruminants", in: Akin, D.E. et al. (eds.), *Microbial and Plant Opportunities to Improve Lignocellulose Utilization by Ruminants*, Elsevier, New York, pp. 137-147.

Waghorn, G.C. and I.D. Shelton (1992), "The nutritive value of Lotus for sheep", *Proceedings of the New Zealand Society of Animal Production*, Vol. 52, pp. 89-92, available at: www.nzsap.org/system/files/proceedings/1992/ab92024.pdf.

Wang, Y. et al. (2003), "Comparison of *in vitro* digestibility of parental and anthocyanin-containing *Lc*-transgenic alfalfa", *Canadian Journal of Animal Science*, Vol. 83, pp. 639 (abstract).

Weller, R.F. et al. (1996), "Animal health", in: Haggar and Padel (eds.), *Conversion to Organic Milk Production*, Technical Review No. 4, Institute of Grassland and Environmental Research, Aberystwyth, United Kingdom.

Wheeler, J.L. (1989), "Variation in HCN potential among cultivars of white clover (*Trifolium repens* L)", *Grass and Forage Science*, Vol. 44, Issue 1, March, pp. 107-109, http://dx.doi.org/10.1111/j.1365-2494.1989.tb01917.x.

Zandstra, I.I. and W.F. Grant (1968), "The biosystematics of the genus Lotus (Leguminosae) in Canada. 1. Cytotaxonomy", *Canadian Journal of Botany*, Vol. 46, pp. 557-583.

Chapter 10

Cultivated mushroom (*Agaricus bisporus*)

*This chapter, prepared by the OECD Task Force for the Safety of Novel Foods and Feeds with Sweden as the lead country, deals with the composition of cultivated mushroom (*Agaricus bisporus*). It contains elements that can be used in a comparative approach as part of a safety assessment of foods derived from new varieties. Background is given on* Agaricus bisporus *domestication, production, consumption, processing, appropriate varietal comparators and characteristics screened by breeders. Then* Agaricus bisporus *nutrients, anti-nutrients and toxicants are detailed. The final sections suggest key constituents for analysis of new varieties for food use (feed use of cultivated mushroom being rare and not requiring additional studies).*

Background

Natural history of Agaricus bisporus

Wild populations of *Agaricus bisporus* (Lange) Imbach, distinct from commonly cultivated strains, are now known from several regions of the world (Mozina et al., 1993; Kerrigan, 1995). Available isolates from these populations lay the foundation for our present knowledge about the species *A. bisporus*, and indicate that wild populations of the species are significant reservoirs of genetic resources presently unexploited by the commercial mushroom industry (Kerrigan and Ross, 1989).

An extensive investigation on the genetic diversity among 342 natural *A. bisporus* isolates from 12 locations around the world, confirmed earlier suggestions by Kerrigan et al. (1993a) that most local *A. bisporus* populations in the United States are made up of two ancestral elements, one indigenous and one being cultivar-like and introduced from Europe (Kerrigan et al., 1993a; Xu et al., 1997). The wide distribution of cultivar-like isolates has been hypothesised to represent recent escape of genotypes from mushroom cultivation, followed by introgression of these genotypes in the wild.

Domestication of Agaricus bisporus

Available records place the cradle of *Agaricus* cultivation in France. A French botanist, Tournefort, first described the primitive method used to grow the mushroom in 1707 (Joly, 1979; Chang and Miles, 2004). Later on, underground caves, where the climate conditions are well suited for growing mushrooms, became popular sites for the cultivation. *Agaricus* cultivation grew rapidly in France and spread to other European countries. By 1870, guidelines on cultural practices of *A. bisporus*, as well as vegetative inoculum ("spawn") was available in England (Robinson, 1870), and around ten years later mushroom cultivation appears to have started in the United States (Kerrigan et al., 1998).

Original cultivars were numerous but needed to be periodically replaced after a few culture cycles because the mycelium weakened or was overrun by pests or moulds (Sinden, 1981). As replacement strains were found more or less randomly in nature (spontaneously appearing on horse manure), the system did not guarantee best quality of the cultivated strains. Furthermore, the diminished use of horses in everyday life threatened the common availability of replacement strains.

The multiplication techniques developed around 100 years ago resulted in a reduction in available genetic variability. Callac (1995) estimated that only seven ancestral European cultivar linages seem to be the origin of all the cultivated strains in the world. However, this situation is changing since wild strains from all of the northern hemisphere are now used in breeding strategies.

Modern methods of cultivation

Mushroom production involves six sequential steps, which consist of: *i)* phase I composting; *ii)* phase II composting; *iii)* spawning and spawn run; *iv)* casing; *v)* pinning; *vi)* cropping. Different types of mushroom require different types of substrates. *A. bisporus* is a compost and, particularly, a leaf-litter degrader, and is able to degrade the major polymers of woody plant materials: cellulose, hemicelluloses and lignin (at least to some extent). This function is to a large extent taken care of by a set of secreted enzymes. The mushroom is cultivated on fermented compost, most commonly based on animal manure and cereal straw, which could be seen as microbial biomass and lignocellulosic residues. The intracellular and secreted proteins of *A. bisporus* actually produce a complete cellulase system, which enables it to grow on cellulose as the sole carbon. The system is induced or repressed depending on the carbon source. Easily metabolised carbon, such as glucose, represses cellulase production whereas more complex molecules induce cellulase production (Chang and Miles, 2004).

The process of mushroom composting, particularly with regard to *Agaricus*, takes place in two distinct phases. Phase I composting (compost preparation) aims at mixing and wetting the raw materials and at starting the composting process during which various microorganisms break down the straw (Chang and Miles, 2004). During this phase, nutrients may also be added (Kurtzman, 1997). There are two purposes of phase II composting (compost conditioning): *i)* to eliminate insects, pests and spores of contaminating microorganisms (pasteurisation) from the phase I composting substrate; *ii)* to bring the substrate to a uniform temperature of 50-55° C, which promotes decomposition of the substrates by thermophilic microorganisms. Through this a more selective medium favouring the growth of the mushroom is produced (Chang and Miles, 2004).

The next step in the cultivation process is spawning, i.e. inoculation of the compost with *A. bisporus* mycelium (spawn) (Kurtzman, 1997). Spawning is generally done by mixing the mycelia throughout the compost, after which the spawn is allowed to grow and produce a thread-like network of mycelium throughout the compost (Vedder, 1978; Van Griensven, 1988). To promote mushroom formation of *A. bisporus*, the compost surface is covered with a surface layer called casing, which is usually a mixture of peat and limestone (Kurtzman, 1997, Volk and Ivors, 2001). Some fungi use light as their signal to form fruiting bodies, but for *Agaricus bisporus*, microorganisms in the casing layer provides the necessary signal to initiate the transition from the vegetative to the reproductive stage in which primordia or "pins", knots of mycelium that eventually develop into mushrooms, appear (Rainey et al., 1990; Rainey, 1991).

The mushroom crop grows in repeating three- to five-day cycles called "flushes" or "breaks". These flushes are followed by a few days when no mushrooms are available to harvest. The individual flushes tend to produce progressively fewer mushrooms. In commercial practice, three to five flushes are picked before the crop is removed to make room for the next. Commonly, mushroom farmers crop their mushrooms for 30-40 days. Most strains of *A. bisporus* (except for the portabella strains) are picked before the veil breaks and the stem elongates (Volk and Ivors, 2001).

There are several excellent reviews on the cultivation of mushrooms available (Chang and Hayes, 1978; Vedder, 1978; Stamets and Chilton, 1983; Van Griensven, 1988; Chang and Miles, 2004).

Breeding of **Agaricus bisporus**

Modern breeding programmes for *Agaricus bisporus* began as recently as about 30 years ago, simply because many techniques routinely used in the breeding of vegetable and cereal crops are not available or have not been adapted to *Agaricus* research. The main factor discouraging breeding work with the mushroom has been its reproduction, which was not understood until the early 1970s.

The majority of *A. bisporus* strains, including cultivated and many wild strains, are predominantly secondarily homothallic (Raper et al., 1972), with a unifactorial mating system comprising multiple alleles (Imbernon et al., 1995). In this reproductive system, the majority of basidia produce only two basidiospores, each of which receives two of the four postmeiotic nuclei (Kerrigan et al., 1993b). Other agarics usually produce uninucleate basidiospores. Most often, the two nuclei in the basidiospores of *A. bisporus* are non-sisters with respect to meiosis II (Summerbell et al., 1989) and carry compatible mating types. The bisporic spores, therefore, give rise to fertile heterokaryotic progeny (n+n) characteristically capable of fruiting (Evans, 1959; Kerrigan et al., 1993b). Virtually all parental heterozygosity is retained in offspring. Homokaryons are rarely produced by most strains and may never be formed by others (Kerrigan et al., 1992). Nuclear fusion (2n) occurs just prior to meiosis in the basidia lining the gills of the mushroom. As meiosis in bisporic *A. bisporus* is accompanied by low recombination frequencies, it is likely that parental heterozygosity will be retained at a frequency much higher than that expected by chance (Summerbell et al., 1989; Allen et al., 1992). This, of course, has hampered breeding of *A. bisporus*.

Identification in 1993 and 2003 of new varieties of the cultivated mushroom, *A. bisporus* (Lange) Imbach var. *burnetti* Kerrigan et Callac and *A. bisporus* (Lange) Imbach var. *eurotetrasporus* Callac et Guinberteau (which resulted in renaming of the traditional variety to *A. bisporus* [Lange] Imbach var. *bisporus*), characterised by tetrasporic basidia and by respectively a heterothallic and a primary homothallic life cycle, in combination with an increased understanding of the molecular genetics of *A. bisporus*, has opened up the possibility to develop new strains of *A. bisporus* with alternative techniques (Callac et al., 1993, 2003).

Production of Agaricus bisporus

The cultivation of mushrooms world wide has increased rapidly over the past 30 years. Even though the actual production of *A. bisporus* increased from 900 000 tonnes in 1981 to just under 2 million tonnes in 1997 (a 2.2-fold increase), its share among cultivated mushrooms decreased during the same period, from 71.6% to 31.8% (Chang and Miles, 2004). However, it is still the mushroom that is produced in the largest quantities. It is cultivated in more than 100 countries all over the world. The major producers of *A. bisporus* mushrooms in 1999-2000 were the People's Republic of China followed by the United States, the Netherlands, France, Poland and Italy (Table 10.1).

Table 10.1. **World production of *Agaricus bisporus* in 1999-2000**

Rank	Country	Production (tonnes)
1	China (People's Republic of)	637 304
2	United States	391 000
3	Netherlands	263 000
4	France	180 000
5	Poland	105 000
6	Italy	102 000

Source: Adapted from Chang and Miles (2004).

Consumption of Agaricus bisporus

The white button mushroom (*A. bisporus*) is the most preferred mushroom in Western Europe and North America (Chang and Miles, 2004). It is consumed fresh, cooked or conserved. However, the available data on *Agaricus* consumption is old. The highest per capita consumption in 1990 was reported in Germany, the Netherlands and Canada (Table 10.2).

Table 10.2. **Annual consumption (kg per capita) of *Agaricus bisporus* in some high-consuming Western countries in 1990**

Country	Fresh mushrooms	Preserved mushrooms	Total
Germany	1.2	2.0	3.2
Netherlands	2.5	0.4	2.9
Canada	1.4	1.3	2.7
France	1.0	1.3	2.3
United Kingdom	2.0	0.20	2.2
United States	0.90	0.90	1.8
Italy	0.70	0.40	1.1

Source: Adapted from Moss and Mitchell (1994).

Processing of Agaricus bisporus

Fruit bodies of *A. bisporus* are not only sold fresh but also processed by the industry before being offered on the market. The technological treatments, which include canning, freezing and various drying processes, might alter the protein, carbohydrate and ash contents of the mushroom (Manzi et al., 2001).

A common preservation method for *A. bisporus* is canning. An important step in the canning process is blanching, which aims both to pre-shrink the mushrooms prior to sterilization and to inactivate the polyphenol oxidase (PPO), an enzyme responsible for browning of the mushroom (Biekman et al., 1997). Blanching may also influence the mineral content of *A. bisporus*, either directly or through ethylene diamine tetraacetic acid (EDTA), an additive in the blanching solution. These changes are probably caused by the elements in the mushroom's tissue binding with chelating agents such as EDTA. The mineral content may also be influenced by excessive amounts of ascorbic acid in the brine of canned mushrooms (Çoşkuner and Özdemir, 1997; 2000). Furthermore, during storage in brine, canned *A. bisporus* may lose some of its protein and fat contents along with some mineral elements, whereas copper, calcium and sodium increase, possibly due to the composition of the brine. Moreover, storage in brine could also lead to increased moisture and ash content (Çaglarlrmak et al., 2001).

Freeze-drying is usually considered to be the drying technique that gives the best quality dried products in terms of nutritional value as well as texture, flavour and colour (Le Loch-Bonazzi et al., 1992). Important for dried mushrooms is the rehydratability of the dried product.

Although freezing might not influence nutritional and mineral content of fruit bodies to any large extent, as long as the product is kept frozen the ice crystals formed in the frozen mushrooms may damage cellular structures and render processes possible after thawing.

Cooking of Agaricus bisporus

Manzi et al. (2001) noted that cooking procedures significantly increase nutrient concentrations by decreasing water content. This finding was confirmed by

Dikeman et al. (2005) who observed increased contents of some carbohydrates, acid hydrolysed fat (AHF), total dietary fibre and insoluble dietary fibre but not of crude protein and soluble dietary fibre in cooked mushrooms. The latter investigators also found reduced levels of chitin in *A. bisporus* after cooking (Dikeman et al., 2005). However, on a dry weight basis, a significant cooking-related loss of protein and fat could be observed in deep-frozen *Agaricus bisporus* (Manzi et al., 2001). The structural damage of the vegetative cells, occurring during the deep-freezing/thawing processes, is suggested to promote the nutrient loss. On the other hand, cooking increased the dietary fibre content. This trend could not entirely be explained by the loss of water during cooking, since it can also be observed on a dry weight basis. According to the authors, these results suggest the occurrence, during severe industrial treatment and/or cooking, of cross-linking reactions among oligosaccharides, monosaccharides and proteins, leading to indigestible products analytically measured in the fibre fraction (Manzi et al., 2001).

Appropriate comparators for testing new varieties

This chapter suggests parameters that *A. bisporus* developers should measure when developing new strains. The morphological, agronomical and chemical data obtained in the analysis of the new mushroom variety should ideally be compared to those obtained from an appropriate near isogenic non-modified variety grown under identical conditions. The evaluation of the extent of equivalence may be enhanced by additional, valid comparisons between the genetically modified mushroom and commercial varieties. These additional data may be generated by the developer and/or compiled from the literature. In the case data are generated by the developer himself, it should be noted that the majority of strains cultivated in the world today originate from only seven ancestral European cultivar lineages, and that it can be useful to have data on both white and brown (including Portabella type of strains) cultivars.

When using literature data, however, they have to be adequately assessed for their quality (e.g. in respect to type of material analysed and analytical method used). Ranges, and when appropriate mean values, should be reported and considered for each parameter investigated. These data would indicate whether the genetically modified lines fall within the natural range in phenotypic expression or critical component concentrations found in non-genetically modified counterparts. The genetically modified and the non-genetically modified varieties tested should be grown under various methods and climates of cultivation.

Critical components include key nutrients, key anti-nutrients and key toxicants for the food in question. Key nutrients are those components in a particular product which may have a substantial impact in the overall diet. These may be major constituents (carbohydrates, proteins and lipids) or minor ones (minerals and vitamins). Key toxicants are those toxicologically significant compounds known to be inherently present in the species, and whose toxic potency and level may impact on human and animal health. Similarly, the levels of known anti-nutrients and allergens should be considered.

Traditional characteristics screened by **Agaricus bisporus** *developers*

The major objective of *Agaricus* breeders remains to create strains giving a good yield of white-capped fruitbodies that are tolerant against *Verticillium*, *Trichoderma*, *Pseudomonas* and other pathogens/diseases. Secondary objects such as having strains with smooth fruit bodies of a suitable form and size are controlled at each step of the selection. Other objectives of mushroom breeding, such as improving the dependence of fruit body development on temperature, the rhythm of flushes or taste are only taken into account at the end of the selection and eventually only at the time of the official description of novel commercial varieties. To a large extent, the foundation for this hierarchy of interests is the areas where basic knowledge about *Agaricus* genetics is available. Thus, knowledge is available on reproduction modes, determinants of cap colour, tolerance against pathogens, form and structure of the cap, and stability of strains. As the understanding of *Agaricus* genetics develops, breeders' interests are likely to widen.

Nutrients in *Agaricus bisporus*

Proximate analysis

The chemical composition of cultivated *A. bisporus* varies between different reports. Observed differences may, to some extent, be explained by the analytical methods being used (Weaver et al., 1977; Cheung, 1997), but are mainly due to several other factors not being controlled. These factors include the genetic constitution – strain (Weaver et al., 1977, Bakowski et al., 1986b), composition of the compost (Maggioni et al., 1968; Bakowski et al., 1986a; Kosson and Bakowski, 1984), flush of the mushroom culture (Bakowski et al., 1986a), developmental stage of fruit body at harvest (Kosson and Bakowski, 1984; Dikeman et al., 2005) and what part of the mushroom was analysed (Kosson and Bakowski, 1984).

Representative data on proximate analysis of fresh *A. bisporus* are presented in Table 10.3. Investigators usually have used standard methods of analysis such as those published by the Association of Official Analytical Chemists (AOAC) to determine moisture, crude protein, fat, fibre, carbohydrate and ash content of the mushroom.

Table 10.3. **Proximate composition of cultivated *Agaricus bisporus***

Category \ Reference	USDA[1]*	Mattila et al.[2]*	Manzi et al.[3]*	Dikeman et al.[4]	Kurasawa et al.[5]	Cheung[6]	CSTJ[7]*	Range
Dry matter (% f.w.)	7.6-8.8	7.7-7.8	7.2	5.5-7.0	7.8-8.4	7.6	6.1	5.5-8.8
Protein[8] (g/100 g d.w.)	28.4-40.8	26.5-27.1	22.7	26.3-31.4	30.4-31.0	26.8	33.3	22.7-40.8
Carbohydrate (g/100 g d.w.)	43.3-57.6	58.4-59.5	61.3		55.1-55.2	61.0	48.7	43.3-61.3
Total dietary fibre (g/100 g d.w.)	7.8-17.0	19.5-20.5	27.5	22.9-30.4		18.2	32.8	7.8-32.8
Fat (g/100 g d.w.)	1.3-4.5 (crude)	4.0-4.3 (crude)	4.6	4.7-5.8 (AHF)**	2.5-3.8 (crude)	1.9	4.9	1.3-5.8
Ash (g/100 g d.w.)	11.2-12.7	10.0-10.1	11.4		10-12	10.3	13.1	10.0-13.1

Notes: * Some values recalculated. Data originally given on a fresh weight basis have been recalculated to dry weight basis. ** AHF: Acid hydrolysed fat.

1. Different strains and growth stages *A. bisporus* (white mushroom, Crimini and Portabella). 2. Different *A. bisporus* strains (brown and white). 3. From local market. 4. Different *A. bisporus* strains (white, Crimini, Portabella) and maturity (immature and mature). 5. Different *A. bisporus* strains (white and brown) from the market. 6. From local market. 7. Average value of *A. bisporus* in Japan. 8. The protein content have been calculated using the conversion factor 4.38 either by the authors or through recalculations, except for data form Mattila et al. (2002b) where the protein content were evaluated by summing the amino acid residues.

Sources: USDA (2005); Mattila et al. (2002b); Manzi et al. (2001); Dikeman et al. (2005); Kurasawa et al. (1982); Cheung (1997); CSTJ (2005).

As shown in Table 10.3, the dry matter is usually between 5.5% and 8.8%, but values as high as 13.7% have been reported (Weaver et al., 1977). The dry matter content is influenced by irrigation (Barden et al., 1990, Beelman et al., 2003), type of compost and strain (Weaver et al., 1977; Bakowski et al., 1986a; Laborde and Delpech, 1991), but it may also vary within and between flushes (Laborde and Delpech, 1991; Bakowski et al., 1986a). Brown strains of *A. bisporus* are usually higher in solids than white strains; Portabella mushrooms harvested fully mature with flat open caps being among the highest (Beelman et al., 2003).

A common way of establishing the crude protein content of a sample has been to calculate it from the nitrogen content using a conversion factor. A factor of 6.25 has frequently been used, based on the assumptions that most proteins contain about 16% nitrogen, that they are digestible by approximately 100% and that only minor amounts of non-protein nitrogen are present in the analysed sample. However, a lower conversion factor may be more appropriate in the case of mushrooms due to the fact that these organisms contain significant amounts of non-protein nitrogen, for example in their chitinous cell walls. Therefore, a conversion factor of 4.38 is advocated for mushrooms, based on 70% protein digestibility (0.7 x 6.25). As some of the references presented in Table 10.3 originally used a conversion factor of N x 6.25, these data have been recalculated using the conversion factor for mushrooms.

The crude fat content is usually determined after solvent extraction, and includes free fatty acids; mono-, di- and triglycerides; sterols; sterol esters; phospholipids; and glycolipids. The fat content is low in *A. bisporus*, usually in the region 1.3-5.8 mg/100 g dry weight (Table 10.3).

Carbohydrates constitute a heterogeneous chemical group and include polyhydroxy aldehydes, ketones, alcohols and acids, as well as their derivatives and polymers. Analytically "carbohydrate" levels are usually calculated by difference. First moisture, protein, fat and ash are determined, and the remainder is named "carbohydrates".

The carbohydrates can be given as total carbohydrate, which includes fibre, or as nitrogen-free carbohydrate (without fibre). Presumably most of the non-protein nitrogen in a mushroom is in the form of chitin contained in the fibre fraction, although small quantities of other nitrogen compounds may also be present. The carbohydrate level reported in *A. bisporus* is between 43.3 and 61.3 g per 100 g dry weight (Table 10.3). Although dietary fibre is included in carbohydrates, some investigators have also reported values on the amount of this chemical group.

Ash is what remains after the organic part of the mushroom has been oxidised through combustion. It is a measure of the total amount of minerals and salts in the mushroom.

Protein

The protein content of *A. bisporus* is influenced by strain (Weaver et al., 1997, Kosson and Bakowski, 1984; Bakowski et al., 1986b), compost composition (Crisan and Sands, 1978; Kosson and Bakowski, 1984, Bakowski et al., 1986a), flush number (Crisan and Sands, 1978; Bakowski et al., 1986a), and time of harvest/developmental stage (Crisan and Sands, 1978; Kosson and Bakowski, 1984; Bakowski and Kosson, 1985; Burton, 1988).

Protein levels are higher in the cup than in the stipe (Kosson and Bakowski, 1984). The protein level in *A. bisporus* is usually in the range 22.7-40.8 g/100 g dry weight

(Table 10.3), the mean of 60 reported values in the literature being 26.2 g/100 g dry weight (median 26.6, minimum 13.6 and maximum 40.8 g/100 g dry weight).[1] This level ranks below most animal meats but above most vegetables, fruits and other foods (Chang and Miles, 2004).

The measurement of crude protein is an indirect but easily performed approximation of total amino acids from the nitrogen level, although vulnerable to the presence of varying levels of non-protein amino acids and other sources of nitrogen in the sample. Therefore, summing up the various amino acids quantified after acid hydrolysis will give more accurate data.

Amino acid composition

The amino acid composition of *A. bisporus* is given in Table 10.4. Essential amino acids are indicated in italics and make up 32-43% of the total amino acid contents in *A. bisporus* (Weaver et al., 1977; Mattila et al., 2002b; USDA, 2005). The relative amount of free amino acids is quite high in *A. bisporus*. Oka et al. (1981) reported around 50%, and Maggioni et al. (1968) between 39% and 46% free amino acids.

The most abundant amino acid in *A. bisporus* is the non-essential amino acid glutamic acid. Not only does glutamic acid occur both as a free amino acid and integrated in proteins, but it is also often covalently linked to other small molecules such as N-(γ-L-glutamyl)-ethanolamine, N-(γ-L-glutamyl)-4-hydroxyaniline and β-N-(γ-L-glutamyl)-4-hydroxymethylphenylhydrazine (agaritine).

The most common essential amino acid is lysine, and the most rare amino acids are those containing sulphur, cysteine and methionine. Therefore, the sulphur amino acids are the limiting amino acids of *Agaricus* proteins.

The amino acid composition of *A. bisporus* fruit bodies is dependent on the strain (genetic factors), type of compost and its composition, the nutrient supplementation (nitrogen or fatty acids), the developmental stage/size of the fruit body and the flush number (Maggioni et al., 1968; Weaver et al., 1977; Kosson and Bakowski, 1984; Bakowski et al., 1985, 1986a, 1986b).

Table 10.4. **Amino acid composition of various strains of cultivated *Agaricus bisporus* expressed as g/100 g protein**

Reference / Amino acid	USDA*[1]			Mattila et al.*[2]		Weaver et al.[3]									Range
	White	Crimini	Portabella	White	Brown	D26	324	310	340	318	348	322	321	341	
Alanine	6.4	7.5	4.2	7.6	7.6	9.7	8.6	11	9.1	11	9.5	11	8.8	9.1	4.2-11
Arginine	2.5	4.9	2.7	5.6	5.2	5.3	6	4.1	5.8	4.3	5.5	5	6.5	6	2.5-6.5
Aspartic acid†	6.3	9.1	6.2	13	13	11	11	11	11	10	11	11	10	10	6.2-13
Cysteine	0.39	0.24	0.48	1.1	1.1	T	T	T	T	T	T	T	T	T	Trace-1.1
Glutamic acid†	11	17	11	21	23	14	19	16	18	15	17	18	17	19	11-23
Glycine	3.0	4.4	2.5	5.1	4.7	4.9	4.9	5.7	5.3	5.0	5.0	4.6	5.0	5.0	2.5-5.7
Proline	2.5	7.0	3.0	5.0	5.0	5.6	5.4	4.8	5.6	10	6.1	5.4	5.7	5.4	2.5-10
Serine	3.0	4.5	2.7	5.2	5.3	5.1	4.8	5.8	5.3	5.4	5.3	5.1	5.6	4.6	2.7-5.8
Tyrosine	1.4	2.2	1.7	14	14	2.6	2.5	2.9	2.3	2.5	2.3	2.1	2.2	2.1	1.4-14
Histidine	1.8	2.7	1.7	2.8	2.6	2.2	2.5	1.6	2.0	1.7	2.2	2.2	2.3	2.3	1.6-2.8
Isoleucine	2.5	4.0	2.0	4.4	4.1	4.7	4.3	4.6	4.3	4.3	4.1	4.0	3.9	4.0	2.0-4.7
Leucine	3.9	6.1	3.2	7.3	6.9	7.2	7.0	7.9	7.5	7.7	7.1	6.7	7.3	6.7	3.2-7.9
Lysine	3.5	10	2.5	6.8	6.1	12	11	8.1	8.4	7.7	9.8	11	11	11	2.5-12
Methionine	1.0	1.9	0.72	1.6	1.4	T	T	T	T	T	T	T	T	T	Trace-1.9
Threonine	3.5	4.5	2.7	5.3	4.9	5.0	4.7	5.6	5.2	5.1	4.7	4.8	4.7	4.3	2.7-5.6
Tryptophan	1.1	2.2	1.2	n.d.	n.d.	n.a.	n.a.	n.a.	n.a.	n.a.	n.a.	n.a.	n.a.	n.a.	1.1-2.2
Phenylalanine	2.8	3.9	2.2	5.1	4.7	4.6	4.4	4.7	4.5	4.7	4.3	4.2	4.3	4.1	2.2-5.1
Valine	7.5	4.6	6.2	5.8	5.8	5.5	4.8	5.9	5.1	5.8	5.3	5.3	5.1	5.2	4.6-7.5
% essential amino acids of total amino acids	43	41	39	34	32	41	38	38	37	37	38	38	39	38	32-43

Notes: T: trace. n.d.: not determined due to degradation in acid hydrolysis. n.a.: not analysed.

* Data originally given on a fresh matter basis have been recalculated to g/100 g protein.

† Asparagine and glutamine cannot be differentiated from aspartic acid and glutamic acid respectively.

1. Protein content calculated using the conversion factor 4.38; white: 40.8 g/100 g d.w.; crimini: 32.5 g/100 g d.w.; Portabella: 28.4 g/100 g d.w. 2. Protein content evaluated by summing the amino acid residues; white: 27.1 g/100 g d.w. and brown: 26.5 g/100 g d.w. 3. Protein content calculated by the authors using the conversion factor 6.25 (thus, the protein contents are not true values but are meant to illustrate differences among strain within a species); D26: 30.2 g/100 g d.w.; 324: 25.6 g/100 g d.w.; 310:19.4 g/100 g d.w.; 340: 29.9 g/100 g d.w.; 318: 38.8 g/100 g d.w.; 348: 24.8 g/100 g d.w.; 322: 37.1 g/100 g d.w.; 321: 33.2 g/100 g d.w.; 341: 26.0 g/100 g d.w.

Sources: USDA (2005); Mattila et al. (2002b); Weaver et al. (1977).

Fat

Crude fat in *Agaricus bisporus* includes all classes of lipid compounds but the total levels are comparatively low, usually in the range 1.3-5.8% of the dry matter (Table 10.3).

The constituents of lipids in the cultivated mushroom *A. bisporus* have been investigated quite extensively (Abdullah et al., 1994). Most studies have been devoted to the fatty acids, which make up around for 0.15% of the fresh *A. bisporus* cap (Cruz et al., 1997). The fatty acid profile reveals a surplus of unsaturated over saturated fatty acids, with ratios usually in the region 3:1-4:1 (USDA, 2006; Mau et al., 1991; Cruz et al., 1997; Aktümsek et al., 1998; CSTJ, 2005; Maggioni et al., 1968), although higher ratios have also been reported (Bonzom et al., 1999). By far the most predominant fatty acid in *A. bisporus*, regardless of the strain and developmental stage of the mushroom, is the unsaturated fatty acid linoleic acid (18:2) (Table 10.5). The data of Cruz et al. (1997) was obtained from mushrooms cultivated on horse manure. For most of the other data in Table 10.5, however, the conditions of cultivation are unknown. Controlling for the developmental stage of fruit bodies when comparing fatty acid profile is important, as opening of the cap has been reported to influence the fatty acid composition depending on the strain. On cap opening, linoleic acid decrease both in the cap and the stem portion of the mushroom (Hira et al., 1990; Cruz et al., 1997). Thus, the fraction of unsaturated fatty acids decreases with advancement of growth stage.

A. bisporus contains neutral lipids (mainly as glycerides, free fatty acids and sterols) and phospholipids. Among the phospholipids, phosphatidylcholine and phosphatidylethanolamine predominate (Holtz and Schisler, 1971; Bonzom et al., 1999). Other identified phospholipids are phosphatidylserine, phosphatidylinositol, phosphatidic acid and cardiolipin. In *A. bisporus* the neutral and polar lipid fraction has been reported to occur in a ratio of 1:2 or 1:1 depending on strain (Holtz and Schisler, 1971).

A. bisporus also contains relatively large amounts of sterols. Unlike plant sterols, which usually contain sterols with one double bond, mushroom sterols are characterised by two double bonds (Parks and Weete, 1991). Sterols are natural components of cell membranes. Mattila et al. (2002a) reported a total amount of 677-789 mg sterols/ 100 g dry weight. In descending order of occurrence the sterols found in *A. bisporus* are: ergosterol (602-654 mg/100 g dry weight; 83-89% total sterols), ergosta-5,7-dienol (47-94 mg/100 g dry weight; 7-12% total sterols), fungisterol (14-26 mg/100 g dry weight; 2-3% total sterols) and ergosta-7,22-dienol (15 mg/100 g dry weight; 2% total sterols) (Mattila et al., 2002a). Several other investigators have also identified ergosterol as the main sterol in *A. bisporus* (Koyama et al., 1984; Huang et al., 1985; Young, 1995).

As vitamin D_2, also called ergocalciferol, is derived by photoirradiation-induced conversion from its precursor ergosterol, it is dependent on the action of sunlight or artificial ultraviolet light (see section *Vitamins* below) (Mattila et al., 2002a). Because of different conditions during *A. bisporus* cultivation, quite variable levels of sterols can be expected in the cultivated mushroom.

Table 10.5. **Fatty acid composition (% of total amount) of fatty acids in cultivated *Agaricus bisporus***

Fatty acid \ Reference	USDA[1]	Mau et al.[2]	Cruz et al.[3]	Koyama et al.[4]	Aktümsek et al.[5]	CSTJ[6]	Maggioni et al.[7]
8:0 Caprylic acid					1.4%		
10:0 Capric acid	0.0-0.9%				0.8%		Trace
12:0 Lauric acid	0.0-1.9%				0.1%		Trace-0.2%
13:0 Tridecanoic acid					2.1%		
14:0 Myristic acid	0.0-1.7%	0.3-0.4%	Trace-1.4%	0.4%	1.0%	0.6%	0.4-0.7%
14:1 Myristoleic acid				(0.5*)	1.2%		
15:0 Pentadecanoic acid					0.2%	1.2%	0.6-1.2%
16:0 Palmitic acid	12.1-19.2%	12.4-12.6%	11.4-16.9%	9.4%	12.6%	12.5%	13.2-18.0%
16:1 Palmitoleic acid				0.7%	3.1%	0.8%	Trace-0.3%
17:0 Eptadecanoic acid		0.5%			0. 7%	0.8%	0.4-0. 9%
18:0 Stearic acid	3.4-4.9%	3.6%	5.1-9.0%	3.0%	2.3%	4.5%	2.9-3.3%
18:1 Oleic acid	0.0-3.4%	1.8%	3.2-7.6%	0.9%	4.9%	2.3%	1.4-2.6%
18:2 Linoleic acid	69.0-76.4%	78.1-78.6%	64.4-78.5%	74.9%	68. 8%	73.4%	70.5-78.3%
18:3 Linolenic acid	0.0-0.9%			n.d[8]	0.7%	0.1%	0.5-0.7%
20:0 Arachidic acid		2.8-3.0%	1.5-2-1%		0.2%	1.9%	
22:0 Behenic acid						1.3%	
24:0 Lignoceric acid						0.7%	
Not specified fatty acids	0.0-8.6%	0	0	10.2%	0		0.6-4.4%

Notes: * Preliminary identification. 1. Different *A. bisporus* strains and growth stages (white, Crimini and Portabella). Trace amounts (< 0.0005 g/100 g) are rounded to 0.0 in the table but may be included in the amount of total fatty acids. 2. Different methods of cultivation, i.e. with or without compost fragmentation at casing. 3. Different strains (AMYCEL 2100 and LION C9) and different developmental stages (button, medium and flat). 4. From local market. 5. From local market. 6. Average value of *A. bisporus* in Japan. 7. From flush 1 and 4, grown on horse manure with N-supplemented compost [$(NH_4)SO_4$ or urea + $(NH_4)SO_4$]. 8. n.d.: not determined

Sources: USDA (2006); Mau et al. (1991); Cruz et al. (1997); Koyama et al. (1984); Aktümsek et al. (1998); CSTJ (2005); Maggioni et al. (1968).

Carbohydrates

Mushrooms are known to contain fairly large amounts of carbohydrates. Including the fibre fraction, *A. bisporus* contains 43.3-61.3 g total carbohydrates per 100 g dry weight (Table 10.3); the mean carbohydrate content of 10 reported values in the literature being 55.4 g/100 g dry weight (median 56.4, minimum 43.3 and maximum 61.3 g/100 g dry weight).[1] The carbohydrates include polysaccharides (such as glucans, glycogen and chitin), monosaccharides (such as ribose, fucose, glucose and mannose), disaccharides (such as trehalose and sucrose), sugar alcohols (such as mannitol and inositol) and sugar acids (such as galacturonic and glucoronic acids) (Crisan and Sands, 1978; Beelman et al., 2003). The fact that starch has been identified in *A. bisporus* (Dikeman et al., 2005) needs confirmation.

As carbohydrates are calculated by difference, data based on a conversion factor of 6.25 for proteins have been recalculated using the mushroom conversion factor 4.38 (true for result from Kurasawa et al., 1982 and CSTJ, 2005). The carbohydrate content from Manzi et al. (2001) has been recalculated subtracting protein, fat and ash from the dry matter.

Dietary fibre

Most of the carbohydrates are structural polysaccharides of the cell walls and are indigestible for humans. Thus, they may be considered as dietary fibre (Beelman et al., 2003). The fibre content as reported in Table 10.3 is measured chemically and varies between 7.8 and 32.8 g/100 g dry weight, the mean of 17 reported values in the literature being 24.4 g/100 g dry weight (median 25.4, minimum 7.8 and maximum 40.1 g/100 g dry weight) (Cheung, 1997; Manzi et al., 2001; Mattila et al., 2002b; Beelman et al., 2003; USDA, 2005; Dikeman, 2005; CSTJ, 2005).[2] As correction for non-protein nitrogen in chitin have been made only in some of the original studies, the calculated average should be taken only as an approximation.

Some compounds claimed to have advantageous and functional properties are present in the dietary fibre fraction. One of these is chitin, a structural polymer of the fungal cell wall occurring in the insoluble fibre fraction. In many strains the chitin fraction will increase as the mushrooms grow and mature (Beelman et al., 2003; Dikeman et al., 2005). Chitin is a nitrogen-containing polysaccharide that consists of monomers of N-acetyl-glucosamine. Around 30% of the total dietary fibre occurs as chitin, and may be detected in the form of glucosamine (Manzi et al., 2001). As chitin contains a significant amount of non-protein nitrogen, this nitrogen will contribute to incorrect crude protein content if determined using the traditional conversion factor for proteins (N x 6.25) after Kjeldahl analysis. A conversion factor of (N x 4.38) is held more appropriate for mushrooms (Crisan and Sands, 1978). Chitin levels in *A. bisporus* vary between 1.8-8.3 g/100 g dry weight (Manzi et al., 2001, Dikeman et al., 2005). β-Glucans, a polymer of glucose, are only found at low levels in the cultivated mushroom (Manzi et al., 2001, Dikeman et al., 2005).

Sugars and sugar alcohols

Among sugars and sugar alcohols in *A. bisporus*, mannitol dominates (Beecher et al., 2001, Tseng and Mau, 1999). In *A. bisporus* it is the main form of storage carbon and may contribute with up to 50% of the fruit body dry weight (Rast, 1965). The mannitol level in the fruit bodies increase during growth and is highest in full grown mushrooms with flat caps (Rast, 1965, Wannet et al., 2000). In full grown mushrooms, the highest amounts occur in the stipe (42.3% of the dry weight) and the cap (33.6% of the dry weight), while lower levels are found in the lamella (8.6% of the dry weight) (Rast, 1965). As the level depends on the growth stage and which part of the mushroom that has been analysed, it is not surprising that different levels (Table 10.6) have been reported (Rast, 1965; Hammond and Nichols, 1976; Ajlouni et al., 1993). In plants, it has been suggested that mannitol accumulates in response to environmental stress, such as to salt (Stoop and Pharr, 1994, Stoop et al., 1996). A similar function may exist in *A. bisporus*. In non-stressed fruit bodies, the concentration of mannitol increases rapidly early during development, then remains relatively constant during fruit body maturation, whereas mushrooms grown under salt stress accumulate larger amounts of mannitol than non-stressed mushrooms (Stoop and Mooibroek, 1998). This observation confirms the notion that mannitol may act as an osmolyte in growing fruit bodies (Jennings, 1984).

Table 10.6. **Mannitol content (% dry weight) in the stipe, cap and gills of cultivated *Agaricus bisporus***

	Stipe	Cap/pileus	Gills	Whole mushroom
Ajlouni *et al.* (1993)	19-28	30	10	26
Hammond and Nichols (1976)	33-52	34-49	12-18	
Rast (1965)	24-42	21-34	9-11	

A. bisporus also contain appreciable amounts of the disaccharide trehalose, usually at fairly constant levels around 1-3% of the dry weight (Hammond and Nichols, 1976, 1979; Ajlouni et al., 1993). However, slightly lower levels have also been reported (Rast, 1965). Occasionally, trehalose levels as high as 18% of the dry weight have been detected in fruit bodies developing between flushes (Hammond and Nichols, 1979).

Very low levels of the organic acids such as fumaric acid, succinic acid and citric acid, and somewhat higher levels of malic acid occur in fresh and stored *A. bisporus* (Le Roux and Danglot, 1972).

Nucleic acids

Only a few studies have measured the nucleic acid content of *A. bisporus*. Nucleic acids occur as RNA and DNA, as well as their precursors. On a dry weight basis mushrooms have been reported to contain 0.17% DNA and 2.49% RNA (Li and Chang, 1982), and 0.11% 5′-nucleotides (Tseng and Mau, 1999). Some of the latter compounds are important for the flavour.

Mineral and trace elements

Mushrooms probably contain every mineral present in their growth substrate (Crisan and Sands, 1978). Therefore, differences in mineral and trace element concentrations in cultivated *A. bisporus* may to a large extent depend on the method of cultivation and the type of compost being used (Vetter, 1989; Ünal et al., 1996; Tüzen et al., 1998; Spaulding and Beelman, 2003). However, the level of minerals and trace elements in mushrooms may also be dependent on strain (Spaulding and Beelman, 2003).

When comparing mineral and trace element levels it is important to compare the same part of the mushroom. According to Vetter (1994) sodium is the only mineral element occurring at higher levels in the stipe than in the cap. Other minerals and trace elements are generally found in the stipe at lower or equal levels to those in the cap (Vetter, 1989; Zródlowski 1995; van Elteren et al., 1998). However, Muñoz and colleagues found higher levels of bismuth, chromium, copper, iron and lead in stalks than in caps (Muñoz et al., 2005).

The uptake of mercury and selenium is much lower in cultivated *A. bisporus* than in wild relatives. Several hypotheses have been suggested to explain this observation, but none of them have hitherto been supported by solid scientific data.

Heavy metal contents in mushrooms grown on non-contaminated composts are usually low. It should be noted that in some studies it has been observed that washing and peeling of mushrooms may reduce the heavy metal content in the consumable parts (Źródlowski, 1995). To what extent this reduction is dependent on the air and soil quality is unknown.

The most common minerals in mushrooms are in general potassium, phosphorous, sodium, calcium and magnesium (Crisan and Sands, 1978, Chang and Miles, 2004). Observed levels are reported on a dry weight basis in Table 10.7. Additional data on the contents of boron, barium, cobalt, chromium, gallium, lithium, molybdenum, strontium, titanium and vanadium are available elsewhere (Vetter, 1989).

Table 10.7. **Mineral contents (on a dry weight basis) in cultivated *Agaricus bisporus***

Mineral element	Reference	USDA*[1]	SLV*[2]	Mattila et al.[3]	Haldimann et al.[4]	Ünal et al.*[5]	Dar[6]	CSTJ*[7]	Various references[8]	Range
Sodium (Na)	g/kg dw	0.66-0.78	0.64	0.42-0.44		0.35-1.0		0.98	0.85-0.96	0.35-1.0
Potassium (K)	g/kg dw	42-58	50	46-47		18-23	21-27	57		18-58
Magnesium (Mg)	g/kg dw	1.2-1.3	1.5	1.3-1.4			1.0-1.4	1.6		1.0-1.6
Calcium (Ca)	g/kg dw	0.40-2.3	0.32	0.13-0.25		0.32-0.49		0.49		0.13-2.3
Phosphorus (P)	g/kg dw	11-16	14	13		8.0-17	12-14	16		8.0-17
Iron (Fe)	mg/kg dw	52-68	42	28-48		90-138	80-146	49		28-146
Zinc (Zn)	mg/kg dw	68-143	64	47-66		467-642	54-77	66		47-642
Selenium (Se)	mg/kg dw	1.2-3.4	0.38	1.4-3.2	1.3-5.7				0.45-1.2	0.38-5.7
Copper (Cu)	mg/kg dw	42-65		29-35		85-110	77-90	52		29-110
Manganese (Mn)	mg/kg dw	6.2-18		5.1-5.5			24-26	6.6		5.1-26
Lead (Pb)	µg/kg dw			35-180	50-490					35-490
Cadmium (Cd)	µg/kg dw			36-96	40-280					36-280
Arsenic (As)	mg/kg dw				0.19-1.5					0.19-1.5
Mercury (Hg)	µg/kg dw				<80-130					<80-130
Chromium (Cr)	µg/kg fw					Trace				Trace
Silver (Ag)	mg/kg dw								0.15-0.62	0.15-0.62

Notes: * Original data recalculated to dry weight basis using the dry matter content stated in the article. 1. Different *A. bisporus* strains and growth stages (white, Crimini and Portabella). 2. From local stores. 3. Different *A. bisporus* strains (white and brown). 4. Different *A. bisporus* strains (white and brown). 5. Different composts, i.e. horse manure or broiler chicken manure or a mix of 70% horse manure + 30% wheat straw. 6. Different developmental stage, i.e. pinhead, button, cup and open, lacking information on *A. bisporus* origin. 7. Average value of *A. bisporus* in Japan. 8. Vetter (2003); Falandysz et al. (1994); Piepponen et al. (1983).

Sources: USDA (2005); SLV (2004); Mattila et al. (2001); Haldimann et al. (1995); Ünal et al. (1996); Dar (1996); CSTJ (2005); Vetter (2003); Falandysz et al. (1994); Piepponen et al. (1983).

Vitamins

As shown in Table 10.8, cultivated *A. bisporus* seems to be a good source of the B-complex vitamins, and of niacin and folate (Beelman et al., 2003; Mattila et al., 2001). For these vitamins, one portion of mushrooms may contribute with close to or more than 10% of the recommended daily intake according to the Nordic Nutrition Recommendations (NNR, 2004).

On the other hand, *A. bisporus* contains very low levels of vitamin A, vitamin D, vitamin E and thiamine (Anderson and Fellers, 1942). The low level of thiamine has been suggested to result from anti-nutritive thiaminases in the mushroom degrading thiamine (Wittliff and Airth, 1970a, 1970b; Wakita, 1976). The low level of D vitamins in *A. bisporus* cultivated indoors contrasts with the comparatively high levels in wild mushrooms. Levels of vitamin D2 are low in spite of relatively high concentrations of the precursor ergosterol. Mattila et al. (1994) found 0.21 µg vitamin D2 (ergocalciferol)/100 g fresh weight, which is one to two orders of magnitude lower than in mushrooms such as *Cantharellus tubaeformis*. These findings are also supported by Teichmann (2005), who found the levels of vitamin D2 in *A. bisporus*/white, *A. bisporus*/brown and *A. bisporus*/Portabella to range between 5.5-6.9, 3.1-3.8 and 4.4-10.1 µg vitamin D2/100 g dry weight, respectively. Whereas the levels of vitamin D2 in *Chantarellus tubaeformis*, *Chantarellus cibarius* and *Boletus edulis* ranged between 209.7-225.7, 138.5-164.0 and 65.2-81.8 µg vitamin D2/100 g dry weight, respectively (Teichmann, 2005). The reason for the low vitamin D2 levels in cultivated mushrooms seems to be that conversion of ergosterol to ergocalciferol (vitamin D2) requires sunlight (or artificial ultraviolet light). Studies have shown that the concentration of vitamin D2 in *A. bisporus* might be increased by as much as 467% by post-harvest UV-irradiation (Mau et al., 1998). No vitamin D3 (cholecalciferol) has been detected in *A. bisporus*.

Table 10.8. **Vitamin content (expressed per kg dry weight) in cultivated *Agaricus bisporus***

Vitamin \ Reference	USDA*[1]	CFC*[2]	CSTJ*[3]	Mattila et al.[4]	Ünal et al.*[5]	Sapers et al.*[6]	Udipi and Punekar*[7]	Range
Retinol (vitamin A) (µg)			0					0
Thiamin (vitamin B_1) (mg)	8.8-12	11	9.8	6	8.1-12		11	6-12
Riboflavin (vitamin B_2) (mg)	53-64	46	48	42-51	49-61	53	26	26-64
Niacin (mg)	476-511	526	492	430-530	580-911	347	507	347-911
Vitamin B6 (mg)	11-14		18					11-18
Pantothenic acid (mg)	170-198		252		223-304			170-304
Folate (mg)	1.8-2.5		4.6	4.5-5.9	6.4-10‡			1.8-10
Vitamin B12 (µg)	5.3-13		0	6.0-8.0				0-13
Vitamin C (mg) (ascorbic acid)	277	263	164	170-210†	635-952		171	164-952
Vitamin D (µg)			98					98
α-Tocopherol (vitamin E) (mg)	1.3–2.3	36	0					0-36
Vitamin K (µg)	0.0		0					0
Carotenoids (mg)		1.3	0					0-1.3

Notes: * Original data recalculated to dry weight basis using the dry matter content stated in the article. † As dehydroascorbic acid. ‡ Folic acid. 1. Different *A. bisporus* strains and growth stages (white, Crimini and Portabella). No data on vitamin C in Crimini and Portabella. 2. Button mushroom. 3. Average value of *A. bisporus* in Japan. 4. Different *A. bisporus* strains (white and brown). 5. Different composts, i.e. horse manure or broiler chicken manure or a mix of 70% horse manure + 30% wheat straw. 6. From commercial packing plant. 7. From the Agricultural College, Pune, India or from local market.

Sources: USDA (2006); CFC (2002); CSTJ (2005); Mattila et al. (2001); Ünal et al. (1996); Sapers et al. (1999); Udipi and Punekar (1980).

Agaricus bisporus *as feed*

In order to explore whether by-products from cultivation and production of *A. bisporus* can be used as animal feed, the lower part of stipes were dried by various methods (freeze-dried, 80°C or 120°C) and ground to powder. The potential usefulness of these products as feed was assessed by comparing their chemical composition with that of freeze-dried samples of commonly consumed mushroom parts (caps and upper parts of the stipe) (Maeda et al., 1993). Drying mushrooms at 120°C seemed to be the most effective treatment to shorten drying time and increase palatability of mushrooms for dairy cattle. Addition of the stipe powder to silage resulted in retained fermentation quality up to an inclusion rate of 5% mushroom material. The pepsin-pancreatin digestibility of the lower part of stipes was around 50% as compared to 80%

for caps and the upper part of stipes. The percentage of soluble nitrogen was around 30% of total nitrogen for stipes as compared to 55% for caps. The crude fat and crude protein content of stipes were half of that in the consumed parts of mushrooms (Table 10.9). Of the minerals investigated, the phosphor content was lower in the by-product than in the consumed part, and the calcium content 15 times higher in the stipe than in the cap.

Table 10.9. **Chemical composition of the commonly consumed part of *A. bisporus* (the cap) as opposed to the by-product (the lower part of the stipe)**

Composition (% of dry matter)	Harvested part (cap)	By-product (stipe)
Crude protein	45.3	24.6-25.7*
Crude fat	2.7	1.0-1.1*
Neutral detergent fiber	41.0	44.5-45.2*
Acid detergent fiber	12.8	20.4-21.3*
Hemicellulose	28.2	23.9-24.2*
Cellulose	11.8	14.1-14.3*
Lignin	0.9	2.4-3.3*
Silica	0.1	3.7-3.9*
Neutral detergent insoluble nitrogen	38.5	64.8-68.2*
Acid detergent insoluble nitrogen	12.7	22.5-30.8*
Crude ash	11.9	18.3-18.5*
Potassium (K)	4.61	3.31-3.33*
Calcium (Ca)	0.02	0.32-0.33*
Magnesium (Mg)	0.16	0.18-0.19
Phosphorus (P)	1.40	0.72-0.76*
Sodium (Na)	0.12	0.14

Note: *Dried at 80°C or 120°C or freeze-dried.

Anti-nutrients in *Agaricus bisporus*

Lectins

Lectins are carbohydrate-binding proteins that are neither enzymes nor antibodies, but which may contain a second type of binding site specific for a non-carbohydrate ligand (Barondes, 1988). Therefore, many lectins have agglutinating activity. Four different isoelectric forms of *A. bisporus* lectins (ABL) have been described (Sueyoshi et al., 1985), possibly differing in glycosylation. They are all tetrameric, composed of four identical subunits (16 kDa), have a total molecular weight in the region around 60 kDa, and have similar specificities for cell-surface carbohydrate receptors. They contain around 4% carbohydrate in the form of glucose, mannose, galactose and glucosamine (Presant and Kornfeld, 1972; Sueyoshi et al., 1985). Crenshaw et al. (1995) isolated and characterised a cDNA clone encoding a lectin gene from *A. bisporus*. Southern blot analysis indicated that at least two lectine genes were present. Deduction of the complete amino acid sequence of ABL has lead to the identification of three potential *O*-glycosylation sites at Ser^5, Thr^{12} and Ser^{85}. The ABL amino acid sequence has been shown to resemble the sequences of saline-soluble fungal lectins in a family of proteins with pesticidal properties (Trigueros et al., 2003). Interestingly, ABL was recently observed to bind to the isolated glucogalactomannan from the cell walls of *Verticillium fungicola*, an *A. bisporus* pathogen causing the "dry bubble" disease (Bernardo et al., 2004).

ABL is not present in the vegetative mycelium of the mushroom. It appears during maturation of the fruit body, indicating that the synthesis of this lectin is developmentally regulated (Bernardo et al., 2004). The ABL mainly binds Galβ1-3GalNAc (Thomsen-Friedenreich antigen, TDF) and holds a particular binding nature different from that of other T-disaccharide specific lectins (Irazoqui et al., 1999). One of the isoforms of the ABL lectin has recently been crystallized and its three-dimensional structure determined by X-ray diffraction (Carrizo et al., 2004, 2005). Binding studies with mono- and disaccharides showed that the lectin has two distinct binding sites per monomer, apparently active independent from the binding sites on the other tetrameres. The specificity of the binding is remarkable as it is able to distinguish two monosaccharides that differ only in the configuration of a single epimeric hydroxyl (Carrizo et al., 2005). As TDF is over-expressed in epithelial cancer cells, the ABL binding has been studied in detail with the intention to develop anti-TDF antibodies with fine carbohydrate-binding for treatment of tumours (Irazoqui et al., 2000). The observation that ABL, in contrast to most lectins which stimulate cell proliferation, is a reversible non-cytotoxic inhibitor of epithelial cell proliferation, has made the *A. bisporus* lectin interesting as a potential agent for cancer therapy (Yu et al., 1993).

Toxicants in *Agaricus bisporus*

Allergens

Allergenicity due to consumption of *A. bisporus* is relatively rare. Such cases have been reported in Germany and India (Pelzer and Freygang, 1997; Pelzer, 1999; Hegde et al., 2002). The latter case was a woman that experienced anaphylaxis (facial oedema and generalised urticaria) minutes after consumption of *Agaricus bisporus* curry (Hegde et al., 2002). The allergen was identified as the low molecular weight compound mannitol.

Other forms of mushroom allergy are related to the cultivation and could be induced by either the compost/culture, and be independent of the mushroom species cultivated, or induced by mushroom tissues, very often spores. Extrinsic allergic alveolitis or hypersensitivity pneumonitis is a disease of the first type occurring in people working with *A. bisporus* and other cultivated mushrooms. Most likely, mushroom lung develops in workers that have worked in sheds in which spawning takes place and where the compost, spawn and organisms living in the media are mechanically mixed. The *Agaricus*-related occupational allergy is rare and mainly manifested as asthma and dermatitis. Basidiocarp and spores have been identified as allergen sources that could elicit occupational asthma (Venturini et al., 2005). The disease may be identified with provocation tests (Kamm et al., 1991). Hand eczema, induced by delayed-type hypersensitivity, and airborne occupational allergic reactions leading to contact dermatitis have been reported in *A. bisporus* workers harvesting mushrooms (Korstanje and van de Staak, 1990; Kanerva et al., 1998). The agent responsible for the contact dermatitis has not been identified.

Biogenic amines

Biogenic amines (histamine, tyramine, putrescine, cadaverine, etc.) is a group of toxins generally produced by carboxylase-positive microorganisms from free amino acids in foods when the food in question is stored under unsuitable conditions. Kalac and Krizek (1997) did not find any of the biogenic amines in fresh or freshly stewed *A. bisporus*. However, after storage of the mushrooms for 48 hours at 20°C,

368 mg putrescine and 37 mg cadaverine per kg dry matter (moisture content of fresh mushrooms approximately 90%) was found in intact fruiting bodies. Neither histamine nor tyramine was found. Putrescine levels were lower in sliced mushroom and stewed sliced mushroom, whereas cadaverine was not found at all. Storage of the mushrooms at 6°C produced no biogenic amines.

Phenylhydrazines

Agaritine (β-N-[γ-L-(+)-glutamyl]-4-hydroxymethylphenyl-hydrazine) was isolated and characterised from the press-juice of *A. bisporus* during the early 1960s (Levenberg, 1961). Mushrooms that contain agaritine also contains a highly active enzyme that catalyses the cleavage of agaritine to 4-(hydroxymethyl)phenylhydrazine and/or the 4-(hydroxymethyl)benzenediazonium ion (Levenberg, 1961). Thus, agaritine is not the single phenylhydrazine in the mushroom. Both the presumed precursors for biosynthesis of agaritine, 4-(carboxy)phenylhydrazine and β-N-[γ-L-(+)-glutamyl]-4-carboxyphenylhydrazine, and the enzymatic degradation product of agaritine, 4-(hydroxymethyl)benzenediazonium ions have been identified in *A. bisporus* (Ross et al., 1982a; Chauhan et al., 1984, 1985). 4-(Hydroxymethyl)phenylhydrazine, which is assumed to be formed on agaritine degradation, has never been found in the mushroom as it is very unstable. Like most synthetic hydrazine derivatives, both agaritine precursors mentioned above and the 4-(hydroxymethyl)benzenediazonium ion have been shown to induce tumours in experimental animals (Toth, 2000). Although agaritine has never been shown to induce tumours in experimental animals, the compound could be regarded as an indicator of the likely presence of potentially toxic phenylhydrazines.

Agaritine

Several factors, including strain-type of *A. bisporus*, the quality of ingredients in the mushroom bed and the cultural practice employed in production, the flush of the cropping cycle and the maturity at harvest may all interact to influence agaritine content of *A. bisporus* fruit bodies. Also, handling of the fruit bodies after harvest (storage time and conditions during storage) have a bearing on the agaritine content. Table 10.10, therefore, shows the amount of agaritine in fresh samples of *A. bisporus* obtained either directly from mushroom growers or purchased at the local market.

As shown in Table 10.10, the range in agaritine content reported in fresh *A. bisporus* is between 80 mg/kg fresh weight and 1 730 mg/kg fresh weight, but usually in the region of 200-500 mg/kg (Andersson and Gry, 2004). The difference in reported agaritine content may to some extent be explained by different parts of the fruit body being studied by different investigators. Agaritine occurs throughout the fruit body at fairly similar levels, but the stipe base and hymenium contain significantly deviating agaritine levels, lower levels in the stipe base and higher levels in the hymenium (Soulier et al., 1993).

Rather small or no differences in mean agaritine content between strain types is apparent in the data of Table 10.10. However, in the study of Speroni et al. (1983), one strain (PSU-351 with brown colour phenotype) had significantly higher agaritine levels than the remaining seven strains, 5 100 mg agaritine per kg mushroom (on a dry weight basis) as compared to 1 700-2 800 mg/kg for other strains. It was hypothesised that PSU-351, which was only recently isolated from nature, contains higher agaritine levels because it had not yet lost its inherited ability to inhibit growth of certain fungi by producing agaritine. It is possible agaritine functions *in vivo* as an antimycotic agent. Also, no relationship could be found between mushroom colour and agaritine level.

As there is an ongoing discussion on whether or not the agaritine levels are higher in young mushrooms than in old ones (smaller vs. larger) (Andersson and Gry, 2004), it is important to register age and size of analysed mushrooms. It should be noted that very few of the studies in Table 10.10 have controlled for the growth stage of the mushroom.

Table 10.10. **Agaritine content of "fresh" cultivated *Agaricus bisporus***

Mushroom sample (g = grower; m = market)	Content (mg/kg fresh weight)[2]		Reference
	Average	Range	
2 fresh samples (m)[1]	440; 720	170-1 170	Ross et al. (1982b)
14 fresh samples of different strains (g)	880	330-1 730	Liu et al. (1982)
2 fresh samples of different strains (g)	304±6.0; 438±2.5		Fischer et al. (1984)
11 fresh samples (m)	368±45	94-629	Fischer et al. (1984)
1 fresh sample[1]	228		Hashida et al. (1990)
2 fresh samples of different strains (g)[1]	~ 180	80-250	Sharman et al. (1990)
5 fresh samples of different strains (m)[1]		160-650	Stijve et al. (1986)
1 fresh sample within a Nordic project (g)[1]	340		Andersson et al. (1994)
2 fresh sample (m)[1]	212; 229		Andersson et al. (1999)

Notes: 1. Size and form of mushroom not given. 2. The dry matter is approximately 10% of the fresh weight.

Other phenylhydrazines

Compared to agaritine, the levels of other phenylhydrazines are low in *A. bisporus*. 4-(Carboxy)phenylhydrazine occurr at around 11 mg/kg fresh weight (Chauhan et al., 1984), β-N-[γ-L-(+)-glutamyl]-4-carboxyphenylhydrazine at 16-42 mg/kg fresh weight (Chauhan et al., 1985; Toth et al., 1997), and the 4-(hydroxymethyl)benzenediazonium ion at 0.6-4 mg/kg fresh weight (Ross et al., 1982a; Toth et al., 1997).

Food use

Identification of Agaricus bisporus *food products*

More than 2 million tonnes of *Agaricus bisporus* arc produced annually worldwide, all destined for human consumption. The mushrooms are either sold fresh or processed by industry into easily stored products (dried and canned mushrooms) or products useful for the food industry (freeze-dried mushrooms). Although mushrooms contain protein, vitamins and minerals, their main role in the human diet is to contribute flavours and enhance the total quality of a dish.

Identification of key nutrients, anti-nutrients and toxicants and suggested analysis for food use

The key constituents suggested to be analysed with appropriate methodology in new varieties of *A. bisporus* intended for human consumption are shown in Table 10.11. As all food products of *A. bisporus* used by consumers and the food industry are derived from the fresh fruit bodies of the mushrooms, it is considered sufficient, in most circumstances, to analyse key constituents only in the fresh mushrooms. It will not be necessary to perform separate analyses of key constituents in commodities such as dried, freeze-dried or canned fruit bodies of *A. bisporus*.

Although it would be nice to have information also on the lectin content of *A. bisporus*, this mushroom constituent could not be suggested in the absence of quantitative data on lectin levels.

Table 10.11. **Suggested constituents to be analysed in fresh fruit bodies of cultivated *Agaricus bisporus***

Constituent	Fruit bodies
Proximates	X
Amino acids	X
Fatty acids	X
Mannitol	X
Trehalose	X
Vitamin B6, riboflavin (B2), niacin, pantothenic acid and folate	X
Agaritine	X

Feed use

It is unlikely that *A. bisporus*, or any other mushroom, will ever be grown to produce animal feed. In the rare cases when by-products of *A. bisporus* cultivation and mushroom processing (mainly stipes) may be used as animal feed, which most probably will be locally in the neighbourhood of the mushroom farms, the verification that the animal feed will have the expected composition would be established by the comparative assessment of composition of fruit bodies required for genetically modified *A. bisporus*. Thus, no additional studies on the chemical composition of by-products of *A. bisporus* cultivation and processing are suggested.

Notes

1. No standard deviation has been calculated due to missing data in the original reports.
2. No standard deviation has been calculated due to missing data in the original reports.

References

Abdullah, M.I. et al. (1994), "Supercritical fluid extraction of carboxylic and fatty acids from *Agaricus* spp. Mushrooms", *Journal of Agricultural and Food Chemistry*, Vol. 42, No. 3, pp. 718-722, http://dx.doi.org/10.1021/jf00039a023.

Ajlouni, S.O. et al. (1993), "Influence of gamma irradiation on quality characteristics, sugar content, and respiration rate of mushrooms during postharvest storage", in: Charalambous, G. (ed.), *Food Flavors, Ingredients and Compostion*, Elsevier Science, Amsterdam, pp. 103-121.

Aktümsek, A. et al. (1998), "*Agaricus bisporus* (Lange) sing.'un Yağ Asidi Bileşimi", *Turkish Journal of Biology*, Vol. 22, pp. 75-79.

Allen, J.J. et al. (1992), "Persistent meiotic arrest in basidia of *Agaricus bisporus*", *Mycological Research*, Vol. 96, Issue 2, February, pp. 125-127, http://dx.doi.org/10.1016/S0953-7562(09)80926-X.

Anderson, E.E. and C.R. Fellers (1942), "The food value of mushrooms (*Agaricus campestris*)", *Proceedings of the Amercian Society for Horticultural Science*, Vol. 41, pp. 301-304.

Andersson, H.C. and J. Gry (2004), "Phenylhydrazines in the cultivated mushroom (Agaricus bisporus)", TemaNord, 2004:558.

Andersson, H.C. et al. (1999), "Agaritine content in processed foods containing the cultivated mushroom (*Agaricus bisporus*) on the Nordic and Czech market", *Food Additives & Contaminants*, Vol. 16, No. 10, October, pp. 439-446.

Andersson, H.C. et al. (1994), "Lack of embryotoxicity and teratogenicity in rats after per oral application of extracts from the cultivated mushroom (*Agaricus bisporus*): A pilot study", in: Kozlowska, H. et al. (eds.), *Proceedings of the International European Food Tox IV Conference*, Polish Academy of Science, Olsztyn, pp. 492-495.

Bakowski, J. and R. Kosson (1985), "Nutritional value and amino acids compostion of the mushroom (*Agaricus bisporus*) at different stages of its development", *Acta Agrobotanica*, Vol. 38, pp. 103-113.

Bakowski, J. et al. (1986a), "The influence of flushes on some constituents of mushrooms (*Agaricus bisporus*) cultivated on different composts", *Acta Agrobotanica*, Vol. 39, No. 1, pp. 99-110, http://dx.doi.org/10.5586/aa.1986.009.

Bakowski, J. et al. (1986b), "Nutritional values of different strains of mushrooms (*Agaricus bisporus*)", *Acta Agrobotanica*, Vol. 39, No. 1, pp. 111-121, http://dx.doi.org/10.5586/aa.1986.010.

Barden, C.L. et al. (1990), "The effect of calcium chloride added to the irrigation water on quality and shelf life of harvested mushrooms", *Journal of Food Protection*, Vol. 53, pp. 759-762.

Barondes, S.H. (1988), "Bifunctional properties of lectins: Lectins redefined", *Trends in Biochemical Science*, Vol. 13, No. 12, December, pp. 480-482, http://dx.doi.org/10.1016/0968-0004(88)90235-6.

Beecher, T.M. et al. (2001), "Water potential and soluble carbohydrate concentrations in tissues of freshly harvested and stored mushrooms (*Agaricus bisporus*)", *Postharvest Biology and Technology*, Vol. 22, Issue 2, May, pp. 121-131, http://dx.doi.org/10.1016/S0925-5214(00)00195-2.

Beelman, R.B. et al. (2003), "Bioactive components in button mushroom *Agaricus bisporus* (J. Lge) Imbach (Agaricomycetideae) of nutritional, medicinal, and biological importance (review)", *International Journal of Medicinal Mushrooms*, Vol. 5, Issue 4, pp. 321-337, http://dx.doi.org/10.1615/InterJMedicMush.v5.i4.10.

Bernardo, D. et al. (2004), "Verticillium disease or 'dry bubble' of cultivated mushrooms: The *Agaricus bisporus* lectin recognizes and binds the *Verticillium fungicola* cell wall glucogalactomannan", *Canadian Journal of Microbiology*, Vol. 50, No. 9, September, pp. 729-735.

Biekman, E.S.A. et al. (1997), "Effects of shrinkage on the temperature increase in evacuated mushrooms (*Agaricus bisporus*) during blanching", *Journal of Food Engineering*, Vol. 33, Issues 1-2, July-August, pp. 87-99, http://dx.doi.org/10.1016/S0260-8774(97)00044-7.

Bonzom, P.M.A. et al. (1999), "NMR lipid profile of *Agaricus bisporus*", *Phytochemistry*, Vol. 50, Issue 8, April, pp. 1 311-1 321, http://dx.doi.org/10.1016/S0031-9422(98)00703-1.

Burton K.S. (1988), "The effects of storage and development on *Agaricus bisporus* proteases", *Journal of Horticultural Science*, Vol. 63, No. 1, pp. 103-108.

Çaglarlrmak, N. et al. (2001), "Determination of nutritive changes of canned mushrooms (*Agaricus bisporus*) during storage period", *Micologia Aplicada International*, Vol. 13, pp. 97-101.

Callac, P. (1995), "Breeding of edible fungi with emphasis on the variability among French genetic resources of Agaricus bisporus", *Canadian Journal of Botany*, Vol. 73, No. S1, pp. S980-S986, http://dx.doi.org/10.1139/b95-348.

Callac, P. et al. (2003), "A novel homothallic variety of *Agaricus bisporus* comprises rare tetrasporic isolates from Europe", *Mycologia*, Vol. 95, No. 2, March/April, pp. 222-231, available at: www.mycologia.org/content/95/2/222.full.

Callac, P. et al. (1993), "Morphological, genetic, and interfertility analyses reveal a novel, tetrasporic variety of *Agaricus bisporus* from the Sonoran Desert of California", *Mycologia*, Vol. 85, No. 5, pp. 335-351, http://dx.doi.org/10.2307/3760617.

Carrizo, M.E. et al. (2005), "The antineoplastic lectin of the common edible mushroom (*Agaricus bisporus*) has two binding sites, each specific for a different configuration at a single epimeric hydroxyl", *Journal of Biological Chemistry*, Vol. 280, No. 11, March, pp. 10 614-10 623.

Carrizo, M.E. et al. (2004), "Crystallization and preliminary X-ray study of the common edible mushroom (Agaricus bisporus) lectin", *Acta Cyrstallographica* Vol. 60, Pt. 4, April, pp. 718-720.

CFC (2002), *China Food Composition*, Yuexin, Y. et al. (eds.), Institute of Nutrition and Food Safety, China CDC, Peking University Medical Press, (in Chinese).

Chang, S.-T. and W.A. Hayes (1978), *The Biology and Cultivation of Edible Mushrooms*, Academic Press Inc., New York.

Chang, S.-T. and P.G. Miles (2004), *Mushrooms. Cultivation, Nutritional Value, Medicinal Effect, and Environmental Impact*, 2nd edition, CRC Press, Boca Raton, Florida.

Chauhan, Y. et al. (1985), "Isolation of N^2-[γ-L-(+)-glutamyl]-4-carboxyphenylhydrazine in the cultivated mushroom *Agaricus bisporus*", *Journal of Agricultural and Food Chemistry*, Vol. 33, No. 5, September, pp. 817-820, http://dx.doi.org/10.1021/jf00065a013.

Chauhan, Y. et al. (1984), "Identification of p-hydrazinobenzoic acid in the commercial mushroom *Agaricus bisporus*", *Journal of Agricultural and Food Chemistry*, Vol. 32, No. 5, September, pp. 1 067-1 069, http://dx.doi.org/10.1021/jf00125a032.

Cheung, P.C.-K. (1997), "Dietary fibre content and compositon of some edible fungi determined by two methods of analysis", *Journal of the Science of Food and Agriculture*, Vol. 73, Issue 2, February, pp. 255-260, http://dx.doi.org/10.1002/(SICI)1097-0010(199702)73:2<255::AID-JSFA723>3.0.CO;2-U.

Çoşkuner, Y. and Y. Özdemir (2000), "Acid and EDTA blanching effects on the essential element content of mushrooms (*Agaricus bisporus*)", *Journal of the Science of Food and Agriculture*, Vol. 80, Issue 14, November, pp. 2 074-2 076, http://dx.doi.org/10.1002/1097-0010(200011)80:14<2074::AID-JSFA746>3.0.CO;2-I.

Çoşkuner, Y. and Y. Özdemir (1997), "Effects of canning processes on the elements content of cultivated mushrooms (*Agaricus bisporus*)", *Food Chemistry*, Vol. 60, Issue 4, December, pp. 559-562, http://dx.doi.org/10.1016/S0308-8146(97)00032-0.

Crenshaw, R.W. et al. (1995), "Isolation and characterization of a cDNA clone encoding a lectin from *Agaricus bisporus*", *Plant Physiology*, Vol. 107, No. 4, April, pp. 1 465-1 466.

Crisan, E.V. and A. Sands (1978), "Nutritional value", in: Chang, S.T. and W.A. Hayes (eds.), *The Biology and Cultivation of Edible Fungi*, Academic Press, Inc., New York, pp. 137-168.

Cruz, C. et al. (1997), "Fatty acid content and some flavor compound release in two strains of *Agaricus bisporus*, according to three stages of development", *Journal of Agricultural and Food Chemistry*, Vol. 45, No. 1, January, pp. 64-67, http://dx.doi.org/10.1021/jf960300t.

CSTJ (2005), *Standard Tables of Food Composition in Japan*, Report of the subdivision of resources, the Council for Science and Technology, Ministry of Education, Culture, Sports, Science and Technology, Japan, Fifth revised and enlarged edition (mainly in Japanese).

Dar, G.M. (1996), "Nutritional changes in cultivated mushroom (*Agaricus bisporus*) during growth and pathogenesis", *Indian Journal of Mycology and Plant Pathology*, Vol. 26, pp. 295-297.

Dikeman, C.L. et al. (2005), "Effects of stage of maturity and cooking on the chemical composition of selected mushroom varieties", *Journal of Agricultural and Food Chemistry*, Vol. 53, No. 4, February, pp. 1 130-1 138.

Evans, H.J. (1959), "Nuclear behaviour in the cultivated mushroom", *Chromosoma*, Vol. 10, Issue 1-6, pp. 115-135.

Falandysz, J. et al. (1994), "Silver uptake by *Agaricus bisporus* from an artificially enriched substrate", *Z. Lebensm. Unters. Forsch*, Vol. 199, No. 3, September, pp. 225-228.

Fischer, B. et al. (1984), "Gehaltbestimmung von Agaritin im Zuchtchampignon (*Agaricus bisporus*) mittels Hochleistungsflüssigchromatographie (HPLC)", *Z. Lebensm. Unters. Forsch.*, Vol. 179, pp. 218-223.

Haldimann, M. et al. (1995), "Vorkommen von Arsen, Blei, Cadmium, Quecksilber und Selen in Zuchtpilzen", *Mitt. Gebiete Lebensm. Hyg.*, Vol. 86, pp. 463-484.

Hammond, J.B.W. and R. Nichols (1979), "Carbohydrate metabolism in *Agaricus bisporus*: Changes in non-structural carbohydrates during periodic fruiting (flushing)", *New Phytologist*, Vol. 83, No. 3, pp. 723-730, www.jstor.org/stable/2433942.

Hammond, J.B.W. and R. Nichols (1976), "Carbohydrate metabolism in *Agaricus bisporus* (Lange) Sing.: Changes in soluble carbohydrates during growth of mycelium and sporophore", *Journal of General Microbiology*, Vol. 93, No. 2, April, pp. 309-320.

Hashida, C. et al. (1990), "Quantities of agaritine in mushrooms (*Agaricus bisporus*) and the carcinogenicity of mushroom methanol extracts on the mouse bladder epithelium", *Nippon-Koshu-Eisei-Zasshi*, Vol. 37, No. 6, June, pp. 400-405.

Hegde, V.L. et al. (2002), "Anaphylaxis caused by the ingestion of cultivated mushroom (*Agaricus bisporus*): Identification of allergen as mannitol", *Allergology Internationam*, Vol. 51, Issue 2, June, pp. 121-129, http://dx.doi.org/10.1046/j.1440-1592.2002.00252.x.

Hira, C.K. et al. (1990), "Biochemical changes in sporophores of two mushroom species before and after opening of caps", *Annals of Biology*, Vol. 6, No. 2, pp. 95-99.

Holtz, R.B. and L.C. Schisler (1971), "Lipid metabolism of (Lange) Sing.: I. Analysis of sporophore and mycelial lipids", *Lipids*, Vol. 6, Issue 3, March, pp. 176-180.

Huang, B.-H. et al. (1985), "The sterol composition of *Volvariella volvacea* and other edible mushrooms", *Mycologia*, Vol. 77, No. 6, pp. 959-963, http://dx.doi.org/10.2307/3793309.

Imbernon, M. et al. (1995), "Allelic polymorphism at the mating type locus in *Agaricus bisporus* var. *burnettii*, and confirmation of the dominance of its tetrasporic trait", *Mushroom Science*, Vol. 14 (part 1), pp. 11-19.

Irazoqui, F.J. et al. (2000), "Novel immunogenicity of Thomsen-Friedenreich disaccharide obtained by a molecular rotation on its carrier linkage", *Glycobiology*, Vol. 10, pp. 781-787.

Irazoqui, F.J. et al. (1999), "Structural requirements of carbohydrates to bind *Agaricus bisporus* lection", *Glycobiology*, Vol. 9, No. 1, January, pp. 59-64.

Jennings, D.H. (1984), "Polyol metabolism in fungi", *Advances in Microbial Physiology*, Vol. 25, pp. 149-193.

Joly, P. (1979), *La culture des champignons*, Dargaud Editeur, Neuilly-sur-Seine.

Kalac, P. and M. Krizek (1997), "Formation of biogenic amines in four edible mushroom species stored under different conditions", *Food Chemistry*, Vol. 58, Issue 3, March, pp. 233-236, http://dx.doi.org/10.1016/S0308-8146(96)00170-7.

Kamm, Y.J.L. et al. (1991), "Provocation tests in extrinsic alveolitis in mushroom workers", *Netherlands Journal of Medicine*, Vol. 38, No. 1-2, February, pp. 59-64.

Kanerva, L. et al. (1998), "Airborne occupational allergic contact dermatitis from champignon mushroom", *American Journal of Contact Dermatitis*, Vol. 9, No. 3, September, pp. 190-192.

Kerrigan, R.W. (1995), "Global genetic resources of *Agaricus* breeding and cultivation", *Canadian Journal of Botany*, Vol. 73, No. S1, pp. S973-S979, http://dx.doi.org/10.1139/b95-347.

Kerrigan, R.W. and I.K. Ross (1989), "Allozymes of a wild *Agaricus bisporus* population: New alleles, new genotypes", *Mycologia*, Vol. 81, No. 3, May-June, pp. 433-443, http://dx.doi.org/10.2307/3760080.

Kerrigan, R.W. et al. (1998), "The indigenous coastal Californian population of the mushroom *Agaricus bisporus*, cultivated species, may be at risk of extinction", *Molecular Ecology*, Vol. 7, Issue 7, January, pp. 35-45, http://dx.doi.org/10.1046/j.1365-294x.1998.00294.x.

Kerrigan, R.W. et al. (1993a), "The California population of *Agaricus bisporus* comprises at least two ancestral elements", *Systematic Botany*, Vol. 18, No. 1, January-March, pp. 123-136, http://dx.doi.org/10.2307/2419792.

Kerrigan, R.W. et al. (1993b), "Meiotic behaviour and linkage relationships in the secondarily homothallic fungus *Agaricus bisporus*", *Genetics*, Vol. 133, No. 2, February, pp. 225-236.

Kerrigan, R.W. et al. (1992), "Strategies for the efficient recovery of *Agaricus bisporus* homokaryons", *Mycologia*, Vol. 84, No. 4, July-August, pp. 575-579, http://dx.doi.org/10.2307/3760324.

Korstanje, M.J. and W.J.B.M. van de Staak (1990), "A case of hand eczema due to mushrooms", *Contact Dermatitis*, Vol. 22, No. 2, February, pp. 115-116.

Kosson, R. and J. Bakowski (1984), "The effect of cultivation methods on the amino acid and protein content of the mushroom (*Agaricus bisporus* Lange/Sing.)", *Molecular Nutrition & Food Research*, Vol. 28, Issue 10, pp. 1 045-1 051, http://dx.doi.org/10.1002/food.19840281011.

Koyama, N. et al. (1984), "Fatty acid compositions and ergosterol contents of edible mushrooms", *Nippon Shokuhin Kogyo Gakkaishi*, Vol. 31, pp. 732-738.

Kurasawa, S.-I. et al. (1982), "Proximate and dietary fibre analysis of mushrooms" (in Japanese), *Nippon Shokuhin Kogyo Gakkaishi*, Vol. 29, pp. 400-406.

Kurtzman, R.H., Jr. (1997), "Production and cultivation of mushrooms in the West, particularly Europe and North America", *Food Reviews International*, Vol. 13, Issue 3, pp. 497-516, http://dx.doi.org/10.1080/87559129709541136.

Laborde, J. and P. Delpech (1991), "Dry matter content of fruitbodies of *Agaricus bisporus* (Lange Sing.): Evaluation during cropping", in: *Science and Cultivation of Edible Fungi*, Maher (ed.), Balkema, Rotterdam, pp. 659-664.

Le Loch-Bonazzi, C. et al. (1992), "Quality of dehydrated cultivated mushrooms (*Agaricus bisporus*): A comparison between different drying and freeze-drying processes", *Lebensm.-Wiss. U. Technol.*, Vol. 25, pp. 334-339.

Le Roux, P. and Y. Danglot (1972), "Composition chimique et valeur alimentaire du champignon cultivé (*Agaricus bisporus*) en fonction des conditions de stockage", *Mushroom Science*, Vol. 8, pp. 471-474.

Levenberg, B. (1961), "Structure and enzymatic cleavage of agaritine, a phenylhydrazine of L-glutamic acid isolated from *Agaricaceae*", *Journal of the American Chemical Society*, Vol. 83, pp. 503-504.

Li, G.S.F. and S.T. Chang (1982), "The nucleic acid content of some edible mushrooms", *European Journal of Applied Microbiology and Biotechnology*, Vol. 15, No. 4, pp. 237-240.

Liu, J.-W. et al. (1982), "Agaritine content of fresh and processed mushrooms [*Agaricus bisporus* (Lange) Imbach]", *Journal of Food Science*, Vol. 47, pp. 1 542-1 544 and 1 548.

Maeda, Y. et al. (1993), "Utilization of heat-dried stipe of mushroom (*Agaricus bisporus* Sing.) for animal feed", *Journal of the Japanese Society of Grassland Science*, Vol. 39, No. 1, pp. 22-27.

Maggioni, A. et al. (1968), "Composition of cultivated mushrooms (*Agaricus bisporus*) during the growing cycle as affected by the nitrogen source introduced in composting", *Journal of Agricultural and Food Chemistry*, Vol. 16, No. 3, May, pp. 517-519, http://dx.doi.org/10.1021/jf60157a037.

Manzi, P. et al. (2001), "Nutritional value of mushrooms widely consumed in Italy", *Food Chemistry*, Vol. 73, No. 3, May, pp. 321-325, http://dx.doi.org/10.1016/S0308-8146(00)00304-6.

Mattila, P. et al. (2002a), "Sterol and vitamin D_2 contents in some wild and cultivated mushrooms", *Food Chemistry*, Vol. 76, No. 3, March, pp. 293-298, http://dx.doi.org/10.1016/S0308-8146(01)00275-8.

Mattila, P. et al. (2002b), "Basic composition and amino acid contents of mushrooms cultivated in Finland", *Journal of Agricultural and Food Chemistry*, Vol. 50, No. 22, October, pp. 6 419-6422.

Mattila, P. et al. (2001), "Contents of vitamins, mineral elements, and some phenolic compounds in cultivated mushrooms", *Journal of Agricultural and Food Chemistry*, Vol. 49, No. 5, May, pp. 2 343-2 348.

Mattila, P.H. et al. (1994), "Vitamin D contents in edible mushrooms", *Journal of Agricultural and Food Chemistry*, Vol. 42, No. 11, November, pp. 2 449-2 453, http://dx.doi.org/10.1021/jf00047a016.

Mau, J.-L. et al. (1998) "Ultraviolet irradiation increased vitamin D2 content in edible mushrooms", *Journal of Agricultural and Food Chemistry*, Vol. 46, No. 12, November, pp. 5 269-5 272, http://dx.doi.org/10.1021/jf980602q.

Mau, J.-L. et al. (1991), "Effect of nutrient supplementation on flavor, quality, and shelf life of the cultivated mushroom, *Agaricus bisporus*", *Mycologia*, Vol. 83, No. 2, March-April, pp. 142-149, http://dx.doi.org/10.2307/3759929.

Moss, S. and D. Mitchell (1994), "Market analysis of the mushroom industry", *Mushroom News*, October.

Mozina, I.A. et al. (1993), "Marking of cultivars and wild strains of cultivated mushroom *Agaricus bisporus* with esterase allozymes", *Biol. Nauki*, Vol. 1, pp. 131-140 (in Russian with English abstract).

Muñoz, A.H.S. et al. (2005), "Subcellular distribution of aluminium, bismuth, cadmium, chromium, copper, iron, manganese, nickel and lead in cultivated mushrooms (*Agaricus bisporus* and *Pleurotus ostreatus*)", *Biological Trace Element Research*, Vol. 106, No. 3, pp. 265-277.

NNR (2004), *Nordic Nutrition Recommendations 2004*, Nordic Council of Ministers, Nord 2004:13.

Oka, Y. et al. (1981), "Quantitative dertermination of the free amino acids and their derivatives in the common edible mushroom, *Agaricus bisporus*", *Journal of Nutritional Science and Vitaminology*, Vol. 27, No. 3, pp. 253-262.

Parks, L.W. and J.D. Weete (1991), "Fungal sterols", in: Patterson, G.W. and W.D. Nes (eds.), *Physiology and Biochemistry of Sterols*, American Oil Chemists' Society, Champaign, pp. 158-171.

Pelzer, A.E. (1999), "Sofrottypallergie auf Champignon", *Allergologie*, Vol. 23, pp. S35-S36.

Pelzer, A.E. and E. Freygang (1997), "Soforttypallergie auf Champignon", *Allergologie*, Vol. 20, pp. 304-306.

Piepponen, S. et al. (1983), "The selenium content of edible mushrooms in Finland", *Z. Lebensm. Unters. Forsch*, Vol. 177, No. 4, pp. 257-260.

Presant, C.A. and S. Kornfeld (1972), "Characterization of the cell surface receptor for the *Agaricus bisporus* hemagglutinin", *Journal of Biological Chemistry*, Vol. 247, November, pp. 6 937-6 945.

Rainey, P.B. (1991), "Effect of Pseudomonas putida on hyphal growth of *Agaricus bisporus*", *Mycological Research*, Vol. 95, Issue 6, June, pp. 699-704, http://dx.doi.org/10.1016/S0953-7562(09)80817-4.

Rainey, P.B. et al. (1990), "A model system for examining involvement of bacteria in basidiome initiation of *Agaricus bisporus*", *Mycological Research*, Vol. 94, Issue 2, March, pp. 191-195, http://dx.doi.org/10.1016/S0953-7562(09)80612-6.

Raper, C.A. et al. (1972), "Genetic analysis of the life cycle of *Agaricus bisporus*", *Mycologia*, Vol. 64, No. 5, September-October, pp. 1 088-1 117, http://dx.doi.org/10.2307/3758075.

Rast, D. (1965), "Zur stoffwechselphysiologischen bedeutung van Mannit und Trehalose in *Agaricus bisporus*", *Planta*, Vol. 64, pp. 81-93.

Robinson, W. (1870), *Mushroom Culture: Its Extension and Improvement*, Frederick Warne and Co., London.

Ross, A.E. et al. (1982a), "Evidence for the occurrence and formation of diazonium ions in the *Agaricus bisporus* mushroom and its extracts", *Journal of Agricultural and Food Chemistry*, Vol. 30, No. 3, May, pp. 521-525, http://dx.doi.org/10.1021/jf00111a028.

Ross, A.E. et al. (1982b), "Occurrence, stability and decomposition of beta-N [gamma-L-(+)-glutamyl]-4-hydroxymethylphenylhydrazine (agaritine) from the mushroom *Agaricus bisporus*", *Food and Chemical Toxicology*, Vol. 20, No. 6, December, pp. 903-907.

Sapers, G.M. et al. (1999), "Structure and composition of mushrooms as affected by hydrogen peroxide wash", *Journal of Food Science*, Vol. 64, Issue 5, September, pp. 889-892, http://dx.doi.org/10.1111/j.1365-2621.1999.tb15934.x.

Sharman, M. et al. (1990), "A survey of the occurrence of agaritine in U.K. cultivated mushrooms and processed mushroom products", *Food Additives & Contaminants*, Vol. 7, No. 5, September-October, pp. 649-656.

Sinden, J.W. (1981), "Strain adaptability", *Mushroom Journal*, Vol. 101, pp. 153-165.

SLV (2004), *Livsmedelsverkets livsmedelsdatabas* (The NFA Food Composition Database), version 04.1.1, The National Food Administration (NFA), Sweden.

Soulier, L. et al. (1993), "Occurrence of agaritine and γ-glutaminyl-4-hydroxybenzene (GHB) in the fructifying mycelium of *Agaricus bisporus*", *Mycological Research*, Vol. 97, Issue 5, May, pp. 529-532, http://dx.doi.org/10.1016/S0953-7562(09)81174-X.

Spaulding, T. and R. Beelman (2003), "Survey evaluation of selenium & other minerals in *Agaricus* mushrooms commercially grown in the United States", *Mushroom News*, Vol. 51, pp. 6-9.

Speroni, J.J. et al. (1983), "Factors influencing the agaritine content in cultivated mushrooms, *Agaricus bisporus*", *Journal of Food Protection*, Vol. 46, pp. 506-509.

Stamets, P. and J.S. Chilton (1983), *The Mushroom Cultivator*, Agarikon Press, Olympia, Washington.

Stijve, T. et al. (1986), "Agaritine, a p-hydroxymethyl-phenylhydrazine derivative in cultivated mushrooms (*Agaricus bisporus*), and in some of its wild-growing relatives", *Dtsch. Lebensm. Rdsch.*, Vol. 82, pp. 243-248.

Stoop, J.M.H. and H. Mooibroek (1998), "Cloning and characterization of NADP-mannitol dehydrogenase cDNA from button mushroom, *Agaricus bisporus*, and its expression in response to NaCl stress", *Applied and Environmental Microbiology*, Vol. 64, No. 12, December, pp. 4 689-4 696.

Stoop, J.M.H. and D.M. Pharr (1994), "Mannitol metabolism in celery stressed by excess macronutrients", *Plant Physiology*, Vol. 106, No. 2, October, pp. 503-511.

Stoop, J.M.H. et al. (1996), "Mannitol metabolism in plants: A method of coping with stress", *Trends in Plant Science*, Vol. 1, Issue 5, May, pp. 139-144, http://dx.doi.org/10.1016/S1360-1385(96)80048-3.

Sueyoshi, S. et al. (1985), "Purification and characterization of four isolectins of mushroom (*Agaricus bisporus*)", *Biological Chemistry Hoppe-Seyler*, Vol. 366, No. 3, March, pp. 213-221.

Summerbell, R.G. et al. (1989), "Inheritance of restriction fragment length polymorphisms in *Agaricus brunnescens*", *Genetics*, Vol. 123, No. 2, October, pp. 293-300.

Teichmann, A. (2005), "Sterol and vitamin D_2 contents in cultivated and wild grown mushrooms", Diploma work Swedish University of Agricultural Sciences, Department of Food Science, Publication No. 196.

Toth, B. (2000), *Hydrazines and Cancer. A Guidebook on the Carcinogenic Activities of Hydrazines, Related Chemicals, and Hydrazine-containing Natural Products*, Harwood Academic Publishers, Australia.

Toth, B. et al. (1997), "Carcinogenesis by the cultivated baked *Agaricus bisporus* mushroom in mice", *Oncology Reports*, Vol. 4, No. 5, September-October, pp. 931-936.

Trigueros, V. et al. (2003), "*Xerocomus chrysenteron* lectin: Identification of a new pesticidal protein", *Biochimica et Biophysica Acta*, Vol. 1 621, No. 3, June, pp. 292-298.

Tseng, Y.-H. and J.-L. Mau (1999), "Content of sugars, free amino acids and free 5′-nucleotides in mushrooms, *Agaricus bisporus*, during post-harvest storage", *Journal of the Science of Food Agriculture*, Vol. 79, Issue 11, August, pp. 1 519-1 523, http://dx.doi.org/10.1002/(SICI)1097-0010(199908)79:11<1519::AID-JSFA399>3.0.CO;2-M.

Tüzen, M. et al. (1998), "Study of heavy metals in some cultivated and uncultivated mushrooms of Turkish origin", *Food Chemistry*, Vol. 63, Issue 2, October, pp. 247-251, http://dx.doi.org/10.1016/S0308-8146(97)00225-2.

Udipi, S. and B.D. Punekar (1980), "Nutritive value of mushrooms", *Indian Journal of Medical Research*, Vol. 72, pp. 241-244.

Ünal, M.K. et al. (1996), "Chemical composition and nutritive value of the cultivated (*Agaricus bisporus*) and wild mushrooms grown in Turkey", *Acta Alimentaria*, Vol. 25, pp. 257-265.

USDA Agricultural Research Service (2006), *National Nutrient Database for Standard Reference*, Release 19 August 2006, available at: http://ndb.nal.usda.gov.

USDA Agricultural Research Service (2005), *National Nutrient Database for Standard Reference*, Nutrient Data Laboratory, Release 18 August 2005, available at: http://ndb.nal.usda.gov.

Van Elteren, J.T. et al. (1998), "Accumulation and distribution of selenium and cesium in the cultivated mushroom *Agaricus bisporus* – A radiotracer-aided study", *Chemosphere*, Vol. 36, Issue 8, April, pp. 1 787-1 798, http://dx.doi.org/10.1016/S0045-6535(97)10064-9.

Van Griensven, L.J.L.D. (1988), *The Cultivation of Mushrooms*, Rustington, Darlington Museum Laboratories, Rustington.

Vedder, P.J.C. (1978), *Modern Mushroom Growing*, Grower Books, London.

Venturini, M. et al. (2005), "Occupational asthma caused by white mushroom", *Journal of Investigational Allergology and Clinical Immunology*, Vol. 15, pp. 219-221, available at: www.jiaci.org/issues/vol15issue03/11.pdf.

Vetter, J. (2003), "Data on sodium content of common edible mushrooms", *Food Chemistry*, Vol. 81, Issue 4, June, pp. 589-593, http://dx.doi.org/10.1016/S0308-8146(02)00501-0.

Vetter, J. (1994), "Mineral elements in the important cultivated mushrooms *Agaricus bisporus* and *Pleurotus ostreatus*", *Food Chemistry*, Vol. 50, Issue 3, pp. 277-279, http://dx.doi.org/10.1016/0308-8146(94)90132-5.

Vetter, J. (1989), "Vergleichende Untersuchung des Mineralstoffgehaltes der Gattungen *Agaricus* (Champignon) und *Pleurotus* (Austernseitling)", *Z. Lebensm. Unters. Forsch.*, Vol. 189, pp. 346-350.

Volk, T. and K. Ivors (2001), "Tom Volk's Fungus of the Month for April 2001", http://botit.botany.wisc.edu/toms_fungi/apr2001.html.

Wakita, S. (1976), "Thiamine-destruction by mushrooms", *Science Reports of the Yokohama National University*, Sec. 2 (23), pp. 39-70.

Wannet, W.J.B. et al. (2000), "HPLC detection of soluble carbohydrates involved in mannitol and trehalose metabolism in the edible mushroom *Agaricus bisporus*", *Journal of Agricultural and Food Chemistry*, Vol. 48, No. 2, February, pp. 287-291.

Weaver, J.C. et al. (1977), "Comparative protein studies (Kjeldahl, dye binding, amino acid analysis) of nine strains of *Agaricus bisporus* (Lange) Imbach mushrooms", *Journal of Food Science*, Vol. 42, pp. 364-366.

Wittliff, J.L. and R.L. Airth (1970a), "[41] Thiaminase I (Thiamine: Base 2-Methyl-4-aminopyrimidine-5-methenyltransferase, EC 2.5.1.2)", *Methods of Enzymology*, Vol. 18, pp. 229-234.

Wittliff, J.L. and R.L. Airth (1970b), "[42] Thiaminase II (Thiamine Hydrolase, EC 3.5.99.2)", *Methods of Enzymology*, Vol. 18, pp. 234-238.

Xu, J. et al. (1997), "Genetic structure of natural populations of *Agaricus bisporus*, the commercial button mushroom", *Journal of Heredity*, Vol. 88, pp. 482-488.

Young, J.C. (1995), "Microwave-assisted extraction of the fungal metabolite ergosterol and total fatty acids", *Journal of Agricultural and Food Chemistry*, Vol. 43, No. 11, November, pp. 2 904-2 910, http://dx.doi.org/10.1021/jf00059a025.

Yu, L.-G. et al. (1993), "Reversible inhibition of proliferation of epithelial cell lines by *Agaricus bisporus* (edible mushroom) lectin", *Cancer Research*, Vol. 53, No. 19, October, pp. 4 627-4 632.

Źródlowski, Z. (1995), "The influence of washing and peeling of mushrooms *Agaricus bisporus* on the level of heavy metal contamination", *Polish Journal of Food and Nutrition Sciences*, Vol. 4/45, pp. 26-33.

Chapter 11

Sunflower (*Helianthus annuus*)

*This chapter, prepared by the OECD Task Force for the Safety of Novel Foods and Feeds with a leading group comprising France (chair), Canada, Germany, Sweden and the United States, deals with the composition of sunflower (*Helianthus annuus*). It contains elements that can be used in a comparative approach as part of a safety assessment of foods and feeds derived from new varieties. Background is given on sunflower origin, production, processing, appropriate varietal comparators and characteristics screened by breeders. Then nutrients in sunflower and its products, anti-nutrients and toxicants are detailed. The final sections suggest key products and components for analysis of new varieties for food use and for feed use.*

Sunflower origins, production and processing of the seeds, appropriate comparators and traditional characteristics

History of the sunflower crop

The sunflower (*Helianthus annuus* L., 2n = 34) is a member of the Compositae (Asteracea) family and the genus *Helianthus*. It originates from North America, where it was traditionally cultivated by the Native Americans. The sunflower was introduced into Spain in the middle of the 16th century, where it was cultivated essentially as an ornamental plant. Its oil-bearing qualities were only discovered in the 18th century. Since then, the sunflower for oil production has been considerably genetically improved. Some of the first improvements, through trait selection and hybridisation, took place in the Russian Federation, then in the United States and aimed at increasing the oil content of the seeds. The breeding resulted in the development of strains with high oleic acid content (Soldatov, 1976). Nowadays, depending on the variety, sunflower contains between 14% and 90% oleic acid (Codex, 2005). Recently, strains with a low content saturated fatty acids have been developed (Delplanque, 2000; Vick et al., 2003).

The introduction of hybrids led to higher yields (Merrien, 1992; Bonjean, 1993). The creation of the first mildew-resistant hybrids also meant that sunflowers could be cropped in areas prone to infection with this disease.

World production of sunflower

The world production of sunflower reached 23.960 thousand tonnes in 2002/03, with an average yield of nearly 1.2 t/ha. The European Union is the world's largest producer, followed by the Russian Federation, Ukraine, Argentina, the People's Republic of China and the United States (Table 11.1).

Table 11.1. **Main sunflower producing countries in 2002/03**

Rank	Country	Production area (1 000 ha)	Production (1 000 t)
1	European Union (25 countries)	2.147	3.718
2	Russian Federation	3.782	3.684
3	Ukraine	2.810	3.510
4	Argentina	2.272	3.340
5	China (People's Republic of)	1.131	1.946
6	United States	0.877	1.112
	World	19.889	23.960

Source: Adapted from Oil World (2005).

Oilseed and non-oilseed sunflowers

There are two types of sunflowers, oilseed and non-oilseed (or confectionery), which are nevertheless of the same species. Oilseed sunflower seeds, constituting the major part of the world production, are characterised by their solid black hulls that are firmly attached to the seed, are used in the crushing industry for oil and for use in wild and domestic bird feeding. Meal resulting from their crushing is mainly used for livestock feeding. The industry has bred high oleic acid oilseed sunflower that has a fatty acid profile similar to Canola oil. The market share for this variety is relatively small, but increasing. It is estimated that 95% of the world production is the traditional oilseed type, and only 5% is the confectionery type (Skrypetz, 2005).

Seeds of non-oilseed sunflowers are characterised by larger, thick, striped, loosely attached hulls that lend themselves to a relative complete dehulling process. These seeds are used for the human food market, as roasted snack foods with shell or as dehulled seeds for the baking industry. Material from non-oilseed sunflowers may be used as both livestock and bird feed (Schild et al., 1991).

Processing of oilseed sunflower seeds

European crushing of sunflower has stabilised at approximately 4.800 thousand tonnes in 2000, after a large increase between 1991 and 1997, due to the high worldwide demand for oils (and particularly sunflower oil) during this period.

The process traditionally used worldwide in crushing plants is described in Figure 11.1.

Figure 11.1. **Classical flowchart for processing oilseed sunflower seeds in the crushing industry**

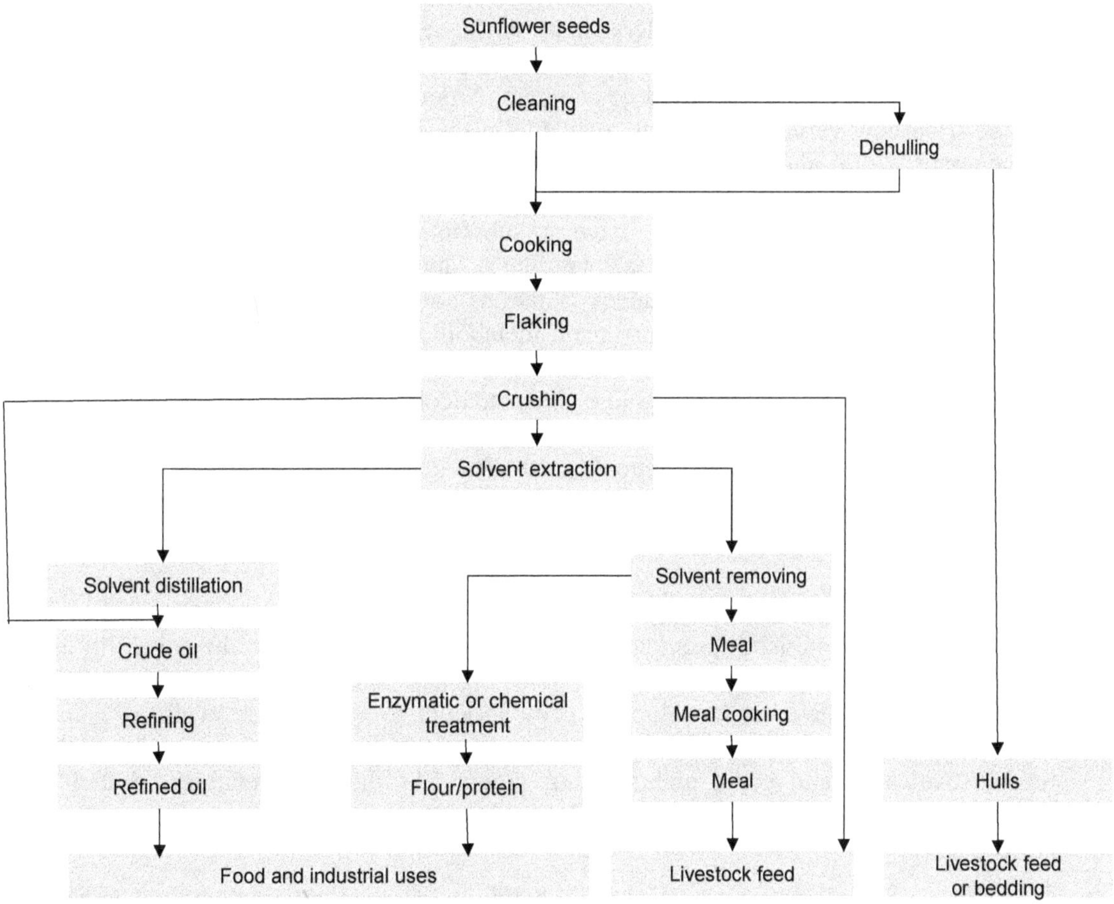

Source: Modified from Laisney (1984, 1992).

Sunflower oil for food consumption is traditionally obtained through two main steps: the crushing of the seeds (mechanical compression followed by solvent extraction) and the refining of the crude oil. The co-product of oil extraction is the meal, which is used in animal feeding as a protein source. In the 1980s, the fibre fraction of the sunflower meal was reduced by dehulling the seeds. Dehulling increases the protein and energy contents of meal and decreases the amount of wax in the crude oil (Evrard et al., 1986). It also has technological benefits: increasing output while decreasing wear and tear on the equipment, although the benefits from dehulled seeds do generally not compensate for the processing costs.

There is no good estimate of the amount of the world's oilseed sunflower seeds that are dehulled or partially dehulled (part-dehulled) prior to crushing. In the United States, the main crushing plants are partially dehulling sunflower seeds prior to crushing to obtain a meal with greater than 30% crude protein and 21% or less crude fiber (Sunflower Technology and Production Agronomy, 1997). In Europe, the dehulling or part-dehulling[1] process is operating in only a few small plants.

Appropriate comparators for testing new varieties

This chapter suggests parameters that sunflower developers should measure. Measurement data from the new variety should ideally be compared to those obtained from the near isogenic non-modified counterpart. A developer can also compare values obtained from new varieties with data available in the literature, or chemical analytical data generated from other commercial sunflower varieties. Components to be analysed include key nutrients, anti-nutrients and toxicants. Key nutrients are those components in a particular product which may have a substantial impact in the overall diet. These may be major constituents (fats, proteins, and structural and non-structural carbohydrates) or minor compounds (vitamins and minerals). They may be complemented with selected secondary plant metabolites for which characteristic levels in the species are known. Similarly, the levels of known anti-nutrients and allergens should be considered. Key toxicants are those toxicologically significant compounds known to be inherently present in the species, i.e. compounds whose toxic potency and levels may impact human and animal health. The key components analysed are used as indicators of whether unintended effects of the genetic modification influencing plant metabolism has occurred or not.

Traditional characteristics screened by sunflower developers

Phenotypic characteristics provide important information related to the suitability of new varieties for commercial distribution. Selecting new varieties is initially based on parent data. Plant breeders developing new varieties of sunflower evaluate many parameters at different stages in the developmental process. In the early stages of growth, breeders evaluate stand count and seedling vigour. As the plant matures, pesticide resistance and disease data is evaluated. The harvested sunflower is measured for yield and seed size.

In some cases, plants are modified for specific increases in certain components, and the plant breeder would be expected to analyse for such components (MAFRD, 2004). A complete review of the biology of sunflower is published in the OECD "Consensus document on the biology of *Helianthus annuus L.* (Sunflower)" (OECD, 2005).

Nutrients in whole sunflower and sunflower products

Composition of oilseed sunflower seeds and non-oil sunflower seeds and kernels

Sunflower seeds are the complete seeds including the hulls, and sunflower kernels are the seeds with the hulls removed. The hulls are difficult to remove from the oilseed type of sunflower and thus the data on oilseed includes the hulls. Hulls are readily separated from the non-oilseed (confectionary) types of sunflower, and thus, the non-oilseed data does not include the hull. According to the USDA, hulls make up 46% of the non-oilseeds, leaving 54% as the kernel. The nutrient content of these two types of sunflowers are shown in Tables 11.2-11.6.

Table 11.2. **Proximate nutrient content of whole sunflower oilseeds and non-oilseeds**

Sunflower seed type	Whole oilseed					Non-oilseed	
						Whole seeds	Kernels
Reference	ST&PA	NRC	Hartman et al.	Ensminger et al.	Kepler et al.	ST&PA	USDA
g/100 g FW							
Dry matter	90.0	91.8	92.6	94.0	91.2	90.0	94.6
g/100 g DM							
Crude protein	19.6	19.2	19.1	22.2	19.0	23.5	24.1
Crude fat	44.0	41.9	43.1	34.4	41.6	25.0	52.4
NDF	24.1	24				32.0	
ADF	16.5	16.7				28.5	
Crude fibre	22.5		15.6	24.1	17.7	24.1	
TDF							11.1
Ash	3.7	5.1	3.9	4.0	4.2	3.8	3.7

Notes: ST&PA: Sunflower Technology and Production Agronomy; FW = fresh weight; DM = dry matter; NDF = neutral-detergent fibre; ADF = acid-detergent fibre; TDF = total dietary fibre.

Sources: ST&PA (1997); NRC (2001); Hartman et al. (1985); Ensminger et al. (1990); Kepler et al. (1982); USDA (2004).

Table 11.3. **Amino acid content (g/100 g dry matter) of whole sunflower oilseeds and non-oilseed kernels**

Sunflower seed type	**Oilseed**		**Non-oilseed kernels**
Reference	ST&PA	Kepler et al.	USDA
Lysine	0.68	0.71	0.99
Histidine	0.49	0.46	0.67
Arginine	1.57	1.79	2.54
Aspartic acid		1.63	2.58
Threonine	0.71	0.68-0.71	0.98
Serine		0.84	1.14
Glutamic acid		4.35	5.89
Alanine		0.53	1.18
Proline		0.93	1.25
Glycine		0.98	1.54
Methionine	0.44	0.33	0.52
Isoleucine	0.79	0.75	1.20
Cystine	0.34	0.34	0.48
Leucine	1.23	1.23	1.75
Phenylalanine	0.89		1.24
Valine	0.95		1.39
Tryptophan	0.23		0.37

Note: ST&PA: Sunflower Technology and Production Agronomy.

Sources: ST&PA (1997); Kepler et al. (1982); USDA (2004).

Composition of sunflower oil

Roughly 80% of the value of sunflower seeds is attributed to their oil content. Like all vegetable oils, sunflower oil is composed of triglycerides (98-99%) and other substances in the unsaponifiable fraction, which are also known as the "minor components" (Evrard et al., 1986; Prévot, 1986).

Table 11.4. **Mineral content (per 100 g dry matter) of whole sunflower oilseeds and non-oilseed kernels**

Sunflower seed type	**Oilseeds**		**Non-oilseed kernels**
Reference	NRC	Beauchemin	USDA
Calcium, Ca (g)	0.71	0.21	0.12
Phosphorus, P (g)	0.51	0.35	0.74
Magnesium, Mg (g)	0.34	0.25	0.37
Potassium, K (g)	1.06	0.72	0.73
Sodium, Na (g)	0.01	0.02	0.003
Sulphur, S (g)	0.21		
Copper, Cu (mg)	2.0	1.6	1.75
Iron, Fe (mg)	14.4	4.7	
Manganese, Mn (mg)	4.5	1.5	2.13
Zinc, Z (mg)	5.3	5.4	5.06
Molybdenum, Mo (mg)	0.18		
Selenium, Se (µg)			59.5

Sources: NRC (2001); Beauchemin (2005); USDA (2004).

Table 11.5. **Vitamin content (per 100 g dry matter) of non-oilseed kernels**

Reference	USDA
Vitamin C, mg	1.48
Thiamine, mg	2.42
Riboflavin, mg	0.26
Niacin, mg	4.75
Pantothenic acid, mg	7.13
Vitamin B6, mg	0.81
Folate, µg	239.86
Vitamin A, IU	52.84
Vitamin E (α tocopherol), mg	36.46
Vitamin K, µg	2.85

Source: Adapted from USDA (2004).

Table 11.6. **Fatty acid composition (g/100 g dry matter) of whole conventional sunflower oilseeds and conventional non-oilseed kernels**

Sunflower type	**Oilseed**	**Non-oilseed kernels**
Reference	Mir	USDA
Myristic acid (C 14:0)	0.02	0.05
Palmitic acid (C 16:0)	2.84	2.95
Palmitoleic acid (C 16:1)	0.03	0.05
Stearic acid (C 18:0)	2.12	2.33
Oleic acid (C18:1)	8.48	9.89
Linoleic acid (C18.:2)	27.8	34.48
Linolenic acid (C18:3)	0.04	0.07
Arachidic acid (C 20:1)	0.06	0.05

Sources: Mir (2005); USDA (2004).

The fatty acid composition is used for the classification of oils. Depending on the predominant type of fatty acid, they can be regarded as either mid-oleic, oleic or high oleic. Sunflower oil from conventionally bred varieties is considered a highly polyunsaturated oil due to its high linoleic acid (C18:2, n-6) content (48.3-74.0%) and its moderate oleic acid (C 18:1) content (14.0-39.4%) (Table 11.7). The level of saturates is 12% on average.

Table 11.7. **Fatty acid profile (% of total fatty acids) of conventional, mid-oleic, high-oleic and high-linoleic sunflower oils**

Sunflower oilseed varieties	**Conventional[1]**	**Mid-oleic[1]**	**High-oleic[1]**
Saturated fatty acids			
Palmitic acid (C 16:0)	5.0-7.6	4.0-5.5	2.6-5.0
Stearic acid (C 18:0)	2.7-6.5	2.1-5.0	2.9-6.2
Monounsaturated fatty acids			
Oleic acid (C 18:1)	14.0-39.4	43.1-71.8	75-90.7
Polyunsaturated fatty acids			
Linoleic acid (C 18:2)	48.3-74.0	18.7-45.3	2.1-17
Linolenic acid (C 18:3)	ND-0.3	ND-0.3	ND-0.3

Sources: 1. Codex (2005). 2. Gunstone (2002); Gibb et al. (2004).

The benefits of the modified varieties (mid-oleic and high oleic) is a higher stability due to their lower content in polyunsaturated acids. Mid-oleic and high-oleic sunflower oils do not need to be hydrogenated when used as frying oil.

Conventional sunflower oil is rich in linoleic acid. Hybrid sunflowers with high oleic acid contents were developed in the United States, France and Spain in the 1980s. They were obtained by mutagenesis breeding of lines in which the desaturase system is blocked to varying degrees. These varieties may contain more than 83% oleic acid and they are considered as high-oleic acid sunflower when their content is higher than 75% (Codex, 2005). The main differences in fatty acid composition between conventional, mid-oleic and high-oleic acid are summarised in Table 11.7.

The oleic acid and linoleic acid contents vary widely according to the variety, the growing conditions and the climate. In oleic sunflowers this is due to:

- Environmental origins: this is essentially a temperature effect. Hot-dry conditions at the end of the crop period lead to an overall decrease in the oil content of sunflower seeds and the fatty acid pattern fluctuates, whereas low temperatures during the maturation phase reduce the oleic acid content and increase the linoleic acid content. Since sunflowers mature in the fall, they produce the most amount of oil and desirable pattern of fatty acids in regions where the autumn is warm.
- Genetic origins: cross pollination also influences the fatty acids patterns of sunflower kernels. When an oleic sunflower field is planted less than 200 metres from a conventional sunflower field, it is quite possible that the oleic sunflower will be pollinated with pollen from conventional sunflowers, reducing the oleic acid content of the harvested seeds. To reduce the chance of cross pollination in certified seed, it has been recommended that the oleic sunflower fields be at least 500 metres from conventional sunflower fields (OECD, 2005).

Between 0.6% and 0.7% of the refined oil is unsaponifiable. This fraction of the oil contains several minor components: waxes, carbohydrates, sterols and antioxidants. It is characterised by high levels of tocopherols (including vitamin E) and phytosterols (Table 11.8). It also has low squalene content.

Table 11.8. **Composition (mg/100 g dry matter) of the minor components of the unsaponifiable fraction of the oil of sunflower varieties**

Sunflower oilseed varieties	Conventional	Mid-oleic	High-oleic
Total sterols	240-500		170-520
Beta-sitosterol [1]	50-70	56-58	42-70
Campesterol [1]	6.5-13.0	9.1-9.6	5-13
Stigmasterol [1]	6.0-13.0	9.0-9.3	4.5-13
Total tocopherols	44-152	50.9-74.1	45-112
alpha (vitamin E)	40.3-93.5	48.8-66.8	40-109
beta	ND-4.5	1.9-5.2	1.0-3.5
Gamma	ND-3.4	0.2-1.9	0.3-3.0

Notes: 1. As a percentage of the total sterol content. 2. ND = non detected.

Source: Adapted from Codex (2005).

Composition of flour and protein concentrate

Sunflower flour and protein concentrate are processed from sunflower seeds for their amino acid content. Table 11.9 provides the essential amino acid content of these products. Sunflower proteins are rich in globulins (55-60%), albumins (17-23%) and glutelins (11-17%) (Canella et al., 1982).

Table 11.9. **Essential amino acid profile of flour and protein concentrates**

g/100 g of crude protein

Amino acid	Flour	Concentrate
Isoleucine	3.7	4.6
Leucine	6.5	6.8
Lysine	3.4	3.4
Methionine + cystine	4.1	3.5
Tryptophan	1.5	1.4
Phenylalanine + tyrosine	8.2	8.7
Valine	4.9	4.6
Threonine	3.3	3.4

Source: FAO (1981).

Composition of sunflower meal

Sunflower meal is a by-product of oil processing as shown in Figure 11.1.

The proximate composition of whole sunflower seed meal and part-dehulled sunflower meal are reported in Tables 11.10 and the amino acid composition in Table 11.11.

Composition of sunflower hulls

Sunflower hulls are derived from the process shown in Figure 11.1. The proximate nutrient composition and the quantity of some minerals are shown in Table 11.12.

Table 11.10. **Composition of sunflower meal derived from whole and part-dehulled seeds**

	Whole sunflower seed meal			Part-dehulled sunflower meal		
Reference	Sauvant et al.		NRC	Sauvant et al.		Preston
	Mean	Standard deviation	Mean	Mean	Standard deviation	Mean
Dry matter g/100 g FW	88.7	1.4		89.7	1.2	92.0
g/100 g DM						
Crude protein	27.7	2.2	28.4	33.4	2.2	38.0
Crude fibre	25.5	2.6		21.2	2.0	20.0
Crude fat	2.0	0.8	1.4	1.7	0.6	2.5
Minerals (ash)	6.2	0.6	7.7	6.7	0.5	8.0
Neutral detergent fibre	41.1	3.7	40.0	35.9	3.6	36.0
Acid detergent fibre	29.3	3.0	30.0	24.7	2.4	24.0
Lignin	10.1	1.4		8.2	1.2	

Notes: FW = fresh weight; DM = dry matter.

Sources: Sauvant et al. (2004; Argentinian type, 2 729 samples whole, 1 141 samples part-dehulled); NRC (2001); Preston (2005).

Table 11.11. **Amino acid and protein content (g/100 g dry matter) of sunflower meal derived from whole and part-dehulled seeds**

	Non-dehulled sunflower meal[1]			Dehulled sunflower meal [2]	
Reference	NRC (2001)	NRC (1998)	NRC (1994)	NRC (1998)	NRC (1994)
Arginine	2.32	2.64	2.56	3.15	3.17
Histidine	0.74	0.73	0.61	0.99	0.97
Isoleucine	1.16	1.43	1.11	1.55	1.59
Leucine	1.82	2.07	1.78	2.48	2.47
Lysine	1.01	1.12	1.11	1.29	1.38
Methionine	0.65	0.66	0.56	0.88	0.89
Cystine	0.50	0.53	0.56	0.71	0.71
Phenylalanine	1.31	1.37	1.28	1.78	1.85
Threonine	1.06	1.16	1.17	1.43	1.44
Tryptophan	0.34	0.42	0.50	0.47	0.46
Valine	1.41	1.66	1.78	1.87	1.94
Glycine					2.26
Tyrosine		0.84		1.11	1.01
Serine			1.11		1.66
Crude Protein	28.4	29.8	25.9	45.4	41.0

Notes: 1. NRC (2001) calculated from percentage of crude protein, using referenced crude protein content of 28.4%; NRC (1994) calculated from referenced dry matter content of 90%. 2. NRC (1998) calculated from referenced dry matter content of 93%; NRC (1994) calculated from referenced dry matter content of 89.8%.

Sources: NRC (1994, 1998, 2001).

Composition of sunflower forage

The green sunflower plant is used as silage and forage. The content of nutrients of the green plant is dependent on its stage of maturity. Sunflower forage has a high amount of moisture at maturity. It is cut and wilted prior to ensiling. The nutrient composition of sunflower silage is shown in Table 11.13.

Table 11.12. **Nutrient content of sunflower hulls**

Parameter	Mean	Range
Dry matter (g/100 g FW)	87.8	85.0-92.0
	In g/100 g DM	
Crude protein	5.0	3.5-9.0
Crude fat	3.0	0.5-3.0
Neutral detergent fibre	70.0	65.0-75.0
Acid detergent fibre	56.0	50.0-63.0
Crude fibre	45.0	40.0-50.0
Ash	2.7	2.0-3.0
Calcium	0.30	0.25-0.35
Phosphorus	0.15	0.10-0.20
Magnesium	0.20	0.15-0.25

Notes: FW = fresh weight; DM = dry matter.

Source: Adapted from Sunflower Technology and Production Agronomy (1997).

Table 11.13. **Composition of sunflower silage**

Source[1]	Putnam et al.	Gregorie	Kling and Wöhlbier			
Stage of maturity	**Mature**		**Before bloom**	**Beginning of bloom**	**In bloom**	**After bloom**
Dry matter (g/100 g FW)		30	12	20	14	15
g/100 g DM						
Crude protein	11-12	12.5	19.3	13.9	14.7	14.0
Crude fat	10-12	10.7	2.7	4.4	2.4	2.8
Crude fibre	31.0		21.1	19.8	23.0	27.4
Acid detergent fibre	32.0	39				
Lignin	10-16	12.3				

Notes: FW = fresh weight; DM = dry matter.

Sources: Putnam et al. (1990); Gregorie (2006); Kling and Wöhlbier (1983).

Anti-nutrients and toxicants in sunflower

Sunflower seeds are not known to contain significant quantities of anti-nutritional factors or toxicants. However, Mulvenna et al. (2005) have detected a precursor of a cyclic trypsin inhibitor in sunflower seeds.

Sunflower kernels and hulls contain phenolic compounds, including chlorogenic and caffeic acids, which are easily oxidised during common processing causing green to brown discoloration in protein isolates and/or concentrates (Sabir et al., 1974a, 1974b). These compounds have been studied both for their additive/synergistic effect on carcinogenesis and their anti-carcinogenic properties (Hirose et al., 1997).

Recent studies have shown that sunflower seeds have been found to contain an allergen, the 2S albumin, which shows homology to the 2S albumins from allergenic nuts (Kelly et al., 2000). However, according to the European legislation, sunflowers and seeds are not required to be labelled as allergens (European Commission, 2005).

The nitrate content of green sunflower plants was found to be almost as high as in oat forage. With concentrations of 0.8% of the dry matter being nitrate, sunflower belongs to the category of plants with a high capacity of nitrate storage (Liebenow, 1971).

Sunflower leaves are known to contain high levels of saponins. The saponin concentration is found to be two to three times higher than in alfalfa or red clover. However, it has not yet been investigated to what extent these substances act as haemolytic agents in farm animals (Koch and Pintácsi, 1969).

Sunflower for food

Whole seeds

Non-oilseed seeds are used for confectionary purposes. Dehulled seeds (i.e. kernels) can be used either to accompany aperitifs (roasted and salted), or in salads or cakes. Whole sunflower seeds (with hulls) are also sold as a snack food. Such seeds are specifically selected and produced for use as snacks (Bonjean, 1993; Agriculture and Agri-Food Canada, 2003). They are bigger, "softer" and contain less oil than oilseed type sunflower seeds. Their hulls are usually striped and relatively thick and can be easily removed.

Sunflower oil

Sunflower oil is traditionally used as salad oil or cooking oil. Due to its low linolenic acid content (< 1%) sunflower oil is sufficiently heat stable to be used for frying.

Sunflower oil is also used in the food industry, mainly in margarines and shortenings as well as in various foods to enhance the composition and physical properties of the food.

Linoleic and linolenic acids cannot be synthesised by humans, and are therefore essential fatty acids in the human diet. A diet either lacking or containing an unbalanced ratio of these fatty acids results in symptoms of deficiency, including reduced growth (IOM, 2002; CNERNA-CNRS, 2001).

Sunflower oil contains sterols, of which the most common are beta-sitosterol, campesterol and stigmasterol. The levels of these compounds are shown in Table 11.8.

Protein concentrates and isolates

Sunflower proteins are produced in the form of flours (55% protein in dry matter), concentrates (70% protein in dry matter) or isolates (85-90% protein in dry matter). Proteins extracted from sunflower seeds have potential nutritional and functional advantages, as they do not contain anti-nutritional substances, do not have a specific taste and contain the two essential sulphur-containing amino acids, methionine and cystine (Table 11.9). Sunflower proteins also have certain physico-chemical properties that could be advantageous for their use in foods (Sosulski, 1984).

The use of sunflower proteins in the food industry has been limited due to the negative effects of certain procedures, such as heat treatment under pressure, commonly used in processing of sunflower seeds into oil and meal. It has been reported that proteins generated from seeds subjected to such procedures have poor solubility and functional properties (Vanktesh and Prakash, 1993). Several researchers are pursuing additional processes, such as enzymatic hydrolysis, as a means of producing more desirable products (Cai et al., 1996; Conde et al., 2005; Parrado et al., 1993). Another deterrent to the use of sunflower proteins in food products is the presence of phenolic compounds, which are easily oxidised into dark green and brown compounds and may cause discoloration of processed and cooked foods. Chlorogenic acid is the principal colour-forming constituent of sunflower kernels, but small quantities of caffeic acid and other phenolic compounds are also present. These compounds bind to proteins and their removal from sunflower products is difficult (Sabir et al., 1974a, 1974b; Saeed and Cheryan, 1989).

Recommendation of key components to be analysed

Roughly 80% of the economic value of oilseed sunflower is attributed to the oil content, which approximately represents half of the seed. Proteins constitute the main economic value of the remaining part of the oilseed which also contains fibre and minerals. As noted earlier, sunflower seeds do not contain any natural anti-nutrients or toxicants. It is recommended that for the oilseed varieties, sunflower seeds and oil be analysed. The key components to be analysed are listed in Table 11.14. For the non-oilseed varieties, only kernels need to be analysed.

For human nutrition, it is important to assess the fatty acid composition of the oil (Table 11.7). Sunflower oil should also be assessed for its tocopherol and sterol content

(Table 11.8). Tocopherol (vitamin E) is an important micronutrient and antioxidant that prolongs the shelf life of the oil and food products containing the oil. Seeds or kernels should be analysed for proximates, amino acids, fatty acids, sterols and tocopherols, minerals and vitamins.

Table 11.14. **Suggested nutrients of sunflower seeds or kernels and oil to be analysed for human food**

Nutrients	Oil	Seeds or kernels
Proximates		X
Amino acids		X
Fatty acids	X	X
Sterols and tocopherols	X	X
Minerals		X
Vitamins		X

Sunflower for feed

Sunflower seeds

Damaged sunflower seeds from the food industry may be fed to beef cattle (Gibb et al., 2004); however, some good quality seeds are also used in dairy cattle rations (Linn, 1990). The indigestible hull of sunflower seeds physically protects the highly degradable unsaturated oil from being released too rapidly, thus giving the by-pass fat effect. The by-pass energy effect is only achieved when cows are fed whole, dried sunflower seeds. The nutrient composition of whole oilseed and non-oilseed (confectionary) sunflowers is listed in Table 11.2.

Sunflower meal

Sunflower meal has two interesting qualities compared with meal derived from other oilseeds:

- it is composed of proteins that are rich in sulphur-containing amino acids (methionine and cystine) compared to soybean meal
- it has not been shown to contain any anti-nutritional factors.

Nevertheless, it has three disadvantages:

- a high cellulose content compared to soybean and rapeseed (low energy value) meals
- a low amount of lysine compared to soybean meal
- the presence of chlorogenic acid that could interfere with the colour of the protein-based products

The nutrient composition of sunflower meal is dependent on the oil content of the seed, extent of hull removal, efficiency of oil extraction and the processing temperature. The fibre in sunflower meal is low in digestibility and may be a disadvantage when balancing rations for non-ruminant and high-producing animals.

The sunflower meal is mainly used for rabbits and ruminants (respective incorporation rates are: 10-12% and 10-20%). Uses for laying birds (less than 5%) or pigs (less than 1%) can be noticed (Burghart and Evrard, 2002).

Dehulled meal, such as "Hipro", in which the protein content has been increased nearly to 40% by an efficient dehulling, can be useful feeds for broilers and fattening pigs. The incorporation rates may then be increased to at least 10% (Evrard et al., 1986).

Sunflower meal is more ruminally degradable (74% of crude protein) than soybean meal (66%) or canola meal (68%). Sunflower meal is high in protein but due to the lack of a sufficient content of lysine, is more suitable for ruminants than non-ruminants.

Sunflower extracted meal is a valuable protein source for the various species/categories in livestock feeding if the feed specific restrictions, i.e. the maximum incorporation rates, are taken into consideration (Table 11.15).

Table 11.15. **Maximum incorporation rates for sunflower extracted meal in rations for livestock feed**

Species/category	Incorporation level (%)
Dairy cows	30
Rearing calves	15
Cattle and bulls	40-50
Sheep, goat	No limitation
Rabbits	30
Growing-finishing pigs	5-10
Poultry	5-10

Source: Adapted from Hoffmann (2001).

Sunflower oil

Sunflower oil has limited applications in livestock feed mainly due to its higher economic value for use with human food preparation for cooking and frying. In small quantities it may be used to reduce dust in animal feeds.

Sunflower silage

Sunflowers can be used as a source of forage by livestock producers. Whole plant sunflower silage has slightly more crude protein and considerably more fat than corn silage on a dry matter basis. The high level of lignin from the fibrous stalk is a disadvantage to sunflower silage.

Sunflower hulls

Sunflower hulls are not a suitable feeding stuff due to the high crude fibre content and the type of binding which causes a negative efficiency of energy utilisation, although they are used as a cheap fibre source to a limited extent in cattle or lamb feeding (Kling and Wöhlbier, 1983). They are used for livestock bedding.

Recommendation of key components to be analysed

When one considers the sunflower products that might be fed to animals, their nutrient content would not be expected to change if the content of whole sunflower seed is not changed. Hence, only the whole sunflower seed and sunflower meal are suggested to be analysed (Table 11.16). However, for amino acids, either whole sunflower or sunflower meal would yield equivalent results. For fatty acids, whole sunflower seeds or sunflower oil would yield equivalent results.

Proximate analysis is used by livestock nutritionists to evaluate feed ingredients and to formulate least-cost rations for livestock. It includes the amounts of crude protein, fat, ash and crude fibre. In the case of ruminants and swine, the traditional analysis for crude fibre is considered not informative and has been replaced by analyses for acid detergent fibre and neutral detergent fibre. For amino acids, the ten essential amino acids (argenine, histidine, isoleucine, leucine, lysine, methionine, phenylalanine, threonine, tryptophan and valine) plus glycine, cystine, tyrosine, serine and proline are the key nutrients. Linoleic acid is the fatty acid of key importance for the seed (OECD, 2002).

Table 11.16. **Suggested nutrients of sunflower seeds and meal to be analysed for animal feed**

Parameters	**Seeds**	**Meal**
Proximates	X	X
Neutral detergent fibre	X	X
Acid detergent fibre	X	X
Amino acids	X	X
Fatty acids	X	
Calcium	X	X
Phosphorus	X	X

Other minerals such as selenium are also important, but their concentration in plants has been shown to reflect the amount of the mineral in the soil. Thus, the minerals other than calcium and phosphorus, and the vitamins are not considered key nutrients.

It has been noted above that sunflower seeds do not contain any natural anti-nutrients.

Note

1. Also referred to as sunflower seed, partially decorticated, extracted, by M4 Council (European) Directive 96/25/EC of 29 April 1996.

References

Agriculture and Agri-Food Canada (2003), "Sunflower seed: Situation and outlook", *Bi-weekly Bulletin*, Vol. 16, No. 16.

Beauchemin, K. (2005), Personal communication, Ruminant Nutrition, Agiculture Agri-Food Canada Research Centre, Lethbridge, Alberta, Canada.

Bonjean, A. (1993), *Le Tournesol*, Les Éditions de l'Environnement.

Burghart, P. and J. Evrard (2002), "Oilseeds: From storage to feed", *Les Points Techniques du CETIOM*, Éditions CETIOM, Paris.

Cai, T. et al. (1996), "Physicochemical properties and yields of sunflower protein enzymatic hydrolysates as affected by enzyme and defatted sunflower meal", *Journal of Agricultural and Food Chemistry*, Vol. 44, No. 11, pp. 3 500-3 506, American Chemical Society, http://dx.doi.org/10.1021/jf9507396.

Canella, M. et al. (1982), "Albumin and globulin components of different sunflower cultivars", *Rivista italiana della sostanze grasse*, Vol. 59, Issue 8, pp. 377-382.

CNERNA-CNRS (2001), *Daily Recommended Amounts for the French Population*, 3rd edition, Lavoisier Tec & Doc, Paris.

Codex (2005), "Standard for named vegetable oils", CODEX-STAN 210 (amended 2003, 2005).

Conde, J.M. et al. (2005), "Effect of enzymatic treatment of extracted sunflower proteins on solubility, amino acid composition and surface activity", *Journal of Agricultural and Food Chemistry*, Vol. 53, No. 20, October, pp. 8 038-8 045.

Delplanque, B. (2000), "Intérêt nutritionnel des huiles de tournesol: Tournesol linoléique et tournesol à haute teneur en oléique", *Ocl*, Vol. 7, No. 6, pp. 467-472.

Ensminger, M.E. et al. (1990), *Feeds and Nutrition*, Second Edition, The Ensminger Publishing Co., Clovis, California.

European Commission (2005), Directive 2005/26/CE of 21 March 2005, *Official Journal of the European Commission*, L 75/33-34, European Union, Brussels.

Evrard, J. et al. (1986), *Tournesol: Conditionnement, transformation, produits*, Cahier technique, CETIOM, Paris.

FAO (1981), *Helia: Scientific bulletin of the F.A.O. Research Network on Sunflower*, Vol. 14, No. 14, July, Food and Agriculture Organization, Rome.

FAO/WHO (1996), "Joint consultation report, Biotechnology and food safety", *FAO Food and Nutrition Paper No. 61*, FAO, Rome, available at: www.fao.org/ag/agn/food/pdf/biotechnology.pdf (accessed 28 January 2015).

Gibb, D.J. et al. (2004), "Value of sunflower seed in finishing diets of feedlot cattle", *Journal of Animal Science*, Vol. 82, No. 9, September, pp. 2 679-2 692.

Gregorie, T. (2006), "Sunflower silage", North Dakota State University Extension Service, available at: www.ext.nodak.edu/extnews/newsrelease/2004/082604/10sunflo.htm.

Gunstone, F. (2002), "Sunflower seed and its products", *Inform*, No. 13, pp. 159-163.

Hartman, A.D. et al. (1985), "Effect of sunflower seeds on performance, carcass quality, fatty acids and acceptability of pork", *Journal of Animal Science*, Vol. 60, No. 1, pp. 212-219, available at: www.animalsciencepublications.org/publications/jas/pdfs/60/1/JAN0600010212.

Hirose, M. et al. (1997), "Carcinogenicity of antioxidants BHA, caffeic acid, sesamol, 4-methoxyphenol and catechol at low doses, either alone or in combination and modulation of their effects in a rat medium-term multi-organ carcinogenesis model", *Carcinogenesis*, Vol. 19, No. 1, January, pp. 207-212.

Hoffmann, M. (2001), Futterspezifische Restriktionen. Arbeitskreis "Futter und Fütterung im Freistaat Sachsen.

IOM (2002), *Dietary Reference Intakes for Energy, Carbohydrate, Fiber, Fat, Fatty Acids, Cholesterol, Protein, and Amino Acids*, National Academies Press, Washington, DC.

Kelly, J.D. et al. (2000), "2s methionine-rich protein (Ssa) from sunflower seed is an Ige-binding protein", *Allergy*, Vol. 55, No. 6, June, pp. 556-559.

Kepler, M. et al. (1982), "Sunflower seeds as a fat source in sow gestation and lactation diets", *Journal of Animal Science*, Vol. 55, No. 5, pp. 1 082-1 086, http://dx.doi.org/10.2134/jas1982.5551082x.

Kling, M. and W. Wöhlbier (Hrsg.) (1983), *Handelsfuttermittel*, Ulmer Verlag Stuttgart.

Koch, E. and M. Pintácsi (1969), "Investigations into some fodder crops for haemolytic saponine", *Agrobotanika*, Vol. XI, pp. 113-117.

Laisney, J. (1992), "Obtention des corps gras", in: *Manuel des corps gras*, Lavoisier Tec & Doc, Paris, pp. 695-730.

Laisney, J. (1984), *L'huilerie moderne: Art et technique*, CFDT, Paris.

Liebenow, H. (1971), "Nitrate und Nitrite in ihrer Beziehung zu Mensch und Tier", *Arch. Tierernährung*, Vol. 21, pp. 635-648.

Linn, J.G. (1990), "Feeding sunflowers to dairy cows", Minnesota Ext. Service, Issue 102, St. Paul, Minnesota.

MAFRD (2004), "Sunflower production and management" webpage, Manitoba Agriculture, Food and Rural Development, www.gov.mb.ca/agriculture/crops/production/sunflowers.html (accessed Feb. 2015)

Merrien, A. (1992), "Source et monographie des principaux corps gras", in: *Manuel des corps gras*, Lavoisier Tec & Doc, Paris, pp. 116-123.

Mir, P.S. (2005), Personal communication. Agiculture Agri-Food Canada Research Centre, Lethbridge, Alberta, Canada.

Mulvenna, J.P. et al. (2005), "Discovery, structural determination, and putative processing of the precursor protein that produces the cyclic trypsin inhibitor sunflower trypsin inhibitor 1", *Journal of Biological Chemistry*, Vol. 280, No. 37, pp. 32 245-32 253.

NRC (2001), *Nutrient Requirements of Dairy Cattle*, Seventh Revised Edition, National Research Council, National Academies Press, Washington, DC, available at: www.nap.edu/catalog/9825/nutrient-requirements-of-dairy-cattle-seventh-revised-edition-2001.

NRC (1998), *Nutrient Requirements of Swine*, Tenth Revised Edition, National Research Council, National Academies Press, Washington, DC, available at: www.nap.edu/catalog/6016/nutrient-requirements-of-swine-10th-revised-edition.

NRC (1994), *Nutrient Requirements of Poultry*, Ninth Revised Edition, National Research Council, National Academies Press, Washington, DC, available at: http://books.nap.edu/catalog/2114.html.

OECD (2005), "Consensus document on the biology of *Helianthus annuus* L. (Sunflower)", ENV/JM/MONO(2004)30, OECD, Paris, available at: www.oecd.org/science/biotrack/46815798.pdf.

OECD (2002), "Consensus document on compositional considerations for new varieties of maize (*Zae mays*): Key food and feed nutrients, anti-nutrients and secondary plant metabolites", ENV/JM/MONO(2002)25 (No. 6), OECD, Paris, available at: www.oecd.org/science/biotrack/46815196.pdf.

OECD (2000), "Report of the Task Force for the Safety of Novel Foods and Feeds", C(2000)86/ADD1, OECD, Paris, available at: www.biosafety.be/ARGMO/Documents/report_taskforce.pdf.

OECD (1997), "Report of the OECD workshop on toxicological and nutritional testing of novel foods", SG/ICGB(1998)1, OECD, Paris, final version available at: www.oecd.org/officialdocuments/publicdisplaydocumentpdf/?doclanguage=en&cote=sg/icgb(98)1/final.

Oil World (2005), "World production arrested area", Statistics update, April.

Parrado, J. et al. (1993), "Characterization of enzymatic sunflower protein hydrolysates", *Journal of Agricultural and Food Chemistry*, Vol. 41, No. 11, November, pp. 1 821-1 825, http://dx.doi.org/10.1021/jf00035a003.

Preston, R.L. (2005), "Feed composition tables. Beef", 1 April, http://beef-mag.com.

Prévot, A. (1986), "L'huile de tournesol aujourd'hui", Journées d'information ITERG et Journées Chevreul AFECG, 14-15 October.

Putnam, D.H. et al. (1990), *Sunflower. Alternative Field Crops Manual*, Co-op. Extension Service, University of Wisconsin, Madison, Wisconsin.

Sabir, M.A. et al. (1974a), "Phenolic constituents in sunflower flour", *Journal of Agricultural and Food Chemistry*, Vol. 22, No. 4, July, pp. 572-574, http://dx.doi.org/10.1021/jf60194a051.

Sabir, M.A. et al. (1974b), "Chlorogenic acid-protein interaction in sunflower", *Journal of Agricultural and Food Chemistry*, Vol. 22, No. 4, July, pp. 575-578, http://dx.doi.org/10.1021/jf60194a052.

Saeed, M. and M. Cheryan (1989), "Chlorogenic acid interaction with sunflower proteins", *Journal of Agricultural and Food Chemistry*, Vol. 37, No. 5, September, pp. 1 270-1 274, http://dx.doi.org/10.1021/jf00089a015.

Sauvant, D. et al. (2004), *Tables de composition et de valeur nutritive des matières premières destinées aux animaux d'élevage*, INRA, Paris.

Schild, J. et al. (1991), "G91-1026 Sunflower production in Nebraska", Historical Materials from University of Nebraska-Lincoln Extension, Paper 772, available at: http://digitalcommons.unl.edu/cgi/viewcontent.cgi?article=1772&context=extensionhist.

Skrypetz (2005), "Sunflower Seed: Situation and Outlook", in *Bi-weekly Bulletin of the Market Analysis Division, Agriculture and Agri-Food Canada, September 20, 2005,* Winnipeg, Manitoba, Canada, http://www.seedquest.com/News/releases/2005/pdf/13559.pdf (accessed 11 Feb. 2015).

Soldatov, K.I. (1976), "Chemical mutagenesis in sunflower breeding", Proceedings of the 7th International Sunflower Conference, Krasnodar, Vol. 7, pp. 352-360.

Sosulski, F.W. (1984), "Food uses of sunflower proteins", in: *Developments in Food Proteins*, Vol. 3, B.J.F. Hudson, Elsevier Applied Science Publishers, London.

Sunflower Technology and Production Agronomy (1997), No. 35, Table obtained from National Sunflower Association website www.sunflowernsa.com/buyers/default.asp?contentID=186.

USDA (2004), *USDA Nutrient Database for Standard Reference*, Release 16.1 Nutrient Data Laboratory homepage, United States Department of Agriculture, Agricultural Research Service, www.nal.usda.gov/fnic/foodcomp.

Vanktesh, A. and V. Prakash (1993), "Functional properties of the total proteins of sunflower (*Heliantus annuus* L.) seed – Effect of physical and chemical treatments", *Journal of Agricultural and Food Chemistry*, Vol. 41, No. 1, January, pp. 18-23, http://dx.doi.org/10.1021/jf00025a005.

Vick, B.A. et al. (2003), "Registration of two sunflower genetic stocks with reduced palmitic and stearic acids", *Crop Science*, Vol. 43, May, pp. 747-748.

WHO (1991), *Strategies for Assessing the Safety of Foods Produced by Biotechnology*, Report of a Joint FAO/WHO Consultation, World Health Organization of the United Nations, Geneva, out of print.

Chapter 12

Tomato (*Lycoperson esculentum*)

*This chapter, prepared by the OECD Task Force for the Safety of Novel Foods and Feeds with Greece as the lead country, deals with the composition of tomato (*Lycopersicon esculentum*). It contains elements that can be used in a comparative approach as part of a safety assessment of foods and feeds derived from new varieties. Background is given on tomato production, processing, uses, appropriate varietal comparators and characteristics screened by breeders. Then nutrients in tomato (of different colours) and its products, toxicants and allergens are detailed. The final sections suggest key constituents for analysis of new varieties for food use and for feed use.*

Background

Production

Tomatoes are cultivated in more than 150 countries around the world on approximately 4 million hectares (ha). The total average annual production over the period 1999-2003 was approximately 108 million tonnes, as shown in Table 12.1. The main producer is the People's Republic of China with approximately 23 million tonnes or 21.8% of the total production. The United States follows, with approximately 12 million tonnes or 10.6% of the total production. Tomato is considered one of the most important vegetables produced in commercial agriculture because of income generated from export.

Table 12.1. **Average annual world tomato production, 1999-2003**

Rank	Country	Production area (ha) (mean value)	Production ('000 tonnes) (mean value)
1	China (People's Republic of)	958 585	23 610.36
2	United States	173 030	11 876.86
3	Turkey	226 000	8 944.20
4	India	494 000	7 564.00
5	Italy	129 728	6 792.40
6	Egypt	185 515	6 417.62
7	Spain	62 539	3 858.83
8	Brazil	60 624	3 318.42
9	Iran, Isl. Rep. of	119 670	3 360.99
10	Mexico	73 219	2 163.64
11	Russian Federation	150 758	1 896.32
12	Greece	41 157	1 849.91
13	Chile	19 413	1 255.53
14	Ukraine	107 832	1 108.50
15	Uzbekistan	28 740	1 028.64
	World	3 985 737	108 365.46

Notes: The production values are the sum of tomatoes grown for industrial use and fresh consumption. Countries are listed in order of production quantities.

Source: FAO (2004).

The commercial tomato belongs to the genus *Lycopersicon.* It is a relatively small genus within the large and diverse botanic family *Solanaceae.* The genus is currently thought to consist of the cultivated tomato, *Lycopersicon esculentum*, and seven closely related wild *Lycopersicon* species. It is worth mentioning that some of the wild species contain valuable genes for disease and pest resistance that can be useful for plant breeders in developing new types of cultivated tomatoes when crossed with *L. esculentum.* All cultivated tomato varieties belong to the species *L. esculentum.*

The most likely region where the tomato was first domesticated is the Puebla-Veracruz area of Mexico, where the greatest varietal diversity of the cultivated form can be found today (Jenkins, 1948). It is thought to have reached this area as a weedy cherry tomato, var. *cerasiforme*, and upon domestication, to have become the large-fruited *L. esculentum* by selection.

Appropriate comparators for testing new varieties

This chapter suggests parameters that tomato breeders should measure. Measurement data from the new variety should ideally be compared to those obtained from the near isogenic non-modified counterpart. Moreover, comparison can be made between values obtained from new varieties and data available in the literature, or chemical analytical data generated from other commercial tomato varieties. Components to be analysed include key nutrients, toxicants and allergens. Key nutrients are those components in a particular product which may have a substantial impact in the overall diet. These may be major constituents (fats, proteins, and structural and non-structural carbohydrates) or minor compounds (vitamins and minerals). Similarly, the levels of known allergens should be considered. Key toxicants are those toxicologically significant compounds known to be inherently present in the species, i.e. compounds whose toxic potency and levels may impact human and animal health. The key components analysed are used as indicators of whether unintended effects of the genetic modification influencing plant metabolism has occurred or not.

Processing

Tomato is processed as shown in Figure 12.1. The most important processing methods are drying (to produce dried tomatoes or a powder) and concentration (to a paste or purée). For each of the processes the tomato should be ripe, red, firm to soft, free of mould growth (by cutting out infected parts) and free of stems, leaves and dirt (by washing) (Gould, 1992). Also common in some places of the world is the use of green tomatoes (normally home-grown) for different recipes e.g. fried green tomatoes, green tomatoes ketchup or chutney or pickles.

Traditional methods in hot, dry regions include sun drying. The tomato halves are placed on clean flat surfaces (e.g. roofs) with the cut side facing up or thread on to strings that are hung in the sun from a branch or beam. In both cases, drying is relatively rapid (depending on the temperature and humidity of the air) but there is a risk that the product may be contaminated by insects, dirt and dust. This risk can be reduced by covering the tomatoes with fine muslin cloth or mosquito netting. The end-product is dark, red, leathery pieces with a strong tomato flavour. Rehydration is relatively slow, but this may be of little importance in cooking applications. Layers of pulp can also be dried to a rubbery consistency and stored in plastic film bags. Alternatively, the post dried pulp can be formed into balls or cubes and then dried in the sun or over a fire. Provided that the humidity is low, the dried product will keep without special packaging for several months. If the humidity rises the product will go mouldy and should be protected either by suitable packaging (e.g. in sealed plastic bags, preferably polypropylene or thick polythene, or in sealed pottery jars covered in oil) or dried slowly over a fire to a low moisture content. The tomatoes should be far enough away from the fire to prevent cooking and contamination with polycyclic aromatic hydrocarbon (PAH). They will be fully dried when they are hard and brittle. Alternatively, artificial drying may be considered.

Tomatoes dried to a low moisture content to become hard (e.g. 5% water) can be pounded or milled to a powder. The most convenient way to store tomato powder is in sealed glass or pottery jars or in sealed thin polypropylene film bags.

Tomatoes can be boiled to evaporate the water. Depending on how much water is removed and what other ingredients are mixed to the pulp, it is possible to obtain a large number of products, the most common of which are shown in Table 12.2.

The basic preservation principle behind all of these products is to remove water by boiling. It results in destruction of enzymes and microorganisms by heat and concentration of the product so that contaminating microorganisms cannot re-grow.

Table 12.2. **Products from tomato pulp**

Product	Other ingredients
Paste	Salt, spice, flavoring, baking soda
Purée	Salt
Jam	(Pectin) sugar, (acid) vinegar, salt, spices
Chutney	
Ketchup	Sugar, vinegar, salt, onions, starches and spices
Soup	Flour, salt, sugar

Note: The solids content is usually measured by refractometer as degrees Brix (°Bx).

There are tomato products for which quality and specification standards are under development by the Codex Committee on Processed Fruits and Vegetables (CCPFV), while for other products no such specifications have been implemented. The main quality parameters of tomato purée and paste are colour, consistency and flavour. However, there are no standardised methods and instruments for defining quality. While colour can be measured objectively, there are currently no standard colour requirements for tomato concentrates. The volatiles responsible for flavour and odour have been identified to the point where the natural odour of tomato paste can be imitated (Hayes et al., 1998).

Processed fruits and vegetables have been long considered to have lower nutritional value than their fresh commodities due to the loss of vitamin C during processing. Studies that have been conducted in order to investigate the effect of thermal processing of ripe raw tomatoes on the quality of the final products in relation to nutrient content and antioxidant activities showed that thermal processing of tomato juice, baked tomatoes, tomato sauce and tomato soup reduced the vitamin C content but increased the amount of total phenolics and the water soluble antioxidant capacity of the tomato products (Gahler et al., 2003; Abushita et al., 2000). Ascorbic acid, alpha-tocopherol quinone and beta-carotene were the most susceptible components to thermal degradation.

Studies on the influence of processing on content of various other antioxidants (phenolics, flavonoids and carotenoids) and total antioxidant activity of tomato sauce show that processing mainly reduce naringenin (a flavonoid) content, and increase the content of chlorogenic acid and all-*trans*-lycopene (Dewanto et al., 2002; Re et al., 2002). The effects of processing on the overall antioxidant activity support the theory of a general increase in bioavailability of individual antioxidants. Thus the total antioxidant activity of both hydrophilic and lipophilic extracts of processed tomatoes were increased.

Processing seems to increase nutrient bioavailability, which could be due to the fact that the nutrients are detached or extracted from their structures. This is particularly true for lycopene (Rao et al., 1998; Shi and Le Maguer, 2000). Moreover, lycopene in raw tomatoes is present mainly as all-*trans*-lycopene. Heat processing of tomato juice enhances its isomerisation to the *cis* isomer, contributing to an increased bioavailability (Stahl and Sies, 1992; Shi and Le Maguer, 2000).

Tomato soup is produced mostly from fresh tomatoes but may also be produced from tomato paste.

Tomato squash is tomato pulp with added sugar syrup to give a concentration of 30-50% total solids (Brix) measured by refractometer. It is not a widespread product as people tend to prefer squashes from other fruits. It is processed in a similar way to juice and may, in addition, contain up to 100 ppm of sodium (or potassium) benzoate preservative in most countries (Gould, 1992).

Figure 12.1. **Processing of tomatoes**

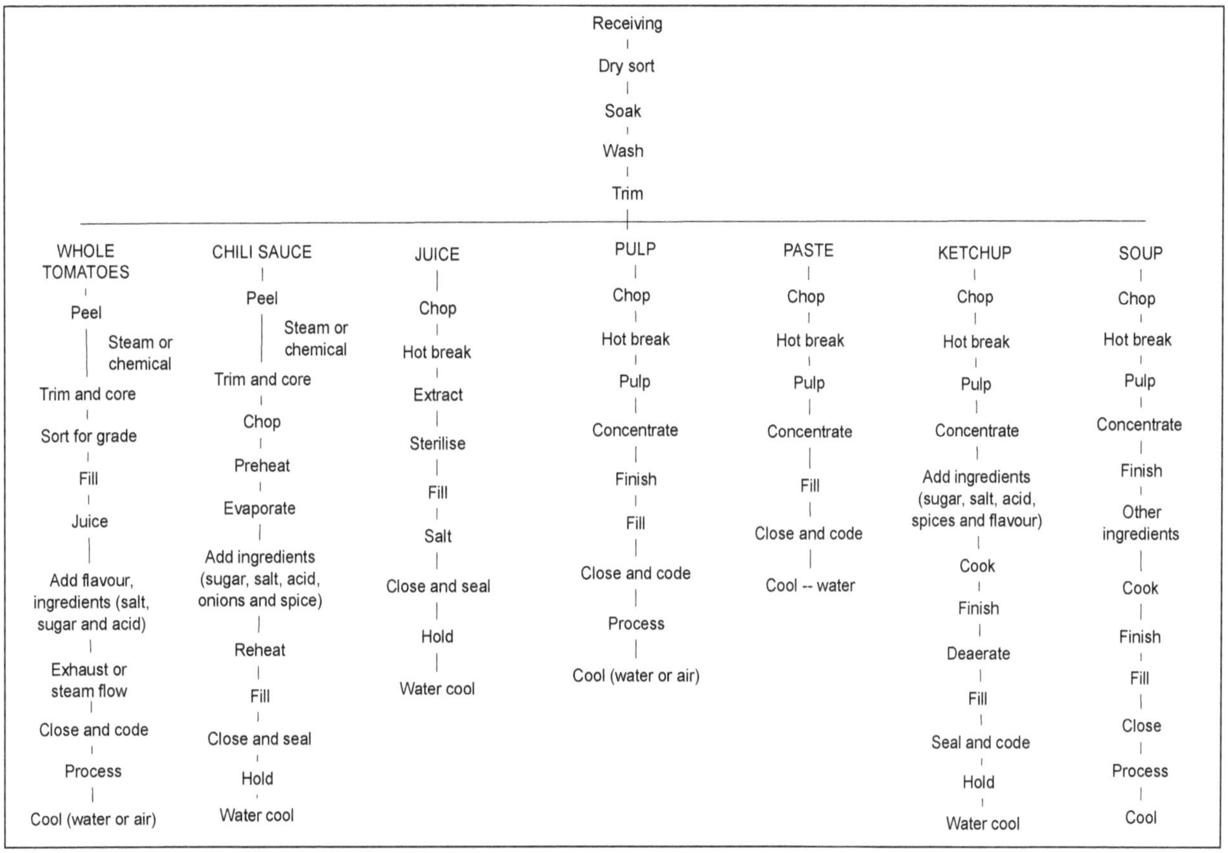

Receiving – Dry sort – Soak – Wash – Trim

- WHOLE TOMATOES: Peel (Steam or chemical) – Trim and core – Sort for grade – Fill – Juice – Add flavour, ingredients (salt, sugar and acid) – Exhaust or steam flow – Close and code – Process – Cool (water or air)
- CHILI SAUCE: Peel (Steam or chemical) – Trim and core – Chop – Preheat – Evaporate – Add ingredients (sugar, salt, acid, onions and spice) – Reheat – Fill – Close and seal – Hold – Water cool
- JUICE: Chop – Hot break – Extract – Sterilise – Fill – Salt – Close and seal – Hold – Water cool
- PULP: Chop – Hot break – Pulp – Concentrate – Finish – Fill – Close and code – Process – Cool (water or air)
- PASTE: Chop – Hot break – Pulp – Concentrate – Fill – Close and code – Cool -- water
- KETCHUP: Chop – Hot break – Pulp – Concentrate – Add ingredients (sugar, salt, acid, spices and flavour) – Cook – Finish – Deaerate – Fill – Seal and code – Hold – Water cool
- SOUP: Chop – Hot break – Pulp – Concentrate – Finish – Other ingredients – Cook – Finish – Fill – Close – Process – Cool

Source: Adapted from Gould (1992).

Uses

Tomato is consumed fresh, in salads, as well as processed. It should be noted that high-quality "salad" tomatoes have the highest value when sold fresh and in good condition. These would not normally be used for processing, unless for home use to save excess at the height of the season. Although tomatoes are commonly consumed fresh, over 80% of the tomato consumption comes from processed products such as tomato juice, paste, purée, ketchup and sauce (Gould, 1992).

Tomato juice is the unconcentrated liquid extracted from mature tomatoes of red or reddish varieties, with or without scalding followed by straining. In tomato juice extraction, heat may be applied without adding water. Tomato juice is strained free from skins, seeds and other coarse or hard substances, but carries finely divided insoluble solids from the flesh of the tomato. The juice may be homogenized, and may be seasoned with salt. When sealed in a container it is processed by heat, before or after sealing, to prevent spoilage (Gould, 1974, 1992).

Tomato purée, tomato pulp is the product prepared by combining at least two of the following optional ingredients:

- the liquid obtained from mature tomatoes of red or reddish varieties
- the liquid obtained from the residue from preparing such tomatoes for canning, consisting of peelings and cores with or without such tomatoes or pieces thereof
- the liquid obtained from the residue from partial extraction of juice from such tomatoes
- salt.

Tomato paste is the product prepared by combining at least two of the following optional ingredients:

- the liquid obtained from mature tomatoes of red or reddish varieties
- the liquid obtained from the residue from preparing such tomatoes for canning, consisting of peelings and cores with or without such tomatoes or pieces thereof
- the liquid obtained from the residue from partial extraction of juice from such tomatoes
- salt (sodium chloride formed during acid neutralizations should be considered added salt)
- spices
- flavouring
- baking soda.

Tomato ketchup is the product prepared by combining at least two of the following optional ingredients:

- the liquid obtained from mature tomatoes of red or reddish varieties
- the liquid obtained from the residue from preparing such tomatoes for canning, consisting of peelings and cores with or without such tomatoes or pieces thereof
- the liquid obtained from the residue from partial extraction of juice from such tomatoes.

The constituents used in the manufacture of ketchup, in addition to tomatoes, are sugar, vinegar, salt, onions, starches and spices.

Chili sauce is of the same general character as ketchup but is made from peeled and cored tomatoes without removing the seeds. It contains more sugar and onions and sometimes is made hotter in flavour than ketchup by the use of more cayenne pepper.

Screening characteristics screened by developers

Domesticated varieties (cultivars) have been developed by selection over the last 10 000 years and inevitably represent a subset of the variation found in their wild ancestors. Unusual or extreme phenotypes, such as large fruit or seed size, intense colour, sweet flavour or pleasing aroma are often selected by humans and maintained in varieties for aesthetic reasons, while synchronous ripening or inhibition of seed shattering (a dispersal mechanism) are selected to facilitate harvest. In the evaluation of tomato varieties, morphological (i.e. stand count, seedling vigour, pesticide resistance and disease resistance), agronomic (i.e. yield, fruit size, fruit uniformity, fruit colour and firmness), and chemical (composition) as well as biochemical parameters (i.e. aroma, flavour) are widely used (FAO-IPGRI, 2005).

At present, great efforts of genetic improvement of tomatoes have focused on the resistance against diseases caused by fungi, bacteria and viruses as well as on the tolerance to stress and pesticide exposure. In some cases, tomato plants are bred for development of varieties having nutritional or health benefits. Research focuses on altering the level of vitamins in order to create a food with enhanced health benefits. Transformation of tomatoes, using molecular techniques, has resulted in transgenic plants with elevated levels of provitamin A, and vitamins C and E, respectively (Herbers, 2003).

Nutrients

The average composition of fresh red tomato is shown in Tables 12.3-12.8, while the average compositions of yellow and orange-coloured ripe tomato fruit varieties and green tomato fruits of non-ripe red-coloured tomato fruit varieties are shown in Tables 12.10-12.13.

Ripe red tomato fruits

Proximate composition

The amount of total solids varies with genetic constitution (tomato variety) and environmental factors such as site of cultivation, soil condition, climate, not least precipitation during the period of fruit development and harvesting. Tomato usually consists of 5.5-6.2% total solids (Table 12.3). However, it has also been reported to be as high as 7.0-8.5% (Gould, 1992).

Table 12.3. **Proximate composition of red ripe tomato**

Nutrient	National Food Institute	USDA	Favier et coll.	Souci et al.	Fineli	Range of mean values
	Mean value, g per 100 g fresh weight[1]					
Water	94.00	94.50	93.80	94.20	94.00[2]	93.80-94.50
	Mean value, g per 100 g dry weight					
Protein	11.67	16.00	12.90	16.38	10.00	11.67-16.38
Fat		3.64		3.62	5.00	3.62-3.64
Ash	8.33	9.09				8.33-9.09
Carbohydrate, by difference	76.67	71.27		44.83	75.00	44.83-76.67
Fiber, total dietary	31.67	21.82	19.35	16.38	23.33	16.38-31.67
Sugars, total	38.67	47.82	56.45	47.84	55.00	38.67-56.45
Sucrose				1.45	1.67	1.45-1.67
Glucose (dextrose)	15.00	22.73		18.64		15.00-22.73
Fructose	23.67	24.91		23.28	33.33	23.28-24.91
Starch	1.50			1.38		1.38-1.50
Pentosan				1.21		1.21
Hexozan				1.90		1.90

Notes: 1. Values calculated based on the percentage water of fresh weight. 2. Varo et al. (1980).

Sources: National Food Institute – Technical University of Denmark (2005); USDA database (2007); Favier et coll. (1995); Souci et al. (1994); Fineli (2004).

Most of the dry matter in tomatoes is carbohydrates (Table 12.3). On a fresh weight basis, the carbohydrate content of tomatoes varies between 2.2% and 3.6%. A substantial fraction is dietary fibre. The major sugars in mature tomato fruit are the hexoses, fructose and glucose, the latter two being derived mainly from the hydrolysis of translocated sucrose (Davies and Hobson, 1981). Of simple sugars, the sucrose level is negligible as it rarely exceeds 0.1% of the fresh weight. The polysaccharide fraction consists of pectins and arabinogalactanes (50%), xylanes and arabinoxylanes (28%), and cellulose (approximately 25%). Reducing carbohydrates comprise approximately 50-65% of the total solids of tomato and consist mainly of glucose and fructose, with fructose usually occurring at higher levels than glucose (Table 12.3; Gould, 1992).

Pectins are polymers of D-galacturonic acid linked together via 1,4-bonds. They are natural constituents of the mature tomato and are responsible for the development of the fleshy red tissue strongly binding the cells together. During the early development of the fruit, an insoluble substance called protopectin is formed and this compound binds firmly to the fruit cells. During maturation of the fruit, protopectin is converted to pectin, which also contributes to stabilising the interaction between cells, but to a lesser extent than protopectin. During the last stages of fruit maturation, when the fruit goes from pink to red, protopectine is converted to pectin. Further maturation allows pectin to be degraded to smaller soluble fragments which show limited binding capacity, leading to soft mature fruits. The modification of pectin occurring during maturation is due to the action of enzymes formed in fruit cells during growth and development. Although these enzymes are formed exclusively during fruit development, their action is continued after harvest. Therefore, these enzymes have an important role in regulating the texture of both fresh and processed tomato products. The total content of pectin in fresh fruit of commercial tomato varieties lies between 0.17% and 0.23% (Goose and Raymond, 1964; Gould, 1974, 1992).

Citric acid is the predominant organic acid in tomato (Table 12.8). Malic acid is the second most important organic acid in the juice of fresh tomato. Processing of tomato juice leads to an increase in the levels of organic acids. Acetic acid level increases by 32% during processing (Gould, 1992), apparently due to the oxidation of aldehydes and alcohols as well as deamination of amino acids. Citric and malic acid levels also increase after processing.

Minerals

The total ash content of red mature tomatoes is a little less than 10% of the dry matter. Tomatoes and tomato products are important sources of potassium, and they also contribute substantially to magnesium and iron intake (Table 12.4). It is worth mentioning that the relatively high ascorbic acid level in tomatoes maintains iron in its reduced form, increasing its potential for being taken up by the body (Gould, 1992).

Fatty acids and phytosterols

Tomatoes have very low fat content (Table 12.3). The most important saturated, monounsaturated and polyunsaturated fatty acids are reported in Table 12.5. However, the level of phytosterols is high (approximately 7 mg/100 g of product).

Proteins and amino acids

Proteins constitute around 11-17% of the dry matter in tomato fruits (Table 12.3). Glutamic acid is the most common amino acid, comprising 48.5% of the total weight of amino acids (Table 12.6). Aspartic acid is the second most abundant amino acid. Proline occurs at the lowest quantity. High temperature processing of tomatoes (e.g. 104.4°C for 20 minutes) increases the level of free amino acids, due to degradation and partial hydrolysis of certain proteins. The greatest increase occurs in the levels of glutamic acid, aspartic acid, alanine and threonine (Goose and Raymond, 1964; Gould, 1992).

Table 12.4. **Mineral composition of red ripe tomato**

Minerals	National Food Institute	USDA	Favier et al.	Souci et al.	Fineli	Range of mean values
	Mean value, per 100 g of dry matter[1]					
Calcium (mg)	166.67	181.82	145.16	162.07	150.00	145.16-181.82
Iron (mg)	8.33	4.91	6.45	5.69	5.00	4.91-8.33
Magnesium (mg)	116.67	200.00	177.42	206.90	183.33	116.67-206.90
Nitrate (mg)				86.21		86.21
Boron (mcg)				1 982.76		1 982.76
Nickel (mcg)	16.67			100.00		16.67-100.00
Phosphorus (mg)	500.00	436.36	387.10	379.31	500.00	379.31-500.00
Potassium (mg)	3 600.00	4 309.09	3 645.16	4 172.41	4 833.33	3 600.00-4 833.33
Sodium (mg)	116.67	90.91	80.65	56.90	41.67	56.90-116.67
Zinc (mg)	1.50	3.09		2.59	3.33	1.50-3.09
Cobalt (mcg)				29.31		29.31
Copper (mg)	0.67	1.07		1.00		0.67-1.07
Manganese (mg)	1.83	2.07		1.88		1.83-2.07
Chromium (mcg)	6.67			327.59		6.67-327.59
Iodin (mcg)	3.33			18.97	16.67	3.33-18.97
Fluoride (mcg)				413.79		413.79
Chloride (mg)				517.24		517.24
Aluminium (mcg)				1 241.38		1 241.38
Silicon (mg)				46.55		46.55
Selenium (mcg)	5.00			16.90	3.33	3.33-16.90

Note: 1. Values calculated based on the percentage water given in Table 12.3.

Sources: National Food Institute – Technical University of Denmark (2005); USDA database (2007); Favier et al. (1995); Souci et al. (1994); Fineli (2004).

Table 12.5. **Fatty acid composition of red ripe tomato**

Fatty acids	National Food Institute	USDA	Favier et al.	Souci et al.	Fineli	Range of mean values
	Mean value, per 100 g dry matter[1]					
Fatty acids, total saturated (g)	0.97	0.51			1.67	0.51-0.97
Palmitic – 16:0 (g)	0.75	0.36		0.55		0.36-0.75
Stearic – 18:0 (g)	0.13	0.15		0.09		0.13-0.15
Fatty acids, total monounsaturated (g)	0.58	0.56			1.67	0.56-0.58
Palmitoleic – 16:1 undifferentiated (g)	0.05	0.02		0.03		0.02-0.05
Oleic – 18:1 undifferentiated (g)	0.53	0.55		0.40		0.40-0.55
Fatty acids, total polyunsaturated (g)	2.50	1.51	2.26		1.67	1.51-2.50
Linoleic – 18:2 undifferentiated (g)	2.17	1.45		1.57	1.38	1.45-2.17
Linolenic – 18:3 undifferentiated (g)	0.22	0.05		0.16	0.12	0.05-0.22

Note: 1. Values calculated based on the percentage water of fresh weight given in Table 12.3.

Sources: National Food Institute – Technical University of Denmark (2005); USDA database (2007); Favier et al. (1995); Souci et al. (1994); Fineli (2004).

Table 12.6. **Amino acid composition of red ripe tomato**

Amino acids	National Food Institute	USDA	Souci et al.	Range of mean values
	Mean value, per 100 g dry matter[1]			
Tryptophan (g)	0.12	0.11	0.10	0.10-0.12
Threonine (g)	0.32	0.38	0.40	0.32-0.40
Isoleucine (g)	0.53	0.36	0.40	0.36-0.53
Leucine (g)	0.47	0.56	0.52	0.47-0.56
Lysine (g)	0.53	0.56	0.50	0.50-0.56
Methionine (g)	0.13	0.13	0.12	0.12-0.13
Cystine (g)	0.10	0.20	0.02	0.02-0.20
Phenylalanine (g)	0.32	0.40	0.41	0.32-0.41
Tyrosine (g)	0.43	0.27	0.21	0.21-0.43
Valine (g)	0.33	0.40	0.40	0.33-0.40
Arginine (g)	0.43	0.38	0.31	0.31-0.43
Histidine (g)	0.18	0.24	0.22	0.18-0.24
Alanine (g)	0.30	0.44	0.45	0.30-0.45
Aspartic acid (g)	1.33	2.15	2.09	1.33-2.15
Glutamic acid (g)	3.17	5.69	5.69	3.17-5.69
Glycine (g)	0.33	0.38	0.31	0.31-0.38
Proline (g)	0.37	0.29	0.28	0.28-0.37
Serine (g)	0.42	0.42	0.48	0.42-0.48

Note: 1. Values calculated based on the percentage water of fresh weight given in Table 12.3.

Sources: National Food Institute – Technical University of Denmark (2005); USDA database (2007); Souci et al. (1994).

Table 12.7. **Vitamin and antioxidant composition of red ripe tomato**

Vitamins	National Food Institute	USDA	Favier et al.	Souci et al.	Fineli	Range of mean values
	Mean value, per 100 g dry matter[1]					
Vitamin C, total ascorbic acid (mg)	250.00	230.91	290.32	327.59	235.00	230.91-327.59
Thiamin (Vitamin B1) (mg)	0.72	0.67	0.97	0.98	1.00	0.67-0.98
Riboflavin (mg)	0.33	0.35	0.81	0.60	0.67	0.33-0.81
Niacin (mg)	11.67	10.80	9.68		13.33	9.68-11.67
Pantothenic acid (mg)	5.50	1.62	4.52	5.34		1.62-5.50
Vitamin B6 (mg)	1.48	1.45	1.29	1.72		1.29-1.72
Folate, total (mcg)		272.73				
Folic acid (mcg)		0.00		379.31		379.31
Folate, food (mcg)	516.67	272.73	322.58		193.33	193.33-516.67
Folate, DFE (mcg DFE)		272.73				272.73
Vitamin A, IU		15 145.45				15 145.45
Vitamin A, RAE (mcg RAE)	1 383.33	763.64			1 113.33	763.64-1 383.33
Retinol (mcg)				1 672.41		1 672.41
Carotene, beta (mcg)	16 533.33	8 163.64	9 677.42	10 206.90		8 163.64-16 533.33
Carotene, alpha (mcg)		1 836.36				1 836.36
Lycopene (mcg)		46 781.82				46 781.82
Vitamin E (alpha-tocopherol) (mg)	18.33	9.82	16.13	14.02	11.67	9.82-18.33
Tocopherol, alfa (mg)				13.79		13.79
Tocopherol, beta (mg)		0.18				0.18
Tocopherol, gamma (mg)		2.18		2.24		2.18-2.24
Vitamin K (phylloquinone) (mcg)	283.33	143.64		98.28	83.33	83.33-283.33
Biotin (mcg)	25.00			68.97		25.00-68.97
Nicotinamide (mg)				9.14		9.14

Note: 1. Values calculated based on the percentage water given in Table 12.3.

Sources: National Food Institute – Technical University of Denmark (2005); USDA database (2007); Favier et al. (1995); Souci et al. (1994); Fineli (2004).

Table 12.8. **Other metabolite composition of red ripe tomato**

Other metabolites	National Food Institute	USDA	Favier et al.	Souci et al.	Range of mean values
	Mean value, per 100 g dry matter[1]				
Malic acid (g)				0.88	0.88
Citric acid (g)				5.66	5.66
Lactic acid (g)				0.10	0.10
Acetic acid (g)				0.14	0.14
Chlorogenic acid (g)				0.17	0.17
Quinic acid (g)				0.14	0.14
Ferulic acid (mg)				12.07	12.07
Fumaric acid (g)				0.03	0.03
Pyruvic acid (mg)				3.28	3.28
Oxaloacetic acid (g)				0.41	0.41
Salicylic acid (mg)				2.24	2.24
Histamine (mg)				34.48	34.48
Carotene, beta (mcg)	16 533.33	8 163.64	9 677.42	10 206.90	8 163.64-16 533.33
Carotene, alpha (mcg)		1 836.36			1 836.36
Lycopene (mcg)		46 781.82			4 6781.82
Lutein + zeaxanthin (mcg)		2 236.36			2 236.36
Cellulose (g)				6.21	6.21
Polyuronic acid (g)				3.97	3.97
Myoinositol (mg)				189.66	189.66

Note: 1. Values calculated based on the percentage water of fresh weight given in Table 12.3.

Sources: National Food Institute – Technical University of Denmark (2005); USDA database (2007); Favier et al. (1995); Souci et al. (1994.

Vitamins and other anti-oxidants

Fresh tomato, tomato juice and other tomato products make a significant contribution to human nutrition due to the concentration and availability of several nutrients in these products and to their widespread consumption. Differences in the amount of nutrients contained in different varieties have been confirmed. In the research of Sahlin et al. (2004) it was shown that between Aranca and Excell (two varieties of tomato), Aranca was found to contain higher levels of ascorbic acid, total phenolics and lycopene, and showed higher antioxidant activity overall.

Levels of vitamins and other antioxidants vary between tomato varieties (Sahlin et al., 2004) and are for the red ripe tomato summarised in Table 12.7. Vitamin C, ascorbic acid, is a vitamin necessary for normal metabolism, wound healing and collagen synthesis. Ascorbic acid levels lie between 12.7 mg and 19.0 mg per 100 g of fresh weight in red ripe tomatoes (Table 12.7). Tomatoes also contain vitamin E, and low amounts of the water-soluble type B vitamins thiamin, niacin and riboflavin (Beecher, 1998).

Lycopene is the most prominent carotenoid in ripe red tomatoes (Table 12.7), where it commonly constitutes around 90-99% of the total carotenoids (Dumas et al., 2003). Other carotenoids in ripe red tomatoes are beta-carotene, gamma-carotene and phytoene as well as several other carotenoids occurring at low levels.

Lycopene is a product extracted from tomato, commonly by the use of solvents. A more environmentally friendly process is the use of supercritical fluid extraction, which minimises the risk of lycopene degradation via isomerisation and oxidation

(Gomez-Prieto et al., 2003). The lycopene content in various tomato products is shown in Table 12.9.

Also, polyphenols contribute to the antioxidant activity of the tomato fruits (Takeoka et al., 2001). The levels of several polyphenols are reported in Table 12.8.

Table 12.9. **Lycopene contents of commonly consumed commercial tomato products**

Group	Product	Lycopene (mg/kg)
I. Products for food preparation	Tomato paste	365.0
	Tomato purée	195.6
	Crushed tomatoes	223.8
II. Sauces	Tomato sauce	130.6
	Spaghetti sauce	191.2
	Pizza sauce	121.7
	Seafood sauce	185.6
	Chili sauce	168.3
III. Condiments	Tomato ketchup	123.9
	Light ketchup	141.0
	Barbecue sauce	42.9
IV. Readily consumed	Tomato juice	101.6
	Condensed soup	72.7
	Ready to serve soup	44.1
	Clam cocktail	43.3
	Bloody Mary mix	42.3

Source: Rao et al. (1998).

Yellow and orange-coloured ripe tomato fruit varieties and green tomato fruits of non-ripe red-coloured tomato fruit varieties

The proximate content of yellow, and orange-coloured ripe tomato fruit varieties and the proximate content of green tomato fruits of non-ripe red-coloured tomato fruit varieties are shown in Table 12.10.

Tomato products

The proximate content of sun-dried red tomatoes is shown in Table 12.10.

Table 12.10. **Proximate composition of yellow, green, orange and sun-dried tomatoes**

Nutrient	Tomatoes, yellow, raw	Tomatoes, orange, raw	Tomatoes, green, raw	Tomatoes, sun-dried
	Mean value, g per 100 g fresh weight			
Water	95.28	94.78	93.00	14.56
Mean value, per 100 g dry matter[1]				
Energy (kcal)	317.80	306.51	328.57	301.97
Energy (kj)	1 334.75	1 283.52	1 357.14	1 265.22
Protein (g)	20.76	22.22	17.14	16.51
Total lipid (fat) (g)	5.51	3.64	2.86	3.48
Ash (g)	10.59	13.22	7.14	14.75
Carbohydrate, by difference (g)	63.14	60.92	72.86	65.26
Fiber, total dietary (g)	14.83	17.24	15.71	14.40
Sugars, total (g)			57.14	44.00

Note: 1. Values calculated based on the percentage water of fresh weight.

Source: Adapted from USDA (2007).

The fatty acid, amino acid, mineral, vitamin and other antioxidants composition of sun-dried red tomatoes is shown in Tables 12.11-12.13. The composition of other tomato products – tomato juice, tomato purée, ketchup, chilli sauce and tomato paste – is shown in Table 12.14.

Table 12.11. **Fatty acid and phytosterol composition of yellow, orange, green and sun-dried tomatoes**

Mean value, per 100 g dry matter[1]

	Tomatoes, yellow, raw	Tomatoes, orange, raw	Tomatoes, green, raw	Tomatoes, sun-dried
Fatty acids				
Fatty acids, total saturated (g)	0.76	0.48	0.40	0.50
16:0 (g)	0.57	0.36	0.29	0.38
18:0 (g)	0.21	0.13	0.10	0.11
Fatty acids, total monounsaturated (g)		0.54	0.43	0.57
16:1 undifferentiated (g)		0.02	0.01	0.01
18:1 undifferentiated (g)	0.85	0.54	0.41	0.56
20:1 (g)	0.04			
22:1 undifferentiated (g)	0.83			
Fatty acids, total polyunsaturated (g)	2.29	1.46	1.16	1.31
18:2 undifferentiated (g)	2.20	1.40	1.11	1.29
18:3 undifferentiated (g)	0.08	0.06	0.04	0.01
Phytosterols (mg)	127.12	76.63		

Note: 1. Values calculated based on the percentage water of fresh weight given in Table 12.10.

Source: Adapted from USDA (2007).

Table 12.12. **Amino acid composition of yellow, orange, green and sun-dried tomatoes**

Mean value, per 100 g dry matter[1]

	Tomatoes, yellow, raw	Tomatoes, orange, raw	Tomatoes, green, raw	Tomatoes, sun-dried
Tryptophan (g)	0.15	0.15	0.13	0.12
Threonine (g)	0.51	0.56	0.43	0.42
Isoleucine (g)	0.49	0.52	0.41	0.40
Leucine (g)	0.76	0.80	0.63	0.61
Lysine (g)	0.76	0.80	0.63	0.61
Methionine (g)	0.17	0.19	0.14	0.14
Cystine (g)	0.28	0.29	0.23	0.21
Phenylalanine (g)	0.53	0.57	0.44	0.43
Tyrosine (g)	0.36	0.38	0.30	0.28
Valine (g)	0.53	0.57	0.44	0.42
Arginine (g)	0.51	0.56	0.41	0.40
Histidine (g)	0.32	0.34	0.26	0.25
Alanine (g)	0.59	0.63	0.49	0.47
Aspartic acid (g)	2.86	3.08	2.37	2.29
Glutamic acid (g)	7.61	8.18	6.31	6.09
Glycine (g)	0.51	0.56	0.43	0.41
Proline (g)	0.38	0.42	0.33	0.31
Serine (g)	0.55	0.59	0.46	0.44

Note: 1. Values calculated based on the percentage water of fresh weight given in Table 12.10.

Source: Adapted from USDA (2007).

Table 12.13. **Mineral, vitamin and carotenoid composition of yellow, orange, green and sun-dried tomatoes**

Mean value, per 100 g dry matter[1]

	Tomatoes, yellow, raw	Tomatoes, orange, raw	Tomatoes, green, raw	Tomatoes, sun-dried
Minerals				
Calcium, Ca (mg)	233.05	95.79	185.71	128.75
Iron, Fe (mg)	10.38	9.00	7.29	10.64
Magnesium, Mg (mg)	254.24	153.26	142.86	227.06
Phosphorus, P (mg)	762.71	555.56	400.00	416.67
Potassium, K (mg)	5 466.10	4 061.30	2 914.29	4 011.00
Sodium, Na (mg)	487.29	804.60	185.71	2 452.01
Zinc, Zn (mg)	5.93	2.68	1.00	2.33
Copper, Cu (mg)	2.14	1.19	1.29	1.67
Manganese, Mn (mg)	2.54	1.69	1.43	2.16
Selenium, Se (mcg)	8.47	7.66	5.71	6.44
Vitamins				
Vitamin C, total ascorbic acid (mg)	190.68	306.51	334.29	45.88
Thiamin (mg)	0.87	0.88	0.86	0.62
Riboflavin (mg)	1.00	0.65	0.57	0.57
Niacin (mg)	24.98	11.36	7.14	10.59
Pantothenic acid (mg)	2.33	3.56	7.14	2.44
Vitamin B6 (mg)	1.19	1.15	1.16	0.39
Folate, total (mcg)	635.59	555.56	128.57	79.59
Folate, food (mcg)	635.59	555.56	128.57	79.59
Folate, DFE (mcg DFE)	635.59	555.56	128.57	79.59
Vitamin A, IU		28 659.00	9 171.43	1 022.94
Vitamin A, RAE (mcg RAE)		1 436.78	457.14	51.50
Carotene, beta (mcg)			4 942.86	613.30
Carotene, alpha (mcg)			1 114.29	
Lycopene (mcg)				47 720.04
Lutein + zeaxanthin (mcg)				1 413.86
Vitamin E (alpha-tocopherol) (mg)			5.43	0.01
Vitamin K (phylloquinone) (mcg)			144.29	50.33

Source: Adapted from USDA (2007).

Tomato pomace is the residue that remains after pressing tomato in the production of ketchup, juice, paste, purée, soup or sauce (NRC, 1983). It is made up of skin, pulp and crushed seed that remain after pressing and some adhering pulp (Ensminger et al., 1990; NRC, 1983). It contains a high amount of water and is usually dried prior to being used in feed. The proximate composition of tomato pomace is shown in Table 12.15, and its amino acid and mineral contents in Tables 12.16 and 12.17, respectively.

Other constituents: Toxicants and allergens

Toxicants

The most important natural toxins in tomatoes are the steroidal glycoalkaloids α-tomatine and dehydrotomatine, possibly produced by the plant as a defense against pathogens and predators including bacteria, fungi, viruses and insects (Andersson, 1999;

Friedman, 2002; Kozukue et al., 2004). Together with phenolic compounds (caffeic acid and naringin) tomatine also contributes to the bitter taste of non-ripe green tomatoes. Tomato glycoalkaloids are synthesised in tomato fruits during early development and then degraded during fruit maturation (Eltyeb and Roddick, 1984, 1985; Kozukue et al., 1994; Friedman and Levin, 1995).

Three factors seem to play a pivotal role in determining changes in glycoalkaloid content in tomatoes: *i)* cultivar (genotype); *ii)* ripening stage; and *iii)* growing conditions (Leonardi, 2000). Non-ripe green fruits contain substantial amounts of α-tomatine. The reported levels vary between negligible and 1 165 mg/kg fresh weight and typically range from 20 mg to 200 mg/kg fresh weight (Andersson, 1999; Friedman, 2004). In contrast, red ripe tomato fruits contain negligible concentrations of tomatine, between nondetectable levels and 23 mg/kg fresh weight, typically around 1 mg/kg fresh weight. The tomato fruit becomes almost tomatine-free if the red fruit is left on the plant for two or three days before being harvested (Kajderowicz-Jarosinska, 1965). A pronounced reduction in the α-tomatine content is obtained also after induction of ripening with artificial techniques (ethylene treatment) (Eltayeb and Roddick, 1984) but not to the same extent as in vine-ripened fruit. Retardation of fruit ripening by treatment with reduced pressure delays the reduction in alkaloid content. Consumer exposure to tomatine may be of toxicological concern mainly in cases where substantial quantities of green non-ripe fruits or red ripe fruits of varieties with naturally high levels of tomatine are consumed. However, tomatoes cultivated in Peru with tomatine content in the range of 500-5 000 mg/kg of dry weight (approximately 30-300 mg/kg of fresh weight) are consumed without apparent acute toxic effects (Rick et al., 1994).

Table 12.14. **Composition of tomato products, per 100 g**

	Canned tomato	Tomato juice				Tomato purée (pulp)	Ketchup	Chili sauce	Tomato paste
		Regular	Concentrated	Dehydrated	Cocktail				
Water %	93.7	93.6	75.0	1.0	93.0	87.0	68.6	68.0	75.0
Food energy (calories)	21.0	19.0	76.0	303.0	21.0	39.0	106.0	104.0	82.0
Protein (g)	1.0	0.9	3.4	11.6	0.7	1.7	2.0	2.5	3.4
Fat (g)	0.2	0.1	0.4	2.2	0.1	0.2	0.4	0.3	0.4
Carbohydrates									
Total (g)	4.3	4.3	17.1	68.2	5.0	8.9	25.4	24.8	18.6
Fiber (g)	0.4	0.2	0.9	3.1	0.2	0.4	0.5	0.7	0.9
Ash (g)	0.8	1.1	4.1	17.0	1.2	2.2	3.6	4.4	2.6
Calcium (mg)	6.0	7.0	27.0	85.0	10.0	13.0	22.0	20.0	27.0
Phosphorus (mg)	19.0	18.0	70.0	279.0	18.0	34.0	50.0	52.0	70.0
Iron (mg)	0.5	0.9	3.5	7.8	0.9	1.7	0.8	0.8	3.5
Sodium (mg)	130.0	200.0	790.0	3 934.0	200.0	399.0	1 042.0	1 338.0	38.0
Potassium (mg)	217.0	227.0	888.0	3 518.0	221.0	426.0	363.0	370.0	888.0
Vitamin A (IU)	900.0	800.0	3 300.0	13 100.0	800.0	1 600.0	1 400.0	1 400.0	3 300.0
Thiamin (mg)	0.05	0.05	0.20	0.52	0.05	0.09	0.09	0.09	0.20
Riboflavin (mg)	0.03	0.03	0.12	0.40	0.02	0.05	0.07	0.07	0.12
Niacin (mg)	0.7	0.8	3.1	13.5	0.06	1.4	1.6	1.6	3.1
Ascorbic acid (mg)	17.0	16.0	49.0	239.0	16.0	33.0	15.0	16.0	49.0

Source: Adapted from Gould (1992).

Table 12.15. **Proximate composition of tomato pomace**

	NRC (1982)	Ensminger et al.	NRC (2001)	Preston	Range
	Grams per 100 g fresh weight[1]				
Dry matter (g)	92.0[2]	25.0[1]-2.0[2]	24.7[1]	92.0[2]	24.7[1]-92.0[2]
	Grams per 100 g dry matter				
Protein (g)	23.5	21.5-22.9	19.3	23.0	19.3-23.5
Ether extract (fat) (g)	10.3		13.3	10.6	10.3-13.3
Ash (g)	7.5		5.5	6.5	5.5-7.5
Neutral detergent fibre			60.0	54.4	54.4-60.0
Acid detergent fibre			47.6	59.8	47.6-59.8
Crude fiber	26.4	27.2-33.7		26.0	26.0-33.7

Notes: 1. Tomato pomace, dehydrated. 2. Tomato pomace, wet.

Sources: NRC (1982; 2001); Ensminger et al. (1990); Preston (2007).

Table 12.16. **Amino acid composition of tomato pomace**

Per 100 g dry matter

	NRC (2001)
Argenine	1.07
Histidine	0.35
Isoleucine	0.62
Leucine	1.52
Lysine	1.43
Methionine	0.09
Cystine	0.09
Phenylalanine	0.80
Threonine	0.62
Valine	0.18
Tryptophan	0.90

Note: 1. Values calculated based on the percentage crude protein of 19.3.

Source: NRC (2001).

Table 12.17. **Mineral composition of tomato pomace**

Value per 100 g dry matter

	NRC (1982)	Ensminger et al.	NRC (2001)	Preston[1]	Range
Calcium, Ca (g)	0.43	0.43	0.22	0.43	0.22-0.43
Magnesium, Mg (g)	0.20		0.28		0.20-0.28
Phosphorus, P (g)	0.60	0.49	0.47	0.59	0.47-0.60
Potassium, K (g)	3.63		0.98		0.98-3.63
Sodium, Na (g)			0.12		0.12
Iron, Fe (mg)	460.00		54.10		54.10-460.00
Zinc, Z (mg)			5.40		5.40
Copper, Cu (mg)	3.30		1.10		1.10-3.30
Manganese, Mn (mg)	5.10		1.10		1.10
Molybdenium, Mo (mg)			0.18		0.18

Note: 1. Values calculated based on percentage dry matter shown for Preston (2007) in Table 12.15.

Sources: NRC (1982; 2001); Ensminger et al. (1990); Preston (2007).

Tomatoes also contain calystegine alkaloids (polyhydroxylated nortropane alkaloids) (Asano et al., 1997, 2001; Andersson, 2002). At higher concentrations, these compounds may inhibit mammalian glycosidases and produce conditions in grazing animals that are phenocopies of inherited deficiencies in various glycosidases leading to lysosomal storage diseases. It is not known whether such diseases can also occur in humans, although hereditary diseases of glycosidase deficiency have been described. The calystegine alkaloids occur as a set of similar compounds, only differing in the number of hydroxyl groups. Tomato fruits contain calystegines A_3 and B_2 at the respective levels of 1.1 mg and 4.5 mg/kg fresh weight (Asano et al., 1997).

Tomatoes, like several other members of the alkaloid-rich nightshade family (*Solanaceae*), contain nicotine, but levels are low and are unlikely to be harmful to consumers (Andersson et al., 2003). The levels reported for red ripe tomatoes range from 2.7 µg to 9.1 µg nicotine per kg fresh weight with only small differences observed between tomato varieties (Domino et al., 1993; Siegmund et al., 1999). The nicotine content is inversely related to the degree of fruit ripening. The highest levels are found in unripe, green fruits and the lowest levels in ripe fruits (Castro and Monji, 1986; Siegmund et al., 1999). Processed tomato products, such as tomato sauce and tomato ketchup, contain slightly higher levels of nicotine than fresh tomatoes (although still very low), probably due to the reduced water content of the processed products (Castro and Monji, 1986; Siegmund et al., 1999).

Allergens

Today, there is limited information available regarding the nature of tomato allergens and rather few attempts have been made to identify and characterise them. Usually, allergy to tomatoes is linked to other types of allergies such as grass pollen and latex. The proportion of food-allergic patients being allergic to tomatoes varies worldwide from 1.5% to 16%, indicating that tomato is a significant allergenic food (Westphal et al., 2003). In Central Europe, tomatoes account for approximately 1.5% of all food allergies, whereas in countries with high tomato consumption it is responsible for approximately 20% of the oral allergy syndromes (Allergopharma Joachim Ganzer KG, 2007). Immunoglobulin E (IgE) cross-reactive profilins have been suggested to account for allergic symptoms in patients suffering from tomato allergy. The most common tomato allergens known to elicit symptoms in food allergic patients are Lyc e 1, Lyc e 2, Lyc e 2.0101, Lyc e 2.0102, Lyc e 3, Lyc e LAT52, and Lyc e NP24 (University of Texas, 2007). Based on the *in vitro* histamine release assays with human basophils, Westphal et al. (2003, 2004) concluded that tomato profilin, Lyc e 1, is a minor human allergen whereas profilin Lyc e 2, beta-fructofuranosidase, is an important human allergen. Additionally, lipid transfer protein (Lyc e 3), which belongs to a family of structurally highly conserved proteins, is a potentially severe food allergen due mainly to its extreme resistance to pepsin digestion and is, therefore, considered a pan-allergen (Asero et al., 2000).

Suggested constituents to be analysed related to food use

Tomato and tomato products are widely consumed by humans all over the world. The popularity of tomato is understandable since the tomato is tasty and is an important source of minerals and vitamins. Tomatoes and tomato products are used as ingredients in many traditional dishes, because of the compatibility with other food ingredients and high nutritional value.

Besides the use of tomatoes and tomato products for direct human consumption, tomatoes and its by-products serve as raw materials for several secondary products. A very valuable constituent of tomato is the red pigment carotenoid lycopene, an exceptionally efficient quencher of singlet oxygen and therefore an important antioxidant. Lycopene, as well as other valuable substances such as beta-carotene, alpha-carotene, alpha-tocopherol, gamma-tocopherol and delta-tocopherol, can be effectively extracted from tomato skins, seeds and other by-products using supercritical fluid extraction technology (Baysal et al., 2000, Rozzi et al., 2002).

Tomato seeds contain high-quality plant proteins that can be supplemented into various food products (Sogi et al., 2005). Studies have revealed that it is economic to utilise protein isolated from tomato seeds due to their higher contents of most essential amino acids compared to the peels, as a substitute for wheat flour used in bakery products, whereas cake made from 10% protein isolate as a substitute for wheat flour had the highest palatability (Attia et al., 2000). Table 12.18 shows suggested nutritional and compositional parameters to be analysed in tomato matrices for food use.

Table 12.18. **Suggested nutritional and compositional parameters to be analysed in tomato matrices for food use**

Parameter	**Tomato (raw)**
Proximate analysis[1]	X
Minerals[2]	X
Vitamins[3]	X
Beta carotene	X
Lycopene	X
Tomatine[4]	X

Notes: 1. Proximate includes protein, fat, total dietary fibre, ash and carbohydrates. 2. Magnesium, potassium. 3. Vitamins include: vitamins C, K, folate. 4. Tomatine includes alpha-tomatine and dehydrotomatine.

Suggested constituents to be analysed related to feed use

Tomato processing waste has been used successfully as animal feed. According to a report on underutilised feedstuffs (NRC, 1983), tomato processing wastes can be divided into three categories according to the type of product that can be recovered. The first is cull tomatoes not accepted for processing. The second is peel residue from whole tomato canning, and about 12% of the original tomato is removed as peel and adhering pulp. The third is tomato pomace, the residue from the manufacture of juice, paste, purée, sauce and ketchup. NRC (1983) reported that 15% of processed tomatoes are processed as whole tomatoes, and 85% are processed as pulp products. USDA (2007) reports that tomato paste is the primary processed product produced worldwide. Thus, the primary residue product available for animal feed is tomato pomace.

NRC (1983) reported that the work of Ammerman et al. (1963, 1965) showed that cull tomatoes have been successfully used as feed for cattle feeder steers, lambs and poultry. Cull tomatoes are reported to contain higher levels of energy and lower levels of fibre than tomato pomace. However, if the tomatoes are green, they could contain glycoalkaloids (see section on toxicants) which have a bitter taste which could restrict animal intake. Also, Ammerman et al. reported reduced carotenoid pigment in skins and shanks of poultry when cull tomatoes were included at 3% of the diet, replacing alfalfa meal.

NRC (1983) reported that the peel residue is limited as animal feed because of the addition of caustic to tomatoes to enhance the mechanical peeling process. The process may increase the pH to 13-14. Also, the moisture remains high at 97-98%.

Tomato pomace has been successfully used as animal feed for many years (NRC, 1983). Because moisture content of the fresh tomato pomace is relatively high (75%; Table 12.15), storage is a problem. Also, its availability is seasonal, mostly in the summer, and requires further processing to make it a useful feed product. Weiss et al. (1997) successfully mixed fresh tomato pomace with corn forage in a 12:88 ratio, respectively and ensiled the mixture as a feed for dairy cattle. A dairy cow feeding study revealed no significant differences between the tomato pomace mixed silage and corn silage on milk yield, milk composition or dry matter intake. A test silo study found that no fermentation occurred when only tomato pomace was included in a silo, but that ensiling it in an air tight plastic bag would provide two months of storage without spoilage. The same researchers found a high lignin level in the tomato pomace which could be of nutritional concern.

Most tomato pomace used for animal feed is dried to about 8% moisture (Table 12.15). NRC (1983) reports that it has been successfully fed to cattle, swine and poultry at a 10-15% dietary level.

The results of feeding tomato pomace on the performance of dairy crossbred steers, showed that: *i)* the average daily gain of the cattle fed dried tomato pomace was higher than the cattle fed with hay and fresh grass; *ii)* total voluntary intake of the cattle fed with tomato pomace was higher than the cattle fed with hay and fresh grass; and *iii)* the economical return of the cattle in the group fed with tomato pomace was the best (Satchaphun et al.,1998). Another study, conducted on ducks, showed that there were no statistical differences in average daily gain, average feed intake and feed conversion ratio, but there was a significant reduction of feed cost per gain (Wanasitchaiwat et al.). Finally, Al-Betawi (2005) reported that tomato pomace has relatively high lysine content, and has been used in feed for poultry at a rate up to 10% of the ration. Ayhan and Aktan (2004) have also shown that tomato pomace can be used in broiler diets at a 5% level.

Table 12.19 shows suggested nutritional and compositional parameters to be analysed in tomato pomace for feed use. As reported, most of the use of tomato processing waste for animal feed is tomato pomace, and most of the tomato pomace is fed to cattle. The nutrients of major concern for cattle are the proximates (crude protein, crude fat [ether extractable], ash, crude fibber and carbohydrates), neutral detergent fibre (NDF), acid detergent fibre (ADF), calcium and phosphorus. Some tomato pomace may be used for poultry where in addition to the aforementioned nutrients, lysine is also important. There are no reports to indicate that any natural toxicants in ripe tomato-based pomace, such as tomatine, are a concern to animals.

Table 12.19. **Suggested nutritional and compositional parameters to be analysed in non-processed tomatoes or tomato pomace for feed use**

Parameter	Non-processed tomatoes	Tomato pomace
Proximate analysis[1]	X	X
Minerals[2]	X	X
Lysine	X	X

Notes: 1. NDF (neutral detergent fibre) and ADF (acid detergent fibre) should be substituted for crude fibre. 2. Calcium, magnesium, potassium, phosphorous and sodium.

For comparative purposes, it is suggested that analysing either the tomato fruit or tomato pomace would suffice. The nutrient content of the pomace would not be expected to change if the nutrient content of the tomato fruit does not change.

References

Abushita, A.A. et al. (2000), "Change in carotenoids and antioxidant vitamins in tomato as a function of varietal and technological factors", *Journal of Agricultural and Food Chemistry*, Vol. 48, No. 6, pp. 2 075-2 081, http://dx.doi.org/10.1021/jf990715p.

Al-Betawi, N.A. (2005), "Preliminary study on tomato pomace as unusual feedstuff in broiler diets", *Pakistan Journal of Nutrition*, Vol. 4, No. 1, pp. 57-63.

Allergopharma Joachim Ganzer KG (2007, *Allergen Database*, www.allergopharma.co (accessed in 2007).

Ammerman, C.B. et al. (1965), "Dried tomato pulp, its preparation and nutritive value for livestock and poultry", *Florida Agricultural Experiment Stations Bulletin*, No. 691.

Ammerman, C.B. et al. (1963), "Tomato by-products as feedstuffs: Nutritive value of dried tomato pulp for ruminants", *Journal of Agricultural and Food Chemistry*, Vol. 11, No. 4, pp. 347-349, http://dx.doi.org/10.1021/jf60128a025.

Andersson, H.C. (2002), *Calystegine Alkaloids in Solanaceous Food Plants*, Nordic Council of Ministers, Food, TemaNord.

Andersson, H.C. (1999), *Glycoalkaloids in Tomatoes, Eggplants, Pepper and Two Solanum Species Growing Wild in the Nordic Countries*, Nordic Council of Ministers, Food, TemaNord.

Andersson, H.C. et al. (2003), *Nicotine Alkaloids in Solanaceous Food Plants*, Nordic Council of Ministers, Food, TemaNord.

Asano, N. et al. (2001), "Dihydroxynotropane alkaloids from calystegine-producing plants", *Phytochemistry*, Vol. 57, No. 5, July, pp. 721-726.

Asano, N. et al. (1997), "The effects of calystegines isolated from edible fruits and vegetables on mammalian liver glycosidases", *Glycobiology*, Vol. 7, No. 8, pp. 1 085-1 088.

Asero, R. et al. (2000), "Lipid transfer protein: A pan-allergen in plant-derived foods that is highly resistant to pepsin digestion", *International Archives of Allergy and Immunology*, Vol. 122, No. 1, May, pp. 20-32.

Attia, E.A. et al. (2000), "Production of protein isolate from tomato wastes", *Egyptian Journal of Agricultural Research*, Vol. 78, No. 5, pp. 2 085-2 096.

Ayhan, V. and S. Aktan (2004), "Potential use of dried tomato pomace in broiler chicken diets", *Journal of Animal Production*, Vol. 45, No. 1, pp. 19-22.

Bacigalupo, M.A. et al. (2000), "Quantification of glycoalkaloids in tomato plants by time-resolved fluorescence using a europium chelator entrapped in liposomes", *Analyst*, Vol. 125, No. 10, October, pp. 1 847-1 850.

Baysal, T. et al. (2000), "Supercritical fluid extraction of beta-carotene and lycopene from tomato paste waste", *Journal of Agricultural and Food Chemistry*, Vol. 48, No. 11, pp. 5 507-5 511.

Beecher, G.R. (1998), "Nutrient content of tomatoes and tomato products", *Proceedings of the Society for Experimental Biology and Medicine*, Vol. 218, No. 2, pp. 98-100.

Castro, A. and N. Monji (1986), "Dietary nicotine and its significance in studies on tobacco smoking", *Biomedical Archives*, Vol. 2, pp. 91-97.

Davies, J.N. and G.E. Hobson (1981), "The constituents of tomato fruit – The influence of environment, nutrition and genotype", *CRC Critical Reviews in Food Science and Nutrition*, Vol. 15, No. 3, pp. 205-280.

Davis, R.A. et al. (1991), "Dietary nicotine: A source of urinary cotinene," *Food and Chemical Toxicology*, Vol. 29, No. 12, December, pp. 821-827.

De Stefani, E. et al. (2000), "Tomatoes, tomato-rich foods, lycopene and cancer of the upper aerodigestive tract: A case-control in Uruguay", *Oral Oncology*, Vol. 36, No. 1, pp. 47-53.

Dewanto, V. et al. (2002), "Thermal processing enhances the nutritional value of tomatoes by increasing total antioxidant activity", *Journal of Agricultural and Food Chemistry*, Vol. 50, No. 10, pp. 3 010-3 014.

Domino, E.F. et al. (1993), "Relevance of nicotine content of common vegetables to the identification of passive tobacco smokers", *Medical Science Research*, Vol. 21, pp. 571-572.

Dorai, T. and B.B. Aggarwal (2004), "Role of chemopreventive agents in cancer therapy", *Cancer Letters*, Vol. 215, No. 2, pp. 219-240.

Dumas, Y. et al. (2003), "Effects of environmental factors and agricultural techniques on antioxidant content of tomatoes", *Journal of Science Food Agric.*, Vol. 83, No. 5, April, pp. 369-382, http://dx.doi.org/10.1002/jsfa.1370.

Eltayeb, E.A. and J.G. Roddick (1985),

Eltayeb, E.A. and J.G. Roddick (1984), "Changes in the alkaloid content of developing fruits of tomato (*Lycopersicon esculentum Mill.*). II. Effects of artificial acceleration and retardation of ripening", *Journal of Experimental Bot*any, Vol. 35, No. 2, pp. 261-267, http://dx.Doi.org/10.1093/jxb/35.2.261.

Ensminger, M.E. et al. (1990*), Feeds and Nutrition*, Second Edition, The Ensminger Publishing Co., Clovis, California.

FAO (2004), *FAOSTAT-Agriculture Data*, www.fao.org/default.htm.

FAO/IPGRI (2005), "Genotypic variation of Kenyan tomato (Lycopersicon esculentum L.) germplasm", *PGR Newsletter*, Vol. 123, pp. 61-67.

FAO/WHO (1996), "Joint consultation report, Biotechnology and food safety", *FAO Food and Nutrition Paper No. 61*, Food and Agriculture Organization, Rome, available at: www.fao.org/ag/agn/food/pdf/biotechnology.pdf (accessed 28 January 2015).

Farlow, C.H. (2004), *Food Additives: A Shopper's Guide to What's Safe & What's Not*, 5th edition, KISS for Health Publishing, United States.

Favier, J.C. et coll. (1995), *Répertoire général des aliments – Table de composition*, Tec & Doc Lavoisier/Cneva-Ciqual, Paris.

Ferrante, V. et al. (2003), "Effect of tomato by-product diet supplementation on egg yolk colour", *Italian Journal of Animal Science*, Vol. 2, No. 1, pp. 459-461.

Fineli (2004), *Finnish Food Composition Database*, Release 6, National Public Health Institute, Nutrition Unit, Helsinki, www.fineli.fi/?lang=en.

Friedman, M. (2004), "Analysis of biologically active compounds in potatoes (*Solanum tuberosum*), tomatoes (*Lycopersicon esculentum*), and jimson weed (*Datura stramonium*) seeds", *Journal of Chromatography A*, Vol. 1 054, Issues 1-2, October, pp. 143-155, http://dx.doi.org/10.1016/j.chroma.2004.04.049.

Friedman, M. (2002), "Tomato glycoalkaloids: Role in the plant and in the diet", *Journal of Agricultural and Food Chemistry*, Vol. 50, No. 21, pp. 5 751-5 780.

Friedman, M. and C.E. Levin (1995), "alpha-Tomatine content of tomato and tomato products determined by HPLC with pulsed amperometric detection", *Journal of Agricultural and Food Chemistry*, Vol. 43, No. 6, pp. 1 507-1 511, http://dx.doi.org/10.1021/jf00054a017.

Gahler, S. et al. (2003), "Alterations of vitamin C, total phenolics, and antioxidant capacity as affected by processing tomatoes to different products", *Journal of Agricultural and Food Chemistry*, Vol. 51, No. 27, pp. 7 962-7 968.

Gómez-Prieto, M.S. et al. (2003), "Supercritical fluid extraction of all-trans-Lycopene from tomato", *Journal of Agricultural and Food Chemistry*, Vol. 51, No. 1, December, pp. 3-7, http://dx.Doi.org/10.1021/jf0202842.

Goose, P. and B. Raymond (1964), *Tomato Puree, Juice and Powder*, Food Trade Press.

Gould, W.A. (1992), *Tomato Production, Processing & Technology*, 3rd edition, CTI Publications Inc., Baltimore, Maryland.

Gould, W.A. (1974), *Tomato Production, Processing and Quality Evaluation*, 2nd edition, AVI Publishing Inc., Westport, Connecticut.

Harwig, J. et al. (1979), "Toxins of moulds from decaying tomato fruit", *Applied and Environmental Microbiology*, Vol. 38, No. 2, pp. 267-274.

Hayes, W.A. et al. (1998), "The production and quality of tomato concentrates", *Critical Reviews in Food Science and Nutrition*, Vol. 38, No. 7, pp. 557-564.

Herbers, K. (2003), "Vitamin production in transgenic plants", *Journal of Plant Physiology*, Vol. 160, No. 7, pp. 821-829.

Jenkins, J.A. (1948), "The origin of the cultivated tomato", *Economic Botany*, Vol. 2, Issue 4, October-December, pp. 379-392.

Kajderowics-Jarosinska, D. (1965), "The content of tomatine in the fruits of tomatoes at different stages of their ripeness", *Acta Agraria Silvestria*, Vol. 5, pp. 3-15.

Kavitha, P. et al. (2004), "Nutrient utilization in broilers fed dried tomato pomace with or without enzyme supplementation", *Indian Journal of Animal Nutrition*, Vol. 21, No. 1, pp. 17-21.

Kavitha, P. et al. (2005), "Nutritive value of dried tomato (*Lycopersicon esculentum*) pomace in cockerels", *Animal Nutrition and Feed Technology*, Vol. 5, No. 1, pp. 101-107.

King, J.A. and G. Zeidler (2004), "Tomato pomace may be a good source of vitamin E in broiler diets", *California Agriculture*, Vol. 58, No. 1, pp. 59-62.

Kiple, K.F. and K.C. Ornelas (2000), *The Cambridge World History of Food*, Vol. 2, Cambridge University Press.

Kozukue, N. et al. (2004), "Dehydrotomatine and a-Tomatine content in tomato fruits and vegetative plant tissues", *Journal of Agricultural Food and Chemistry*, Vol. 52, No. 7, April, pp. 2 079-2 083.

Kozukue, N. et al. (1994), "Alphatomatine purification and quantification in tomatoes by HPLC", *Journal of Food Science*, Vol. 59, Issue 6, pp. 1 211-1 212, http://dx.doi.org/10.1111/j.1365-2621.1994.tb14678.x.

Leonardi, C. et al. (2000), "Antioxidative activity and carotenoid and tomatine contents in different typologies of fresh consumption tomatoes", *Journal of Agricultural and Food Chemistry*, Vol. 48, No. 10, October, pp. 4 723-4 727.

Linseisen, J. et al. (1998), "Effect of a single oral dose of antioxidant mixture (vitamin E, carotenoids) on the formation of cholesterol oxidation products after ex vivo LDL oxidation in humans", *European Journal of Medical Research*, Vol. 21, No. 1-2, pp. 5-12.

Moulton, K. et al. (1994), "Dimensions of the global processing tomato industry", University of California at Berkeley, Department of Agricultural and Resource Economics and Policy, *CUDARE Working Paper Series*, No. 692.

Müller, N. et al. (2003), "Tomato products and lycopene supplements: Mandatory components in nutritional treatment of cancer patients", *Current Opinions in Clinical Nutrition and Metabolic Care*, Vol. 6, No. 6, November, pp. 657-660.

National Food Institute – Technical University of Denmark (2005), "Pour la tomate danoise produite in situ", *Danish Food Composition Databank*, www.foodcomp.dk/fcdb_details.asp?FoodId=0790.

NRC (2001), *Nutrient Requirements of Dairy Cattle*, Seventh Revised Edition, National Research Council, National Academies Press, Washington, DC, available at: http://www.nap.edu/openbook.php?record_id=9825.

NRC (1983), *Underutilized Resources as Animal Feedstuffs*, National Academies Press, Washington, DC, available at: www.nap.edu/catalog/41/underutilized-resources-as-animal-feedstuffs.

NRC (1982), *United States – Canadian Tables of Feed Composition: Nutritional Data for United States and Canadian Feeds*, Third Revision, National Academies Press, Washington, DC, available at: www.nap.edu/catalog/1713/united-states-canadian-tables-of-feed-composition-nutritional-data-for.

OECD (2003), "Considerations for the safety assessment of animal feedstuffs derived from genetically modified plants", ENV/JM/MONO(2003)10, OECD, Paris, available at: www.oecd.org/science/biotrack/46815216.pdf.

Ozcelik, S. et al. (1990), "Toxin production by *Alternaria alternata* in tomatoes and apples stored under various conditions and quantification of the toxins by high-performance liquid chromatography", *International Journal of Food Microbiology*, Vol. 11, No. 3-4, pp. 187-194.

Preston, R.L. (2007), "Feed composition tables, Beef", 1 May, Beef Magazine.

Rao, A.V. and S. Agarwal (2000), "Role of antioxidant lycopene in cancer and heart disease", *Journal of the American College of Nutrition*, Vol. 19, No. 5, pp. 563-569.

Rao, A.V. et al. (1998), "Lycopene content of tomatoes and tomato products and their contribution to dietary lycopene", *Food Research International* Vol. 31, No. 10, pp. 737-741.

Re, R. et al. (2002), "Effects of food processing on flavonoids and lycopene status in a Mediterranean tomato variety", *Free Radical Research*, Vol. 36, No. 7, pp. 803-810.

Rick, C.M. et al. (1994), "High α-tomatine content in ripe fruit of Andean Lycopersicon esculentum var. cerasiforme: Developmental and genetic aspects", *Proceedings of the National Academy of Science*, Vol. 91, No. 26, December, pp. 12 877-12 881.

Rozzi, N.L. et al. (2002), "Supercritical fluid extraction of lyopene from tomato processing byproducts", *Journal of Agricultural and Food Chemistry*, Vol. 50, No. 9, pp. 2 638-2643.

Sahlin, G.P. et al. (2004), "Investigation of the antioxidant properties of tomatoes after processing", *Journal of Food Composition and Analysis*, Vol. 17, No. 5, pp. 635-647.

Satchaphun, B. et al. (1998), "Use of tomato pomace as animal feed: 3) Use of dry tomato pomace as roughage for steers", Research Project No. 36, pp. 1 308-1 354.

Scientific Committee on Foods (1999), Opinion on a request for consent to place on the market a tomato fruit genetically modified to down-regulate the production of polygalacturonase (PG), and solely intended for processing.

Shi, J. and M. Le Maguer (2000), "Lycopene in tomatoes: Chemical and physical properties affected by food processing", *Critical Reviews in Food Science and Nutrition*, Vol. 40, No. 1, pp. 1-42.

Siegmund, B. et al. (1999), "Determination of the nicotine content of various edible nightshades (*Solanaceae*) and their products and estimation of the associated dietary nicotine intake", *Journal of Agricultural and Food Chemistry*, Vol. 47, No. 8, August, pp. 3 113-3 120.

Sogi, D.S. et al. (2005), "Biological evaluation of tomato waste seed meals and protein concentrate", *Food Chemistry*, Vol. 89, No. 1, pp. 53-56.

Souci, S.W. et al. (1994), *Food Composition and Nutrition Tables*, Medpharm – CRC Press, Stuttgart & Boca Raton.

Stahl, W. and H. Sies (1992), "Uptake of lycopene and its geometric isomers is greater from heat-processed than from unprocessed tomato juice in humans", *Journal of Nutrition*, Vol. 122, No. 11, November, pp. 2 161-2 166.

Takeoka, G.R. et al. (2001), "Processing effects on lycopene content and antioxidant activity of tomatoes", *Journal of Agricultural and Food Chemistry*, Vol. 49, No. 8, pp. 3 713-3 717.

Universidad de Almeria, Area de Tecnologia de Alimentos (2005), *Guide to Food Additives*, www.ual.es/~jlguil/Aditivos/colors100-181.htm.

University of Texas (2007), *Structural Database of Allergenic Proteins*, University of Texas Medical Branch, http://Fermi.utmb.edu/cgi-bin.SDAP (accessed in 2007).

USDA (2007), *National Nutrient Databases for Standard Reference*, Release 20, United States Department of Agriculture, www.ars.usda.gov/Services/docs.htm?docid=17476 (accessed in 2007).

Varo, P. et al. (1980), "Mineral element composition of Finnish foods. VII. Potato, vegetables, fruits, berries, nuts and mushrooms", *Acta Agriculturae Scandinavia*, S22, pp. 89-113.

Wanasitchaiwat, V. et al., "Use of tomato pomace as animal feed. 5) For Muscovy", Research Project No. 34 (5/37)-0513-036.

Weiss, W.P. et al. (1997), "Wet tomato pomace ensiled with corn plants for dairy cows", *Journal of Dairy Science*, Vol. 80, No. 11, pp. 2 896-2 900.

Westphal, S. et al. (2004), "Tomato profilin Lyc e 1: IgE cross-reactivity and allergenic potency", *Allergy*, Vol. 59, No. 5, pp. 526.

Westphal, S. et al. (2003), "Molecular characterization and allergenic activity of Lyc e 2 (beta-fructofuranosidase), a glycosylated allergen of tomato", *European Journal of Biochemistry*, Vol. 270, No. 6, pp. 1 327-1 337.

List of OECD consensus documents on the safety of novel foods and feeds, 2002-14

CONSENSUS DOCUMENT	LEAD COUNTRY(IES)	YEAR ISSUED	VOLUME
CROPS			
Alfalfa (*Medicago sativa*) **and other temperate forage legumes**	Canada and the United Kingdom	2005	Vol. 1
Barley (*Hordeum vulgare*)	Finland, Germany and the United States	2004	Vol. 1
Cassava (*Manihot esculenta*)	South Africa	2009	Vol. 2
Cotton (*Gossypium hirsutum* and *G. barbadense*)	United States	2009	Vol. 2
Cultivated mushroom (*Agaricus bisporus*)	Sweden	2007	Vol. 1
Grain sorghum (*Sorghum bicolor*)	United States and South Africa	2009	Vol. 2
Low erucic acid rapeseed (Canola)	Canada	2011	Vol. 2
Maize (*Zea mays*)	Netherlands and the United States	2002	Vol. 1
Oyster mushroom (*Pleurotus ostreatus*)	Sweden	2013	Vol. 2
Papaya (*Carica papaya*)	Thailand and the United States	2010	Vol. 2
Potato (*Solanum tuberosum* ssp. *tuberosum*)	Germany	2002	Vol. 1
Rice* (*Oryza sativa*)	Japan*	2004*	Vol. 1
Sugar beet (*Beta vulgaris*)	Germany	2002	Vol. 1
Sugarcane (*Saccharum* ssp. hybrids)	Australia	2011	Vol. 2
Soybean (*Glycine max*)	United States	2012	Vol. 2
Sunflower (*Helianthus annuus*)	Canada, France, Germany and the U.S.	2007	Vol. 1
Sweet potato (*Ipomea batatas*)	South Africa and Japan	2010	Vol. 2
Tomato (*Lycopersicon esculentum*)	Greece	2008	Vol. 1
Wheat (*Triticum aestivum*)	Australia	2003	Vol. 1
FACILITATING HARMONISATION			
Animal feedstuffs derived from genetically modified plants	Canada and the United Kingdom	2003	Vol. 1
Unique Identifier for transgenic plants (revised version) (guidance document)	Working Group on Harmonisation of Regulatory Oversight in Biotechnology	2006	Vol. 1
Molecular characterisation of plants derived from modern biotechnology	Canada, *joint publication of the Biosafety Working Group and the Food/Feed Safety Task Force*	2010	Vol. 2

* Rice document under revision, new issue expected in 2015.

Published in the Series on the Safety of Novel Foods and Feeds, by number

1	Consensus Document on Key Nutrients and Key Toxicants in Low Erucic Acid Rapeseed (Canola) (2001) – ***REPLACED with revised Consensus Doc. No. 24 (2011)***
2	Consensus Document on Compositional Considerations for New Varieties of Soybean: Key Food and Feed Nutrients and Anti-Nutrients (2001) – ***REPLACED with revised Consensus Doc. No. 25 (2012)***]
3	Consensus Document on Compositional Considerations for New Varieties of Sugar Beet: Key Food and Feed Nutrients and Anti-Nutrients (2002)
4	Consensus Document on Compositional Considerations for New Varieties of Potatoes: Key Food and Feed Nutrients, Anti-Nutrients and Toxicants (2002)
5	Report of the OECD Workshop on the Nutritional Assessment of Novel Foods and Feeds, Ottawa, Canada, February 2001 (2002)
6	Consensus Document on Compositional Considerations for New Varieties of Maize (*Zea mays*): Key Food and Feed Nutrients, Anti-Nutrients and Secondary Plant Metabolites (2002)
7	Consensus Document on Compositional Considerations for New Varieties of Bread Wheat (*Triticum aestivum*): Key Food and Feed Nutrients, Anti-Nutrients and Toxicants (2003)
8	Report on the Questionnaire on Biomarkers, Research on the Safety of Novel Foods and Feasibility of Post-Market Monitoring (2003)
9	Considerations for the Safety Assessment of Animal Feedstuffs Derived from Genetically Modified Plants (2003)
10	Consensus Document on Compositional Considerations for New Varieties of Rice (*Oryza sativa*): Key Food and Feed Nutrients and Anti-Nutrients (2004) – ***Under revision***
11	Consensus Document on Compositional Considerations for New Varieties of Cotton (*Gossypium hirsutum* and *Gossypium barbadense*): Key Food and Feed Nutrients and Anti-Nutrients (2004)
12	Consensus Document on Compositional Considerations for New Varieties of Barley (*Hordeum vulgare* L.): Key Food and Feed Nutrients and Anti-Nutrients (2004)
13	Consensus Document on Compositional Considerations for New Varieties of Alfalfa (*Medicago sativa*) and Other Temperate Forage Legumes: Key Feed Nutrients, Anti-Nutrients and Secondary Plant Metabolites (2005)
14	An Introduction to the Food/Feed Safety Consensus Documents of the Task Force for the Safety of Novel Foods and Feeds (2006)
15	Consensus Document on Compositional Considerations for New Varieties of the Cultivated Mushroom *Agaricus Bisporus*: Key Food and Feed Nutrients, Anti-Nutrients and Toxicants (2007)
16	Consensus Document on Compositional Considerations for New Varieties of Sunflower: Key Food and Feed Nutrients, Anti-Nutrients and Toxicants (2007)
17	Consensus Document on Compositional Considerations for New Varieties of Tomato: Key Food and Feed Nutrients, Anti-Nutrients, Toxicants and Allergens (2008)
18	Consensus Document on Compositional Considerations for New Varieties of Cassava (*Manihot esculenta* Crantz): Key Food and Feed Nutrients, Anti-Nutrients, Toxicants and Allergens (2009)
19	Consensus Document on Compositional Considerations for New Varieties of Grain Sorghum [*Sorghum bicolor* (L.) Moench]: Key Food and Feed Nutrients and Anti-Nutrients (2010)
20	Consensus Document on Compositional Considerations for New Varieties of Sweet Potato [*Ipomoea batatas* (L.) Lam.]: Key Food and Feed Nutrients, Anti-Nutrients, Toxicants and Allergens (2010)
21	Consensus Document on Compositional Considerations for New Varieties of Papaya (*Carica papaya* L.): Key Food and Feed Nutrients, Anti-Nutrients, Toxicants and Allergens (2010)
22	Consensus Document on Molecular Characterisation of Plants Derived from Modern Biotechnology (2010)
23	Consensus Document on Compositional Considerations for New Varieties of Sugarcane (*Saccharum* spp. hybrids.): Key Food and Feed Nutrients, Anti-Nutrients and Toxicants (2011)
24	Revised Consensus Document on Compositional Considerations for New Varieties of Low Erucic Acid Rapeseed (Canola): Key Food and Feed Nutrients Anti-Nutrients and Toxicants (2011)
25	Revised Consensus Document on Compositional Considerations for New Varieties of Soybean [*Glycine max* (L.) Merr.]: Key Food and Feed Nutrients, Anti-Nutrients, Toxicants and Allergens (2012)
26	Consensus Document on Compositional Considerations for New Varieties of Oyster Mushroom [*Pleurotus ostreatus*]: Key Food and Feed Nutrients, Anti-Nutrients and Toxicants (2013)

Note: **The individual documents composing the Safety of Novel Foods and Feeds Series, latest version, are available online at the OECD BIOTRACK website: www.oecd.org/biotrack.**
The Series of Biosafety Consensus Documents (environmental safety), issued by the OECD Working Group on the Harmonisation of Regulatory Oversight in Biotechnology,as well as the OECD *Biotech Product Database*, are also available at the same address.

ORGANISATION FOR ECONOMIC CO-OPERATION AND DEVELOPMENT

The OECD is a unique forum where governments work together to address the economic, social and environmental challenges of globalisation. The OECD is also at the forefront of efforts to understand and to help governments respond to new developments and concerns, such as corporate governance, the information economy and the challenges of an ageing population. The Organisation provides a setting where governments can compare policy experiences, seek answers to common problems, identify good practice and work to co-ordinate domestic and international policies.

The OECD member countries are: Australia, Austria, Belgium, Canada, Chile, the Czech Republic, Denmark, Estonia, Finland, France, Germany, Greece, Hungary, Iceland, Ireland, Israel, Italy, Japan, Korea, Luxembourg, Mexico, the Netherlands, New Zealand, Norway, Poland, Portugal, the Slovak Republic, Slovenia, Spain, Sweden, Switzerland, Turkey, the United Kingdom and the United States. The European Commission takes part in the work of the OECD.

OECD Publishing disseminates widely the results of the Organisation's statistics gathering and research on economic, social and environmental issues, as well as the conventions, guidelines and standards agreed by its members.

OECD PUBLISHING, 2, rue André-Pascal, 75775 PARIS CEDEX 16
(97 2012 11 1 P) ISBN 978-92-64-18013-0 – 2015-02

www.ingramcontent.com/pod-product-compliance
Lightning Source LLC
LaVergne TN
LVHW081401110826
845149LV00010B/1634

* 9 7 8 9 2 6 4 1 8 0 1 3 0 *